PRIMATE
PATTERNS

PRIMATE PATTERNS

PHYLLIS DOLHINOW, Editor

University of California, Berkeley

HOLT, RINEHART AND WINSTON, INC.

New York Chicago San Francisco Atlanta

Dallas Montreal Toronto London Sydney

PREFACE

The number of laboratory and field studies of nonhuman primate behavior has increased steadily during the last ten years and many more projects are planned. Research results, most of which are written for the specialist in primate behavior, are published in a wide range of professional journals and texts. The general reader who wishes to find an introduction to the primates has a problem in common with the teacher who desires to use a brief, non-technical but authoritative reference on primates in the classroom: both have great difficulty in finding such a book. An increasing awareness of the need for an introduction to the field of primate studies prompted the editing of this volume. *Primate Patterns* is designed to be used in a variety of courses including psychology, zoology, and physical, cultural, and social anthropology, as well as to be read by anyone who is interested in how man's closest living relatives behave.

It can be argued that there are as many sets of primate patterns as there are different kinds of primates. Indeed, it can be argued further that even a single kind of nonhuman primate, depending upon the circumstances of its life, can show different patterns. How, then, does one recognize a primate pattern? How large or small a segment of behavior is a primate pattern? Good questions – and not at all easily answered. That is what this book is about, even though not all the authors address this problem in so many words. Each agrees, however, that the aim of investigation into primate behavior is not description alone, that ultimately we all seek to understand the multiple factors that are responsible for each pattern and the relationship of patterns to each other.

This book represents only a very small sample of all the primate behavior studies of the last decade and it is intended as an introduction rather than a comprehensive survey of what is known of primate ways of life. It was originally intended to make selections from two previously published volumes *Primate Behavior: Field Studies of Monkeys and Apes* (1965) and *Primates: Studies in Adaptation and Variability* (1968), but it soon

v

became apparent that the field of primate studies had far outgrown the confines of topics contained in these two volumes. As a result, three chapters were taken from each book, three were reprinted from other sources, and three were written new for this book.

Primate Behavior: Field Studies of Monkeys and Apes, edited by I. DeVore, was written in 1963 by a group of scholars who were members of a project on primate behavior at the Center for Advanced Study in the Behavioral Sciences during 1962–1963. The three selections from this volume are papers on individual species of Old World primates, the "Behavior of the Mountain Gorilla" by George Schaller, "Baboon Social Behavior" by K. R. L. Hall and I. DeVore, and "The North Indian Langur" by P. Jay Dolhinow (substantially rewritten). All three served to introduce a species of Old World primate and each was based on fieldwork undertaken by the author. Each primate has since been, or is being, studied by other investigators; it was the realization these subsequent studies brought of the variability within the behavior of the two species of monkeys that was in part responsible for the attention devoted to variability and adaptation in the 1968 volume.

Primates: Studies in Adaptation and Variability, edited by P. Jay, was based on the results of a symposium on primate social behavior sponsored by the Wenner-Gren Foundation for Anthropological Research and held at the European Conference Headquarters of the Foundation, Burg Wartenstein, Austria, in 1965. One chapter reprinted here, van Lawick-Goodall's "A Preliminary Report on Expressive Movements and Communication in the Gombe Stream Chimpanzees," is based on more than five years of continuous observations of the Gombe chimpanzees and is devoted to detailed analysis of one aspect of chimpanzee behavior. (The reader can get a good notion of what a difference several additional years of fieldwork can make by reading van Lawick-Goodall's earlier chapter on chimpanzee behavior in the 1965 DeVore volume.) "Social Learning in Monkeys" by the late K. R. L. Hall states very clearly that for an animal to be normal, most of its learning must take place in a social context. This is a point of view shared by the majority of fieldworkers and a view that is influencing the investigation of primate behavior in captivity. "Aggressive Behavior in Old World Monkeys and Apes" by S. L. Washburn and D. A. Hamburg remains an important statement on the social functions of aggressive behavior in nonhuman primates. A great deal more work has been done on aggression, its biological basis, functions, and control, and much of it has been along lines suggested in this paper.

Each of the three chapters reprinted from other sources discusses aspects of primate behavior. "Field Studies of Old World Monkeys and Apes" by S. L. Washburn, P. Jay, and J. B. Lancaster was written some five years ago but it asks questions about primate behavior and suggests lines of research that have, in fact, been pursued by many researchers since 1965.

"The Development of Primate Social Relations and Motor Skills through Play" by P. Dolhinow and N. Bishop concentrates on one of the most important aspects of the social development of mammals. This chapter is far from being a definitive statement, but rather, is an introduction to some of the problems involved in the study of one of the most universal and important activities of young mammals. "Evolution of Primate Behavior" by S. L. Washburn and R. S. Harding brings together many diverse lines of research and directs their results to the study of the evolution of primate behavior patterns.

The remaining three chapters were written for this book. In "Social Communication in Some Old World Monkeys and Gibbons," J. Shirek-Ellefson summarizes many attributes of social communication common to Old World primates. She emphasizes *Macaca irus* and reports her fieldwork results from Singapore. The remaining two new chapters start and close the volume. The book opens with "The Nonhuman Primates: An Overview" as an introduction to the primates—their substantial complexity of forms and their history. The concluding chapter "Primate Patterns" looks at the great diversity of forms of primate behavior and seeks to draw out from the variability some common threads that may be said to characterize the lives of a majority of the living Old World monkeys and apes. This chapter also touches upon some of the important problems in the analysis of primate behavior.

All the chapters listed above are arranged in two sections: the first "Field Studies" and the second "The Analysis of Behavior: Special Topics of Primate Behavior." The emphasis throughout is on field studies, and the contributors represent many diverse disciplines and backgrounds of experience. The following remains true, as was stated in the preface to the 1968 book: "Many areas of research are involved because social behavior is multidimensional. It has a biological basis; animals interact as they are predisposed by factors that include anatomy and physiology. The sum of these factors is the outcome of interactional processes between evolutionary heritage and environment. In few areas of research has it become clearer than it is in primate behavior studies that properly framed questions and valid answers will require the crossing of traditional academic boundaries."

It was a difficult task to select the few studies reprinted here from among the many excellent reports that have been published in the last decade. The reader is referred to the bibliography for a notion of the quantity and variety of information that has been gathered about the nonhuman primates. It is also apparent from even a scanning of the bibliography that this book is, indeed, far from a representative sample of what is available.

For those readers who wonder to what extent and in what ways the behavior of the living nonhuman primates is useful to the study and understanding of man, past and present, several chapters suggest ways in which patterns of living primate behavior are relevant to human history. We shared a common ancestor with the African apes far more recently than we suspected until modern methods of dating and assessing evolutionary distances were developed. The quantum distance as measured in culture and language that man has traveled since his identity with his ape ancestors makes a comparison of his behavior with that of the living primates exceedingly complex.

I wish to thank the Wenner-Gren Foundation for Anthropological Research for an editing grant that helped make this book possible. S. L. Washburn has been of constant assistance throughout the planning and execution of this book, and his help is gratefully acknowledged. Anne Brower, as with both earlier volumes, has provided crucial editorial assistance and has made innumerable suggestions all along the way. Many people assisted in the production of each manuscript by reading and commenting upon it. To all I express my thanks, to Lynda Muckenfuss who compiled the bibliography for this book, to Dana Olson for the name and subject indexes, and to Becky who has given me many new notions about primate behavior.

P. D.

Berkeley, Calif.
February 1972

CONTENTS

ix

PRIMATE
PATTERNS

chapter one

THE NONHUMAN PRIMATES: AN OVERVIEW

Phyllis Dolhinow

Since the emergence of Primates some 70 million years ago, there have been many major fluctuations in their varieties and numbers, and at times in the past their success—even their mere survival—must have been in the balance. Today, although there are far fewer kinds than there were at certain times in the past, the living primates are a large and highly diversified group of mammals, encompassing more than 50 living genera in which at least 200 species[1] are well defined. Among living primates are animals as diverse as the tree shrews (if they are included in the order), the lemurs, and men.

Modern nonhuman primates found in Central and South America are designated New World, and those in Asia and Africa are designated Old World. In general they are limited to tropical and subtropical forests, although some have adapted to life in savannas, in arid regions, and in cold climates as far north as Japan. A small population of macaques, probably originally from North Africa, survives on the Rock of Gibraltar. In spite of this evidence of adaptability, many modern primate genera are greatly reduced, in comparison with their former distribution and numbers, both in the geographical areas they occupy and in the size of their populations.

Early in primate history prosimians dominated the trees for millions of years and gradually spread to cover North and South America, Europe, Asia, and Africa. We know of at least 60 prosimian genera that did not survive to the present, and it is likely that there were still others which left no trace in the fossil record. Radiations that eventually developed into the living primates—including, of course, man himself—originated from only a few of the early kinds of prosimians.

Although an ever increasing number of fossils has been unearthed in recent archeological investigations, the fossil record of primate evolution is not as complete as that of some mammals, such as the horse. Fortunately

---

[1] For discussions of primate classification the reader is referred to Simpson 1945, 1949 (Chapter 7), 1962, and 1963; Piveteau 1957; Fiedler 1956; Clark 1959; and Napier and Napier 1967.

we are not limited only to fossil evidence in the study of primate evolution; field studies of the living primates and research in molecular biology are also valuable aids.

The first primates were inconspicuous, small, insectivorous mammals living in trees where they probably escaped the notice of most larger animals. In the evolution of primates the earliest and most basic structural changes were adaptations in the hands and feet that facilitated climbing by grasping. This very important and efficient locomotor adaptation included the replacement of claws by nails and the development of hands and feet as grasping organs, so it became possible for the animal to hold onto a branch or twig by the skin on the ends of the fingers and toes (Clark 1936; Bishop 1962). The shape of the terminal bone of each digit tells whether it bore a nail or a claw, making it possible to know by inspection which fossil forms had nails, claws, or a combination of both.

Primitive mammals and many living ones, such as rodents, climb trees by digging their claws into the bark for purchase. In these forms each digit ends in a claw that is firmly attached to the terminal bone of the finger or toe. This pad of specialized, weight-bearing skin is relatively immobile because it is firmly attached, and it forms a stable and unmoving base for the claw. In primate evolution the claw became thinner as the digits elongated and the terminal phalanx flattened. Pads of skin supporting the claws reduced in size as did the claw itself. Eventually the specialized thickened weight-bearing skin of the palm and sole extended to cover the ventral surfaces of the digits, and these extensions onto the digits became highly sensitive tactile surfaces.

Many factors undoubtedly influenced the development of nails, one being the relative inefficiency of climbing with claws as early primates increased in size. There were also advantages in being able to move farther out onto the ends of branches to reach food. An animal that could move from one limb to another, grasping each handhold and foothold securely, could get to food that would otherwise be unobtainable. The ability to move out to the ends of limbs perhaps also increased the number of kinds of foodstuffs available to the animal.

As Le Gros Clark (1959) describes the events in primate history, along with climbing by grasping there was an elaboration of the visual apparatus and gradually the development of three-dimensional or binocular vision in which the eyes are rotated to look forward and the visual fields overlap. With larger eyes and improved retinas came color vision and the development of an encircling bony protection for the eyes. None of these features was essential to living in trees, since other mammals without them succeed in doing so, but they must have given the primates some advantage in their competition for arboreal niches. In general, as visual efficiency increased, there was a reduction in the importance of smell. Sight-oriented monkeys in moving around in their environment and interacting with each other depended much less than did prosimians on scent markings and other olfactory-based behaviors. Of the nonhuman primates, only prosimians depend heavily on specialized vibrissae to gain tactual information of their world.

Other changes in the course of evolution affected most parts of the primate body and behavior, from reproducing to eating. The gestational process became increasingly efficient in the nourishment of the fetus, and this coupled with an increase in the length of infant dependency had important implications for the life of primates. As in the discussion of the development of climbing by grasping, it is not sufficient merely to detail structural changes that have taken place in primate evolution; the importance of each complex of change to the individual's behavior and reproductive success must be assessed as well. For every feature that changed over time the question has to be asked: What effect did this change have on the behavior of the individual? It is seldom possible to answer with certainty, but our guesses become more educated as we learn more about the behavior of living primates and the relationships between structures and behavior in them. Our appreciation of what an animal does, given the morphological structures it possesses, is extremely useful in reconstructing the probable or possible behavior of extinct forms with the same or similar morphological features.

It is essential to know how closely related different forms are if we are going to generalize from the living primates to those in the past. Many claims have been made as to which primates are most closely related to each other and to us; the candidates for our closest relative range from a tarsier to a chimpanzee. It is extremely important to understand the implications of these differing points of view in trying to answer questions about the evolution of primate behaviors. Some of the most important of these questions ask what behaviors are basic to all primates, what patterns characterize different major radiations of the primates, which did man's ancestors share with other primates, and which were characteristic of early man after he was distinct from the other primates. The answers to these questions will vary according to the length of time that is judged to have elapsed during different stages of primate evolution and according to the characteristics attributed to the immediate precursor population.

On the one hand, if it is maintained that the human lineage (Hominidae) diverged from the lineage of the ape (Pongidae) and of the monkey (Cercopithecidae) in the Oligocene period, then there would have been three separate but parallel evolutionary lines persisting for approximately 30–35 million years, and man would have developed from a quadrupedal monkey, not an ape. All the similarities man shares with the living apes, especially the African apes, would have had to develop in a parallel fashion, a most improbable series of evolutionary events.

If on the other hand, as some have suggested, man diverged from an ancestral ape soon after the divergence of monkeys and apes and after at least the beginning of the modifications of the ape shoulder and arms associated with the locomotion pattern of brachiation, then man's relationship to the apes would be much closer than his relationship to the monkeys. In this case the length of time since the divergence would be in the order of 15–20 million years. The small quadrupedal fossils from the Oligocene and early Miocene do not possess the anatomical features of brachiation, and thus man's separation was from an ancestral ape population, and it must have occurred in the late Miocene or early Pliocene after at least the start of the development of brachiation as a mode of locomotion and feeding. By that time *Dryopithecus* and *Pliopithecus* displayed structural features characteristic of the ape shoulder even though the remaining parts of the skeleton resembled that of a quadrupedal monkey. Obviously, the later the separation of early man from apes, the more features they would hold in common. Washburn has suggested strongly that one such common feature may have been knuckle-walking specializations of the African apes (Washburn 1968*a*). The Asiatic apes had meanwhile developed other anatomical specializations for locomotion in the trees. If man had a time of knuckle-walking in his history then the time span since the divergence from apes would be close to 5 million years. This is a substantially shorter time than that estimated by proponents of a prebrachiation ancestry for man, and a great deal shorter still than that estimated by those who maintain man's most immediate ancestor was a quadrupedal monkey.

Molecular data (Goodman 1968; Sarich and Wilson 1967*a*, 1967*b*) offer strong support for the position that the Hominidae evolved from the Pongidae very recently and that the sequence of divergence in Old World primate evolution from earliest to most recent is: monkeys, Asian apes, African apes, and finally, man. It is likely, according to the changes in blood albumins studied by Wilson and Sarich, that after the separation of the Pongidae from the Cercopithecidae, the apes shared a long period of common ancestry before the Asian and African apes separated from a common ancestral form. Man's divergence from the African apes, then, would be relatively recent.

Bipedalism as a favored means of locomotion did not appear until the early hominid forms of *Australopithecus,* and there are indications that the mode of bipedalism in these forms was relatively inefficient, especially slow walking over long distance (Zihlman 1967). The many major differences between the head and brain of hominids and of apes developed only after the divergence of the hominid from the pongid lineage. It was not until relatively very late in human evolutionary history that substantial increases in brain size and in the complexity of culture occurred, and most likely in the development of human language as well. Early man was at an *Australopithecus* stage of development, a bipedal and simple tool-using creature with a brain approximately the size of that of a living great ape, for a period of approximately 2.5 million years, or for about four times as long as the rest of human history.

Points of view as to the course and timing of human evolution differ according to the kinds of evidence used. However, it now appears that biochemical analysis offers clocks, or measures, of evolutionary distances that are compatible with sequences already outlined in the fossil record. Eventually, with the collaboration of specialists in many fields, we will be able to reconstruct the paths of primate evolution in much more detail than is possible at present.

A brief survey of the major kinds of living primates as summarized in the outline below will give an impression of the variability in form and adaptation to be found among them. It is traditional to begin with the most primitive, the tree shrews, although there is some question whether they are the most primitive insectivorelike primate or the most primatelike insectivore. Their exceedingly primitive structure — the absence of any climbing-by-grasping adaptation of either hands or feet, the presence of claws on all digits, and the many features of skull, teeth, and brain that are unlike those of primates — would seem to exclude them from the order Primates (Van Valen 1965). Behaviorally they are also very dissimilar from the vast majority of primates.

The Prosimii suborder of Primates today includes lemurs and tarsiers (in addition to the tree shrews if they are admitted to the order). During the first half of the age of mammals there were many different varieties of prosimians, most of which became extinct. Not only was there tremendous diversity among them, but also they were widely distributed over a large portion of the world. Now they are completely extinct in all northern and temperate regions and in the entire New World. With the exception of the diverse forms in Madagascar, where there are no monkeys, most Old World prosimians are small and nocturnal and in close competition with monkeys.

The island of Madagascar has been separated geographically from the mainland of Africa for more than 30 million years, and during this time there has been a tremendous local radiation of lemurs. There have been almost no natural predators and a complete lack of competition from other primates on Madagascar, in contrast to the mainland where there have been a great many predators and many kinds of monkeys. On the African mainland and in Asia the prosimians have developed nocturnal habits and thereby greatly reduced the competition with monkeys for food and living space.

I. PROSIMIANS

A. *Distribution.* Living forms are limited to the Old World (Asia and Africa including Madagascar). Fossil forms were very numerous in both the New and Old Worlds, but became extinct in all northern and temperate regions and in the New World. The prosimian period of primate evolution lasted from about 65 to 40 million years ago.

B. *General characteristics.* Prosimians retain much of the special senses of a primitive mammal; most are nocturnal.

C. *Kinds*.

	Rapid leapers	*Very slow movers*
1. Asia:	Tarsiidae	Lorisidae (*Loris, Nycticebus*)
2. Africa:	Galagidae	Lorisidae (*Perodicticus*)

3. Madagascar: Many kinds of lemurs, no monkey competition so some forms are diurnal. During the Pleistocene many Madagascar forms became extinct.
4. Tree Shrew: Very primitive form that retains claws and other features that suggest it may be a member of Insectivora rather than of Primates.

II. NEW WORLD MONKEYS

A. *Distribution*. These forms live in areas from southern Mexico into South America. Fossil forms had a wider distribution. The New World monkeys evolved independently from a New World prosimian, and similarities with the Old World monkeys are due to parallel evolution.

B. *General Characteristics*. The New World forms are much less uniform than are the Old World monkeys. There are many very ancient lines and the degree of relatedness among some of them is unclear. Some forms have prehensile tails, and one is nocturnal. All are strictly arboreal.

C. *Kinds*. There are many major types including the following:
1. *Alouatta*, the howler monkey, has a prehensile tail and is the largest of the New World monkeys.
2. *Aotus*, the owl monkey, is the only nocturnal monkey.
3. *Ateles*, the spider monkey, has a prehensile tail.
4. *Cacajao* are closely related to the *Pithecia;* the former are bright crimson faced small monkeys with long shaggy red coats, and the latter, the sakis, are short-tailed smaller forms.
5. *Callicebus*, a small form, lives in small social groups and defends territories.
6. *Callitrichidae*, or marmosets, are numerous with at least two genera and many species.
7. *Cebuella*, or the pygmy marmosets; this is the smallest of the New World monkeys.
8. *Cebus*, is very common, the "organ grinder" monkey; it has a prehensile tail.
9. *Lagothrix*, the wooly monkey, has a prehensile tail.
10. *Saguinus*, are the various kinds of tamarins.
11. *Saimiri*, the squirrel monkey, lives in large groups.

III. OLD WORLD MONKEYS

A. *Distribution*. Old World monkeys are abundant in Asia and Africa, but there are only a few in Europe (probably from North Africa). Living forms occupy habitats from very dry savanna to dense tropical rain forest, Fossil forms are found in Europe, Asia, and Africa. The Old World monkeys became separate from the prosimians in the Oligocene.

B. *General characteristics*. All Old World monkeys are quadrupedal forms adapted to climbing by grasping with their hands and feet. All are diurnal.

C. *Kinds*. There are two major types, which may be distinguished at the subfamily level and are separated primarily on the basis of visceral specializations.
1. Colobinae have been distinct since at least the Miocene; these forms are specialized to enable them to eat mature leaves. Most forms are arboreal and only a few kinds spend much time on the ground.
    a. Asia: Langurs. Many kinds; most arboreal. In southeast Asia there are two species groups of *Presbytis* (*Trachypithecus* and *Presbytis*) and three closely related genera *(Nasalis, Pygathrix, and Rhinopithecus).*
    b. Africa: A single genus, *Colobus,* which are arboreal. There are a few kinds.
2. Cercopithecinae include many different types. Many of these forms are ground-living.
    a. Asia: Macaques. Many species living in different habitats. Most of these monkeys spend at least part of the day on the ground and some have very widespread geographical distributions.
    b. Africa: There are many kinds, some of which are highly specialized in habitat preference. Major forms are:
        (1.) *Cercopithecus* are very diverse genera with many arboreal forms. One of the major forms, the vervet, also spends much time on the ground.
        (2.) *Erythrocebus,* the patas, are primarily open-ground-living in small one-male groups. They are very swift and adapted to running and hiding from predators.
        (3.) *Papio,* the baboons, are very widely distributed, living in habitats ranging from savanna to forest.
        (4.) *Cercocebus,* the mangabeys, are primarily arboreal.
        (5.) *Theropithecus* are restricted in distribution to northeast Africa; the gelada has a very distinctive social organization.

IV. APES

A. *Distribution.* Africa and Asia. The living forms are few in number with only one small form, the gibbon, surviving. In the fossil record there are many smaller forms in Africa during the Miocene. The ape period of primate evolution lasted from about 25 to 5 million years ago. Man has been separate from the apes for from 4 to 7 million years.

B. *General characteristics.* All the apes have a basic brachiating anatomy, although only the small ape, the gibbon, brachiates as an adult. African apes walk quadrupedally on their knuckles.

C. *Kinds.*
1. Asia: Forms are strictly arboreal.
    a. Gibbon. There are two kinds, the smaller gibbon and the larger siamang. Both live in small groups of a mated pair and their recent offspring. Adults of the same sex are hostile and the small groupings actively defend small territories.
    b. Orangutan, a large ape, is arboreal, living either in pairs or alone; groupings of more than three are extremely rare.
2. Africa: there are two forms; both are large and spend much time on the ground as adults.
    a. Chimpanzee, a knuckle-walking large ape, with exceedingly complex social behavior. Restricted in geographical distribution, in forest or edge of forest habitats.

    b. Gorilla, the largest of apes; adults seldom move much in the trees
       because of their size. Even more restricted than chimpanzees in their
       geographical distribution, gorillas are found only in forest habitats.

The tarsier, a nocturnal specialized jumping primate about the size
of a two-week-old kitten, has been regarded as closer than any other prosimian to the line of evolution leading to monkeys (Clard 1959). At one time
the genus *Tarsius* was anatomically diversified and widespread in distribution, in contrast to the one kind surviving today in Borneo, the Philippines,
and the Celebes.

Lemurs are divided into two groups, the Lemuriformes and the Lorisiformes. The former are confined to the island of Madagascar and represent
a striking variety of types, from a mouse-sized lemur through larger ones
about the size of a small dog to the bizarre and improbable aye-aye. The
lorisiformes include the lorises of Asia and the remaining lemurs found on
the mainland of Africa—the galagos, pottos, and angwantibos.

All lemurs and the tarsier have a grasping hand and foot but they differ
in the way they combine nails on some digits and claws on others. The
lemurs have a claw on the second toe and the tarsier only retains "toilet"
claws on the second and third toes. The aye-aye has elongated digits but
its nails are histologically similar in structure to claws. Prosimians in general
depend much more on smell for information about their environment than
do monkeys, and in this they resemble the primitive mammals more than
they do the rest of the living primates. Related to the importance of olfaction
are the scent glands many prosimians have that are used in marking behaviors within and among social groups. Specialized tactile hairs are present
on all forms. Vision is not so highly developed in the prosimians as in the
monkeys or apes; they perceive little or no color and, further, lack stereoscopic vision.

The New World monkeys, the Ceboidea, are an exceedingly diverse
and very ancient set of groups. All of these New World forms evolved from
New World prosimians, and it does not appear that there was a common
monkey ancestor with Old World monkeys. The many similarities between
New and Old World monkeys are due to parallel evolution since the Oligocene some 30 million years ago when the separate forms developed. The
Old World forms are not nearly so diversified and many of the differences
among them are of much more recent origin than are the differences among
New World forms. Often known as the Platyrrhine monkeys, the Ceboidea
comprise a large number of genera and are distributed through Central and
South America as far south as Argentina.

Unlike many African and Asian forms that spend a great deal of time
on the ground, all the Ceboidea are strictly arboreal. Some have developed
a prehensile tail, which can act almost as another hand and is used to hang
by as well as to grasp. The squirrel, spider, wooly, and capuchin (or organ
grinder) monkeys are the most familiar since they are commonly used as

pets. The largest of all New World monkeys is the howler and the smallest forms are some of the many different kinds of marmosets. The only nocturnal monkey in the world is the *Aotus*. There are others with quite distinctive features and habits, but much less is known of the New World forms than of those in Asia and Africa. Because of their tropical and arboreal habitats the Ceboidea are, in general, exceedingly difficult to study. However, although Old World monkeys have been studied much more thoroughly than New World monkeys, representatives of only relatively few genera have been observed in free-ranging conditions.

The Cercopithecidae, or Old World monkeys, have been separate from other primates for about 30 million years, or since the Oligocene. All are quadrupedal forms with long, narrow, deep trunks related to their basic adaptation of climbing by grasping (Schultz 1936). The two subfamilies of Old World monkeys, the Cercopithecinae and the Colobinae, have been separate since about the middle Miocene. Both subfamilies include arboreal and terrestrial types, although most of the Colobinae spend the greater part of their days as well as their nights in trees. The major division among the monkeys of Asia and Africa is based on visceral specializations. The Colobinae have greatly enlarged stomachs or colons, so that in some the stomach and stomach contents comprise one quarter of the weight of the animal (Washburn 1944). Their specializations include the ability to digest large quantities of mature leaves, and at least some have developed special gut bacteria to accomplish this (Bauchop and Martucci 1968).

The Colobinae are especially abundant in Asia. *Presbytis entellus* (see Chapter 5), the common Indian langur, is found in the drier parts of both India and Ceylon. This form spends more time on the ground than do any of the other Asian langurs. In Southeast Asia there are two species groups of *Presbytis*, called *Trachypithecus* and *Presbytis* by Pocock (1934), and three genera: *Nasalis, Pygathrix,* and *Rhinopithecus* (Washburn 1944). It has been suggested that they are possibly all one genus characterized by oddshaped noses and their preference for living in very wet habitats. The African Colobinae are one genus, *Colobus,* with many fewer species than the Asian Colobinae.

	Africa	*Asia*
Colobinae	*Colobus*	*Presbytis* (incl. 4 species groups)
		Nasalis
		Pygathrix
		Rhinopithecus
Cercopithecinae	*Papio*	*Macaca*
	Cercopithecus	
	Cercocebus	
	Theropithecus	
	Erythrocebus	

The Cercopithecinae contain the following major genera: the *Macaca* (macaques) of Asia, and *Papio* (baboons), *Cercopithecus* (vervets, etc.), *Cercocebus* (mangabeys), *Theropithecus* (gelada), *Erythrocebus* (patas)— all of African distribution. The baboons and macaques, on the one hand, are very similar in morphology and behavior and probably should be contained within one genus. *Cercopithecus*, on the other hand, is a large and complex genus, with at least three major species groups occupying many adaptive niches. Of this subfamily, most spend the majority of their time in trees but there are a few, such as the savanna baboon, the gelada, hamadryas, and patas, that are morphologically and behaviorally adapted to spending much of their days on the ground, often at a considerable distance from the safety of trees.

The Pongidae, or apes, include the African chimpanzee and gorilla (which Simpson in 1963 suggested be placed in a single genus *Pan*), the Asian orangutan, and the gibbons of southeastern Asia (at least five species of small gibbon plus the one large siamang). It is believed that the hominoid divergence from the cercopithecoid ancestral populations occurred approximately 24 million years ago (Sarich 1969). In addition to striking behavioral differences between monkeys and apes, there are major differences in modes of locomotion and in the morphology of body parts related to locomotion. There has been a relative elongation of the arms and concomitant changes in the trunk structure in the apes (Schultz 1936). The ape trunk is short, wide, and very shallow compared to the monkey trunk, and the ape lumbar region is short with small back muscles which correspond to a set of movements of the ape back different from those used by typical quadrupedal monkeys. This general type of ape structure is termed brachiation and originally, as mentioned earlier, it was a specialization or adaptation to feeding at the ends of branches, permitting the animal to swing or hang beneath them.

Both the chimpanzee and the gorilla are knuckle-walkers; that is, when moving on flat horizontal surfaces they progress on all fours, placing their weight on the middle bone of the fingers. The organutan, in contrast, is much better adapted to remaining in trees and progressing on all fours only under unusual circumstances. The orang's limbs are extraordinarily mobile, and for long periods of time the animal can assume postures that to us appear most uncomfortable. Orangs rarely come to the ground, and when they do, it is obvious by the way they place their hands and feet in walking that this is a strange and unnatural mode of locomotion for them. Tuttle (1967) and Lewis (1969) have described the anatomy of the hand and wrist for the great apes. The gibbons are much less at home and efficient on the ground than are gorillas and chimpanzees, but compared to the orang, the gibbon seems to have relatively few problems in moving on a flat horizontal surface. Because the gibbon's arms are so long relative to its trunk and legs, it generally walks upright and uses its long slender arms for balancing, held either out to the side or slightly above the head.

Clearly, the only two apes that spend much time on the ground are the chimpanzee and the gorilla. These African forms had been classified as brachiators on the basis of anatomical structures, and only when they were observed in the wild was it realized that adult gorillas do not brachiate and adult chimpanzees seldom do. The young of both will brachiate so long as they are small enough that the trees easily support their weight. Categorizing an animal on the basis of structure alone can be exceedingly misleading. It can be equally misleading to group together animals because they do the same thing. For example, both the African chimpanzee and the New World spider monkey sometimes hang under branches, but each does it with substantially different anatomical structures (Washburn, personal communication). Animals with very similar anatomical structures may not use the same means of locomotion, as in the example of the adult gorilla, that seldom if ever brachiate, and the gibbon, that uses brachiation as its major means of moving from one place to another. It is imperative to see exactly what the animal does in nature to determine the full range of movement it uses and how much of the time and under what circumstances each kind of movement is used. It is also essential to study locomotion developmentally in each life stage.

Estimates of the date of divergence between the pongid and hominid lines depend on the evidence considered. According to recent biochemical data (Sarich and Wilson 1967a, 1967b), the separation took place about 4–5 million years ago. Man is most closely related to the chimpanzee and gorilla of all the living apes, and behaviorally the chimpanzee of all non-human primates is by far the most similar to man.

Although the Hominidae are approximately 5–10 million years old (that is, the length of time from the emergence of earliest ape-men or man-apes until the present), the earliest bipedal, toolmaking, hominid genus represented in the fossil record is *Australopithecus*, an early man with a brain about the size of that of a modern great ape. It seems reasonable to assume that *Australopithecus* had at least as sophisticated a system of communication and at least as much toolmaking skill as the living chimpanzee. However, the archeological record reveals that it was not until the middle of the Pleistocene, about 600,000 years ago, with the emergence of the genus *Homo*, that man's history of cultural proliferation and skillful toolmaking and use began. The rudiments of speech also probably first began at the time of *Homo*.[2] *Homo erectus* had a brain approximately twice the

[2] Lenneberg (1967) estimates that all modern languages had a common stock from 30,000 to 50,000 years ago. This is substantially later than the appearance of *Homo*, but there was doubtlessly a long period of time during which the brain and behavior were becoming more complex, leading to the emergence of language as we think of it late in the Pleistocene.

size of that of an ape or of an earlier *Australopithecus*. Thus, human communication systems, skillful tool manufacture and use, and complex cultural traditions evolved in only the last few hundred thousand years of the much longer hominid history.

The range of behavior of the primates forms a very general gradient from the less social and simpler patterns of behavior characteristic of the prosimians, through the monkeys, to the complex behaviors of apes, and finally to human behavior. Prosimians have been studied by several workers (Petter 1962*a*, 1962*b*, 1965), and lemur behavior has been summarized recently by Jolly (1966). All prosimians mature rapidly, most are nocturnal, some are raised in litters in nests, and few form even relatively permanent complex social groups of any size.

In contrast, monkeys mature more slowly, all except one are diurnal, single births are usual, and they have very social lives. Often individuals are born, live, and die within the same troop. Social groups may vary in size from a mated pair with their recent offspring (as the *Callicebus* observed by Mason 1968) to troops of more than 100 individuals of both sexes and all ages (Hall and DeVore, this volume). Even within a single species there may be considerable variation in the composition, stability, and structure of the troop, and in the relationships of individuals to each other and to conspecifics in other social groups. In general, social behavior is much richer among monkeys than among prosimians; the relatively immobile and expressionless face of a prosimian is simply incapable of expressing many of the nuances of emotion that most monkeys can signal without difficulty.

Apes are intermediate between monkeys and man in many respects, but apes, especially African apes, are much closer to man than monkeys are to apes. Apes are slow to mature and have a lengthened period of infant dependency. This allows more time for learning, practicing, and experimenting when the animals are immature, and there is much more for an ape to learn than for a monkey. The female monkey is socially and sexually mature at approximately four years of age, whereas the female ape is not mature until at least eight, and the human female's age of maturity is twice that of the ape's. For the female nonhuman primate there is an approximate coincidence of sexual and social maturity since when she has an infant she assumes typical adult social roles within the troop.

Both the great apes and man are far more object-oriented than are any of the monkeys. Although several New World monkeys have been observed to manipulate objects and sometimes use sticks in food getting (Thorington 1967), there is by no means the same degree of object use that one sees among chimpanzees (van Lawick-Goodall 1968*b*). In the Gombe National Park chimpanzees make and use several different kinds of tools to get food, and objects are used also in cleaning and in aggression and defense. Chimpanzees and gorillas construct nests in which they sleep at night and sometimes during the day. Few of these apes have ischial callosities, the toughened patches of skin on which an animal can sit for a long time without cutting off the blood flow to the legs with the result that their legs "fall asleep." Thus the African apes, in a sense, have replaced a morphological feature with the behavioral trait of nest building and now sleep on their sides rather than sitting up in monkey fashion (Washburn 1957).

The communication systems of the chimpanzee is by far the richest of any nonhuman primate. Included within its repertoire are gestures of placation, begging, and reassurance. It is likely that monkeys do more of this than we have noted but with far less frequency and without the subtleties characteristic of chimpanzee communication.

The most persistent and intense social bond in the life of the monkey or ape is that between a mother and her offspring. It is difficult to say just how long this bond lasts, but certainly among chimpanzees at the Gombe National Park it is a persistent and important relationship into at least the offspring's young adulthood, a period of many years (van Lawick-Goodall 1968*b*). Among rhesus monkeys on Cayo Santiago some mother–young bonds also persist for years (Sade 1965, 1966, 1968; Wilson 1968). A difficulty in making cross-species comparisons of this characteristic is that without long-term field studies it is impossible to determine uterine kin lines. In contrast to the importance of the mother–offspring bonds there is no "social father" among monkeys and apes, except possibly in those very few kinds such as gibbons and *Callicebus* where the biological father is known because the social groups are composed of a mated pair. It is not known whether the relationships between these group males and their offspring are different from those found in troops where many males may mate with each receptive female. The importance of nonhuman primate sibling groups is not clear, except that among chimpanzees siblings continue to have more to do with one another than they do with unrelated chimpanzees, and the offspring of a sibling will be played with when no other young will be.

Many species of Old World monkeys show a surprising degree of behavioral variability in their adaptation to different ecological areas, although, of course, not all species show an ability to survive in different habitats. In general, the more arboreal the species the greater is the tendency for it to be restricted in geographical distribution and to occupy a narrower range of habitat types. Some monkeys, such as the rhesus, are remarkable for their extreme hardiness and ability to survive in an amazing variety of habitats, from high deciduous forests where men are seldom encountered to the teeming heart of an Old Delhi bazaar.

We know remarkably little about the geographical distribution of many species of both Old and New World monkeys, and even less about the relationships and interactions among closely related species that share the same area. Few field studies have been directed to interactions among primate species living in the same location. Distributions of closely related macaque species are frequently not known; for example, the distribution of the North Indian rhesus and the South Indian bonnet macaque has never been investigated systematically. In some areas there are actually hybrid forms of monkeys where two varieties of a single, usually widespread species come together or overlap (Booth 1968), or where two species' distributions overlap, as between the anubis and hamadryas baboons.

More information about closely related living species and their present distributions and interactions will help in piecing together the evolution of behavior patterns as well as of morphological features. These data will clarify past taxonomic differentiations and shed light on the events in primate evolution that led to the diverse forms of the living primates.

part 1

FIELD STUDIES

Only the complex factors of a natural environment can elicit the full range of behavior patterns that are potential in most species, if not in all. Neither behaviors nor the causes of modification of behavior patterns are simple, and we are far from understanding the complexities of behavioral adaptation to any given environment. Primate behavior was not always understood as particularly variable or complex, no doubt because so few species had been studied and because comparative studies were lacking of the same kind of monkey living in different habitats. The early views of behavior influenced laboratory work as well as fieldwork. Since early guidelines for measuring "normal" behavior were not based on field studies, a laboratory monkey that was able to survive with its cagemates and to reproduce was viewed as a "normal" monkey, or at least normal enough for the purposes of research.

A great deal has been written contrasting field and laboratory research, including many tiresome recriminations as to the merits of one versus the other. Until quite recently very few individual scientists had had any experience working in both field and laboratory, with the result that there was a general lack of communication between workers in each area and an almost complete lack of appreciation of respective goals and methods. It is not a question of the laboratory versus the field; the significant issue is which problems are most appropriate to which method of investigation. Without a constant exchange of the information gained from each, understanding of primate behavior will not be as complete as it might be. It is absurd for the field researcher to maintain that the field is the only place any "normal" pattern of behavior develops, or that the processes of behavioral development can most profitably be investigated in the field. It is equally absurd for the laboratory researcher to maintain that the only way to investigate behavior is to control all save one or two variables, or that many kinds of social deprivation during maturation do not significantly alter an animal's behavior.

16

On the one hand, the field is the appropriate place to investigate such problems as the relationships of patterns of behavior to population pressures, to ecological or climatic changes, to the presence of predators, or to events in the history of a social group; it is also the place to investigate the social structure and dynamics of groups with more than a few members. On the other hand, specific questions about some aspect of ontogeny or the factors that affect perception require for their answers the controls possible only in a laboratory, or (and this may be the better alternative) in a captive colony or social group.

Some attempt should be made to assess the extent and kinds of behavior deprivation and alterations of behavior patterns that occur when animals are reared in captivity. Under these necessarily artificial and highly controlled conditions the animal is usually subjected to surroundings and handling that may have exceedingly long-lasting effects on its behavior and physical well-being. No documentation is available on the exact degree and kinds of behavioral changes that occur under differing captive conditions. It is not possible at this time to say precisely what percentage of behaviors or responses vary in clearly defined ways in captivity. Some species are affected much more than others; some simply do not survive or reproduce in captivity. Many laboratory traumas and stresses are obvious, as for example those associated with crowding, catching, transporting, and medication. The experienced fieldworker has little or no difficulty in identifying laboratory-reared animals, especially if they have been caged in pairs or singly during most of the formative months. Stereotyped motor patterns and other forms of neurotic behavior are well documented among laboratory animals and may appear, along with normal species-typical patterns, in the behavior repertoires of captive animals.

Many behavior patterns depend upon a rich social context for their development, and the processes of that development are often obscure. Even in the wild a young chimpanzee does not automatically learn how to termite or to build a nest to sleep in at night. If a young chimpanzee is orphaned before the time it has mastered these skills, it does not learn them even though there are many adult chimpanzees to observe (van Lawick-Goodall 1968b). In the field it is possible to see the results of accommodation and learning but not the processes by which they occur. It is next to impossible to sort out the meaningful variables among those that comprise an animal's experiences and to say with any certainty that some are more important than others or that certain of them must occur at specific ages of development.

Artificial colonies are limited in their usefulness for investigating some problems. Social groups in captivity, although they may be composed of approximately the same age-sex structure as a free-ranging group, seldom if ever replicate the generational and sibling ties that form part of social experience for a wild monkey or ape. Space is a severe problem in most captivity situations, including space for the group, a home range, as well as the more immediately important social space for each member of the group. Many and possibly most dominance fights are avoided in the wild by animals moving out of sight of each other, either by running away once a chase starts or by moving away cautiously in the early stages of tension build-up. Often in the wild, animals hostile to one another tend to avoid each other by staying in different parts of the troop, and nature is full of trees, shrubs, or tall grasses behind which an animal may move to escape being seen.

How much energy, anger, or tension directed by a captive animal into an attack on a cagemate that cannot escape in flight might be dissipated by running 100 yards? This is one of several important questions that deserve attention before there can be an assessment of the extent to which a laboratory colony can provide adequate surroundings for the expression and development of specific kinds of behavior. Species differ in their requirements, as for example with respect to space, and open-area, ground-living forms probably need much more space than does a group of arboreal monkeys that inhabits dense forest treetops. Designing colony space in ways that maximize psychological distances and provide hiding places or complex surfaces may go a long way toward increasing the square feet of space usable by the animals.

From the fieldworker's point of view it is a great advantage to observe a species first in captivity in order to familiarize oneself with at least some of the animals' behavior patterns before undertaking a field study. Some field studies have been preceded by brief observations of captive animals (Shirek-Ellefson 1967), and several species have been studied in captive social groups prior to any field investigation, for example *M. nemistrina* (Kaufman and Rosenblum 1966). Ideally, a fieldworker has experience observing before going to the field, including practice in recording behavioral events, in photographing, and in other related skills that are essential to a field study. The more advanced training a researcher has, the less initial difficulty he will encounter in trying to acquire these abilities in the field where conditions are often very difficult.

A behavior repertoire prepared before going to the field, based on observations of captive animals, should be done only with caution and a willingness to revise drastically, if necessary, after seeing the animals in the wild. Any predetermined list of items to check off creates problems for the researcher that are not always apparent at first. A catalog at one's elbow may serve to reduce the observer's sensitivity to variations on the items and to other possibly more important units of behavior. An observer may tend to record only what the categories call for; once the repertoire is defined, revisions become unlikely. Behavior is extremely variable, especially during the early months of life when an individual animal's patterns are developing. The forms and changes of gestures and vocalizations during ontogeny are least amenable to rigid categorization and easy checklist recording. Until a species is known well, nothing replaces a continuous running set of notes that attempts to record as much of what the observer notices as possible. Later, as the animals become known and the worker is thoroughly familiar with their behavior patterns, the items recorded may appropriately be dictated by the kinds of questions the observer seeks to answer. The more specific the inquiries, the more specific the items to be recorded.

A brief survey of the literature of the field in the last 40 years reveals the major trends in primate studies. In the late 1920s and early 1930s a three-months study was considered respectably long. Social relationships were pictured as basically dyadic interactions, and social organization was described simply and with respect to only a few social roles. Very few species were observed, but they included monkeys and apes from the Old and the New World. Starting in the 1930s C. R. Carpenter prepared a series of excellent monographs based on very short field studies, continuing work that had been started earlier by Yerkes and Yerkes (1929) and their colleagues. Carpenter stands out as responsible in large part for the theoretical orientations that continue to influence studies of the monkeys and apes.

During the years spanning the Second World War, field studies were set aside and it was not until the mid-50s that students once again started naturalistic observations. Two long-term studies were begun before the 1950s, but they were exceptions rather than the rule. In 1938 Carpenter had established a monkey colony on Cayo Santiago, a short distance off the coast of Puerto Rico, with rhesus monkeys imported from India, but work there was intermittent and there was little continuity in detailed behavioral observations. The scientific study of Japanese macaques was started in 1948 and the Japan Monkey Centre was formed in 1956. Studies have continued until the present on several troops of Japanese macaques.

When studies were only three months long not much time could be devoted to locating a study site, but this changed when longer periods of research became the norm. In the 1950s and early 1960s a fieldworker would go to great lengths in order to locate a suitable spot for a field study: an area in which conditions were as ideal as possible, that is, where the animals were as they had been for centuries past and observation conditions optimal. Since field studies were still limited in time and since few primates had been observed in the wild, each study attempted to concentrate on the most "typical" troop in as untouched an area as possible. In fact, expediency sometimes dictated the choice of observation area, locating studies in areas that were far from pristine or unaffected by man — such as much-visited game parks.

Troops, then as now, were carefully selected for observation only after a survey of as large an area as possible to determine the limits of group size and composition. We still wish to observe in as unaltered a situation as possible, but we now realize that some species are able to thrive in many different habitats, some of them much modified by human activity, and manage to survive in still others that appear much less than optimal. Because this variability in habitats makes it much less clear where to observe, one's choice of troop depends ultimately upon the problem orientation of the observer.

Some notions of primate behavior have been exceptionally simplistic, for example, typological descriptions of behavior. Fieldworkers returned to civilization with reams of notes that were digested eventually into reports purporting to describe the species that had been observed. Norms of behavior were neatly laid out and the reader was left with the notion that he had read the definitive behavior catalog of that animal. It was comfortable, indeed, to return from the field and check off, as finished, another species from the hundreds of living kinds of primates, but this state of affairs did not last long.

Field studies in the 1950s were undertaken with the assumption that it was necessary to record at least an annual cycle of activity, although not all studies did. Since some behaviors vary seasonally (for example, births may occur only or mostly during certain months), it was felt that even a minimum amount of data should include all possible variations in behavior during a 12-month period, and several studies did include at least part of a second annual cycle. Now, in contrast to our earlier satisfaction with a one-year study, we recognize the need for substantially longer periods of observations, using teams of observers, alternating to assure continuous recording of events over years. This has produced a wealth of data never before available. Questions can be asked and answered on genealogical relationships, on changes in behavior over time, and on social development that is not limited to the first year of life. With long-term studies the crucial time dimensions of past individual experience and group history can be added to the dynamics of a group's behavior.

As of 1969 only relatively few species have yet been studied, but a most important development in the last five years has been the recognition and description of variability in the behavior of groups within species that live in different habitats. Vervets at Chobi behave differently from those at Lolui (Gartlan and Brain 1968), and city rhesus are different from forest. Now the emphasis is on dimensions of variability, of adaptations to local social and ecological conditions, and effort is directed toward unraveling the relationships among the many factors of the environment and life in social groups. Notions concerning these relationships will remain tentative until there are more studies, but sufficient data are available so that it has become possible to begin to speculate (Crook 1970b).

Still and motion photography have become an indispensable part of the documentation of behavior in the field and allow detailed analysis of actions and interactions that the observer's eyes alone might miss altogether. Experimental work is being started in the field, although techniques (as for example, for brain surgery) have not been perfected for many of the tasks at hand. Telemetry is just beginning to be used in the field and will provide some of the most needed data concerning an individual animal's reactions and participation in social interactions. On a simpler level, increasing use is being made of the trapping and releasing of animals to study social dominance and other aspects of behavior.

The following chapters are based on fieldwork during the last 15 years. They report on only a very small number of the species studied, the methods used, and the results gained from field studies during that time. Two chapters concern apes and two concern Old World monkeys. All four primates have been studied at least once again since these investigations, and the Gombe chimpanzees have been continuously observed by van Lawick-Goodall and her staff from 1960 to the present.

The chimpanzee chapter represents a very small portion of the wealth of data on most phases of chimpanzee development and behavior in the Gombe[1] population. The photographic record alone, taken by Hugo van Lawick, comprises a vast store of information that has hardly been touched. Chimpanzees have been observed in other habitats than the Gombe. A team of Japanese scientists has worked south of Kigoma at Kabongo Point and the Kasakati Basin on the eastern shore of Lake Tanganyika south of the Gombe National Park (Izawa and Itani 1966; Azuma and Toyoshima 1962, 1965). Their research indicates chimpanzees in that area live in more stable social groups of larger size than is characteristic of Gombe chimpanzees. Recently Nishida (1968) has reported on observations of six large groups of chimpanzees in a study area in the northeastern part of the Mahali Mountains in western Tanzania. He states

> The chimpanzees live in a clear-cut social unit which consists of adult males, adult females, and immature animals. The permanency, stable membership, and integrative nature of the unit-group were confirmed during the course of this study. The size of the unit-groups ranges from 30 to 80.
>
> The unit-group generally splits up into temporary subgroups that repeat joining and parting. The size of the subgroups of the baited population ranges from one to 28 head, the mean being 8.1 head . . . subgroups are usually composed by social bonds on the basis of similar age, sex, blood relationship, and/ or sexual attraction (1968:167).

Since Schaller's study of the mountain gorilla another project has been started on gorilla behavior that is producing a great deal of new information (L. S. B. Leakey, personal communication). The new study confirms Schaller's descriptions of the lethargy and generally relaxed nature of gorillas. Their behavior continues to form a sharp contrast to that of chimpanzees whose volatile and demonstrative natures result in much greater

[1] It is interesting to note that since 1968, the original publication date of the chimpanzee article, the status of the Gombe has changed from a Reserve to a National Park, in large part owing to the efforts and continuing research of the van Lawicks. All too few natural populations of primates are being protected against human incursions through farming, hunting, and careless planning.

social activity and expressiveness. It is possible that a comparison of the results from the gorilla study now under way with the chimpanzee research in Tanzania will clarify some of the effects of habitat and diet on a species' behavior.

Today the Hall and DeVore chapter on baboons represents only a brief stage in the history of baboon studies. Since the 1950s, when the fieldwork for this chapter was undertaken, there have been several major studies of baboons living in different habitats. It is clear that additional studies in savanna habitat are needed, specifically in areas where there is a full complement of natural predators and minimum interference by man. Neither Hall nor DeVore was able to spend a continuous annual cycle of observation with the same troop. In both study areas in South and East Africa the habitat of the animals was definitely affected by man. Perhaps as a result of this interference and the short observation periods the baboon troop emerges from these studies as a "closed society" with almost no interchange of members among different troops.

Studies by Rowell (1966a, 1966b, 1967a, 1967b) and Ransom (personal communication) indicate that the organization of social groups of forest-living baboons may be substantially different from that of the savanna baboons observed by Hall and DeVore. Rowell describes the forest baboons of Ishasha in Queen Elizabeth Park, Uganda, as being without a rigid dominance hierarchy, and she documented frequent exchange of males among troops. Ransom also reports a relatively less rigidly structured troop and the exchange of troop members in the forest troops of the Gombe, Tanzania. Subtle gestures of appeasement form an important part of the repertoire of the Gombe baboons, unlike any gestures reported by either Hall or DeVore for the savanna baboons they observed. Studies are needed of baboons in areas with minimal human intervention and maximum natural predator pressure. Only then will it be possible to determine the effect of predation on the baboon social group and to assess the extent to which it is a major factor in baboon social life as emphasized by both Hall and DeVore.

In the years since the research for the langur monkey chapter was undertaken there have been several intensive studies of langurs in other locations of India and Ceylon—with remarkably different results. It may be that there is greater diversity in the behavior of this genus, *Presbytis*, than in any other genus studied to date (Yoshiba 1968). Clearly, additional studies are needed if we are to sort out the factors influencing the diversity of behavior among langurs.

In addition to those reported on here, many other species have been observed in the wild, including both New and Old World monkeys and the other two apes. Some species have been observed in a variety of habitats. The common African vervet monkey *(Cercopithecus aethiops)* has been studied by Gartlan and Brain (1968), and on the basis of their work they have contrasted the social behavior of vervets in two areas of East Africa. Struhsaker (1967a, 1967b) has observed vervets in Amboseli Park in East Africa, and more recently Lancaster (personal communication) has observed vervets in Zambia. Each of these studies offers still further evidence of behavioral variability in different habitats.

Research scientists from the Japan Monkey Centre and from Kyoto University have observed several Old World species including the Japanese macaques (Imanishi and Altmann 1965; Imanishi 1957; Itani 1959; Tsumori 1967), Southeast Asian langurs (Furuya 1961), bonnet macaques in South India, and Indian langurs (Sugiyama 1964, 1965a, 1965b; Yoshiba 1967, 1968), as well as the chimpanzees mentioned earlier. Not all these studies have been as long in duration as those on the native Japanese macaques, but the tendency has been for teams to continue observations for as many years as possible. Provisioning, or artificially feeding the animals, has been an important part of this fieldwork, although it is reported that not all species accept food readily, or at all.

Rhesus macaques have been observed intermittently since 1938 on Cayo Santiago (Altmann 1962; Koford 1963a, 1963b, 1965, 1966; Sade 1964, 1965; Wilson 1968) and in northern India by Lindburg (1967) Neville (1966), Southwick, Beg, and Siddiqi (1961a, 1961b), and others.

An excellent study of the Hamadryas baboons was undertaken by Kummer (1967, 1968a, 1968b) and Kurt (Kummer and Kurt 1963, 1965). Geladas have been observed by Crook (1966). Both these African monkeys live in exceedingly arid country where there is a sparse food supply and few sleeping trees, and their behavior provides many contrasts with that of baboons observed by Hall, DeVore, Ransom, Rowell, and others. Instead of having a social system of troops with many members of both sexes and all ages, gelada and Hamadryas group in one-male units with several females and their offspring. Contact among many one-male units occurs at night when groups as large as several hundred animals gather at sleeping cliffs.

The number of studies of New World monkeys is increasing, although they present formidable problems in observation. All New World monkeys are arboreal and most live in extremely dense forests, where it is very difficult to see the animals or to follow them for long. Mason (1968) has reported on the behavior of the Callicebus monkey, and more recently Klein has returned from a study of the spider monkey in Colombia. Other studies have been completed and additional ones are under way so that in the future a good deal more information on New World primates will be available.

Prosimians have received their share of attention in field studies and outstanding among them are those by Jolly (1966) and Petter (1962a, 1962b, 1965) on Madagascar lemurs.

In the studies reported here and in others not included, it is often difficult to tell precisely what field techniques and observation methods were used or what assumptions concerning behavior were made. Techniques and methods may have varied, even with a single observer, depending on the assistance available and on local conditions. Many studies have not included systematic sampling of behaviors, or of the behavior of specific individuals or classes of animals. An observer has limitations as to how much he can see and record accurately. If the troop is large the observer may not be able to go back through notes and pull out data on frequencies of a given behavior or for a specific animal with any certainty that the record is at all complete, unless the observations were made specifically to include these data. Systematic attention directed to representative animals and with respect to specific activities will help assure that data are gathered to answer as many questions as possible. However, followed too rigidly or exclusively, this method may also direct the attention of the observer so narrowly that he will miss the unusual event or important interaction if animals not on the schedule for that day or hour are involved. There are gains and losses inherent in each system. The important factor is to select that method or combination of methods most appropriate to the goals of the study and to the situation.

In summary, the following four chapters represent only a small portion of primate studies to date, but they do sample four very differently adapted kinds of nonhuman primates. Each is suggestive of further research problems in the field and in the laboratory.

chapter two

A PRELIMINARY REPORT ON EXPRESSIVE MOVEMENTS AND COMMUNICATION IN THE GOMBE STREAM CHIMPANZEES[1]

Jane van Lawick-Goodall

INTRODUCTION

Primates have a wide range of expressive movements and calls, many of which have a communicative function; that is, they appear to have some effect on other individuals that see or hear them and thus they may help to maintain interindividual relationships within the social structure as a whole. Each of these movements and calls is likely to be affected, both in structure and in probable function as a signal, by a number of variables: the environmental and behavioral situation at the time, and the motivational state and individuality of the sender and of other members of the group. Moreover, primate signals invariably occur in clusters:

> In most situations it is not a single signal that passes from one animal to another but a whole complex of them, visual, auditory, tactile, and sometimes olfactory. There can be little doubt that the structure of individual signals is very much affected by this incorporation in a whole matrix of other signals (Marler 1965:583).

The data to be presented here were obtained during 45 months of observation (between 1960 and 1965) of free-ranging chimpanzees at the Gombe Stream Chimpanzee Reserve, Tanzania, East Africa. This reserve consists of about 30 square miles of rugged mountainous country, stretching

[1] The work was carried out in Tanzania, East Africa, and I am grateful to the Tanzania government officials in Kigoma and to the Tanzania Game Department for their cooperation. The research was financed initially by the Wilkie Foundation and subsequently by the National Geographic Society, and I express my gratitude to both organizations.

I should also like to thank Dr. L. S. B. Leakey, Honorary Director of the National Museum Center for Prehistory and Palaeontology, who initiated the expedition; Baron Hugo van Lawick for his valuable photographic and film record and for his assistance in all aspects of the research; my assistants, Miss Edna Koning and Miss Ivey; and the African research. Finally, I am indebted to Professor R. A. Hinde, Sub-department of Animal Behaviour, Cambridge University, for his helpful criticisms during the early stages of this manuscript.

along the eastern shores of Lake Tanganyika and supporting a small population (probably between 100 and 150) of *Pan troglodytes schweinfurthi,* the Eastern or long-haired chimpanzee.

The animals may be divided into the following main age classes on the basis of behavioral characteristics: infant, juvenile, adolescent, and adult.

The infant (0 to 3–3½ years) suckles, is transported by the mother (either continuously or occasionally), and sleeps in the same nest with the mother at night. During its first six months the infant is almost completely dependent on the mother for food, transport, and protection. At about six months it makes its first movements away from her, but remains extremely dependent for the following six months. Locomotor patterns develop rapidly, and the infant commences to eat some solid foods during the second half of its first year. Social interactions with other individuals are frequent; the infant may be patted or groomed by adults, and played with, groomed, and carried by older infants, juveniles, and adolescents. An infant frequently approaches individuals joining its group to "greet" them. The white tail tuft is fully developed and at maximum length by the end of the first year. In the second year the infant begins to walk beside the mother for short distances, and the proportion of solids eaten increases. Other individuals continue to show tolerance. During this year most if not all of the gestures used by adult chimpanzees during interindividual interactions appear. During its third year, the infant rides less and less frequently on the mother's back. By the end of the year, its diet differs little from that of the adult. Older individuals are still generally tolerant, but the infant receives a number of gentle rebuffs and behaves with increasing caution. It is still protected by the mother.

The juvenile (3–3½ to 6–7 years) is no longer dependent on the mother for food or transport and makes its own nest at night. The mother still protects her offspring on occasions, and the juvenile continues to move around with her for most of the time. Rebuffs from older individuals become increasingly severe. The white tail tuft gradually becomes less conspicuous.

The adolescent period is from 6–7 to 11–13 years. In the male chimpanzee the period of adolescence is fairly well defined and can be said to commence with puberty (about 7 years) and to continue until the individual is socially mature and becomes integrated into the adult male hierarchy. In the female adolescence is not easy to define, but probably commences between 6 and 7 years. During the first year or so of adolescence the female may show a very slight swelling of the anogenital region, but this in no way approximates a normal sexual swelling and does not arouse attention from the males. At a later period of adolescence the female shows the normal external features of the estrous cycle.

These chimpanzees are nomadic in that they normally sleep in a different place each night and, although for the most part they keep within the same general area, they follow no regular circuit in their daily search for food. The distance and direction of this daily wandering varies with the availability of food; thus when a tasty fruit is ripe in one part of the reserve only, some chimpanzees may move far beyond their normal home range in order to feed there (see also Goodall 1962, 1963, 1965).

The entire chimpanzee population of the reserve can best be described as one loosely organized group or community within which all or most individuals are familiar with each other and may from time to time move about together. With the exception of infants and young juveniles, which normally move around with their mothers, each individual within this community is an independent unit since it may move about on its own from time to time. In fact, mature and adolescent males are frequently encountered alone, and mature and adolescent females are also, but less frequently. Individual chimpanzees often join to form temporary associations that may remain stable for a few hours or a few days. These may consist of any combination of age-sex classes and may number from 2 to more than 30 (the average being 6). Membership within such temporary associations is continually changing as individuals or numbers of individuals move off, either to travel about on their own or to join with other chimpanzees.

There are certain individuals that may associate with each other on many occasions (thus an adolescent moves around more often with its mother than with any other single individual), but the only association that is stable over a number of years is that of a mother and her younger offspring (van Lawick-Goodall 1968b).

The temporary nature of chimpanzee associations results in an *apparently* loose social structure (Goodall 1965; Reynolds and Reynolds 1965). However, when regular observations became possible on the interactions between the various individuals it gradually became evident that the social status of each chimpanzee was fairly well defined in relation to each other individual. In other words, it was often possible to predict, when for example two chimpanzees met on a narrow branch, which animal would gain right of way; that animal could then be described as the dominant one of the two.

At the beginning of the study it was decided that for close-range observation it would be necessary to habituate the chimpanzees to the presence of an observer. This was a lengthy process (Goodall 1965), but finally most of the apes continued their normal activities even when the observer was within 30 to 40 feet of them.

In 1962 a mature male, David Greybeard (all chimpanzees known individually were named) visited my camp to feed on the fruit of a palm tree; he returned each day until the fruit was exhausted. During one such visit he took some bananas from my tent, and after this my African staff left these out for him. When fruit on another palm in camp ripened he returned, again took bananas, and subsequently visited camp from time to time for the bananas alone; on such occasions he was sometimes accompanied by other chimpanzees. Gradually, more and more individuals followed him to camp,

TABLE 2-1   Individuals observed visiting the feeding area, April 1964 – March 1965

						Infants			
Adults	No. of Observations	Adolescents	No. of Observations	Juveniles	No. of Observations	0–1	No. of Observations	1–3	No. of Observations
Males									
David Graybeard	383	Faben	641			Flint	666	Merlin	272
Mr. McGregor	377	Figan	631			Goblin	201	Sniff	20
Mr. Worzle	365	Evered	431						
Mike	323	Pepe	340						
J. B.	292	McD	221						
Leakey	287	Charlie	206						
Humphrey	283								
Hugo	277								
Goliath	244								
Huxley	205								
Hugh	121								
Totals (11)	3157	(6)	2470			(2)	867	(2)	292
Females									
Flo	666	Pooch	319	Fifi	673	Cindy	49	Gilka	309
Melissa	390	Gigi	91	Miff	283	Jane	37	Little Bee	20
Olly	309	Sally	21						
Marina	272								
Mandy	218								
Circe	123								
Madam Bee	20								
Sophie	20								
Totals (8)	2018	(3)	431	(2)	956	(2)	86	(2)	329

and in 1963 I decided to set up an artificial feeding area on a permanent basis.[2] It then became possible to make fairly regular observations on the different chimpanzees, since every time they were in the area they detoured to have a meal of bananas.

The number of chimpanzees visiting the feeding area gradually increased. Table 2–1 lists the 38 individuals that came most frequently between April 1964 and March 1965, together with the number of "observation periods" when it was possible to observe each one for longer than 10 minutes at a time.[3] Although the total number of these observation periods for each individual is not an exact measure of the amount of time the animal was observed, the totals for various age-sex classes provide an approximate index to the relative amounts of time during one year that I was able to observe males as compared with females, and so forth. These measures were used as a means of determining the relative frequency in each age-sex class of some of the gestures and postures that form part of the communicatory system of chimpanzees.

CALLS, FACIAL EXPRESSIONS, AND AUTONOMIC BEHAVIOR

Here I shall briefly discuss three aspects of behavior that will be referred to throughout the chapter. I have not yet commenced a detailed analysis of calls and facial expressions, but the more obvious of these are listed in Table 2–2, together with the behavioral contexts in which they were most commonly recorded and the observed responses of other individuals. Each specific call or facial expression referred to in the text is described in this table.

Calls

The chimpanzee has a wide range of readily distinguishable calls. These express his "emotions" at the time, and although, as Andrew (1963b) says, they are not given "in order, for example, to warn fellows of impending danger, in a way a man might cry 'Look out'," nevertheless they often do communicate the mood of the individual to his fellows.

---

[2] After a good deal of experimenting, the feeding area now comprises some 30 cement boxes half sunk into the ground, each containing from 10 to 15 bananas; these boxes can be opened by handles situated not less than 10 yards away. Thus when a large number of chimpanzees arrive at the same time each one can have a box opened for him. The boxes are widely spread out, and the area has been designed to stimulate as little abnormal aggressive behavior as possible.

[3] Individuals sometimes remained at or near the feeding area for several hours; the average mean length for observation periods during one year was calculated for three females and was between 45 and 55 minutes for each.

TABLE 2–2 Some common calls and facial expressions

Call or Sound	Expression of Face
A. Between relaxed individuals.	
1. *Soft grunt.*	Typical relaxed or alert face.
2. Very *soft groan,* "hm-hmmm."	Typical relaxed or alert face.
B. In connection with feeding.	
1. Loud barking sounds, with a wide range of variation in tone and pitch.	Mouth slightly opened at each sound; lips may be slightly retracted to show teeth.
2. Short, high-pitched, single-syllable *shrieks.*	Mouth wide open; lips retracted from teeth; corners of the mouth drawn back.
3. Soft or loud *grunts.*	Various jaw and lip positions during eating.
C. During nonaggressive physical contact with other individuals.	
1. "*Panting*," soft.	Lips may be very slightly parted.
2. "*Copulatory pants*," louder and hoarser than C.1.	Lips may be very slightly parted.
3. "*Laughing*," soft panting sounds which may become jerky panting grunts.	"*Play face.*" At a low intensity the lower lip is retracted to show the lower teeth; at higher intensities the mouth is opened and lip retraction increases until all forward teeth may be exposed. Laughing does not always accompany "play face."
D. During various types of social activity (not obviously aggressive).	
1. Quiet "huu."	Lips pushed slightly forward.
2. Series of "*panting hoots and calls.*" There are many variations; the calls may rise or fall in pitch, and are sometimes long and drawn out toward the end.	"*Hoot face.*" Lips pushed forward into trumpet, but corners of mouth may be retracted during inhalation. At end of series mouth may open wider, lips covering teeth.
3. "*Panting hoots* with *shrieks or roars.*" Same as above, but always rise in pitch and volume.	Same as above, but during shrieks and roars mouth is wide open and, in some individuals, teeth showing.
E. During aggressive behavior.	
1. (*No sound.*)	"*Glare.*" Lips compressed, animal stares fixedly at another individual.
2. "*Soft bark*," single syllable breathy exhalation, like a cough.	Mouth half open; lips pushed slightly forward but covering teeth.
3. "*Waa bark*," loud single syllable bark.	Similar to E.2 but mouth wider open.

Situation in Which Normally Elicited	*Observed Response of Other Individuals*
1. Relaxed situation in group during resting, grooming, traveling, and so forth.	May repeat similar sound.
2. Same as above.	Same as above.
1. Eating or approaching desirable food.	May look toward or approach calling animal.
2. Arriving close to desirable food, when commencing to feed.	May look toward or approach calling animal.
3. Feeding.	May look toward or approach calling animal.
1. Social grooming. Also when approaching another animal prior to and during "kissing," "bowing," and so forth.	
2. Copulation.	
3. Social play. The extreme form with wide open mouth was only observed during violent tickling or wrestling.	When one individual approaches with a "play face" another may respond by playing.
1. As another approaches; response to group calling in the distance.	
2. As another approaches; response to distant calls; when arriving in group; during meat eating; when lying in nest at night.	Frequently may join in and also hoot. May reach out to touch calling individual.
3. During charging displays, drumming, and so forth; sometimes when males cross ridge between valleys.	May join in; if subordinate to the calling animal, may climb a tree, or "hide."
1. Sometimes prior to chase or attack. Prior to copulation.	Approaches and presents for copulation in this context.
2. When threatening subordinate or other species of which it is not afraid.	Subordinate normally shows submissive behavior.
3. Similar to E.2. Also when threatening superior from a distance; when another chimpanzee is being attacked.	Looks in direction of disturbance; may make similar call.

TABLE 2–2 – Continued

Call or Sound	*Expression of Face*
4. "*Wraaah*" call, long and drawn out, clear "savage sounding" call.	Similar to E.3.
5. High pitched *scream calls;* may be short or long and drawn out, and may lead to glottal cramps.	Mouth half or wide open; lips retracted from teeth and gums; corners of mouth drawn back. May occur without sound.
F. During "anxiety" or "frustration" situations.	
1. "*Hoo call*," soft, low pitched single syllable sound; less breathy than D. 1.	Lips pushed quickly forward, animal stares at worrying stimulus.
2. "*Hoo whimper*," similar but lower pitched sound.	"*Pout face*," lips pursed and pushed right forward; eyes wide open and staring at mother or other individual.
3. "*Whimpering*," series of "hoo" sounds rather like whimpering of puppy at times.	Initially similar to above but then the corners of the lips are drawn back and lower lip retracted from teeth; "*whimper face.*"
4. "*Crying*," loud hoarse yells and screams; may lead to glottal cramps.	Mouth wide open; teeth show.
G. During submissive behavior and when individual is frightened.	
1. "*Bobbing pants*," loud hoarse panting sounds.	Mouth half open or wide open; lips normally pulled tightly over teeth. Lips may be pushed forward slightly at end of call.
2. "*Panting shrieks*," loud screamlike sounds.	Mouth open; lips pushed slightly forward.
3. "*Squeak calls*," short high-pitched, squeaky screams.	"*Grinning.*" Initially the lips are parted, the corners drawn back, and an oblong expanse of closed teeth shown. In more extreme form the mouth is opened and the lips retracted fully from teeth and gums.
4. (*No sound.*)	"*Silent grin.*" Same as above without calling.
5. "*Screaming*," loud screaming with many variations; some screams have a rasping quality; some are long and drawn out. May lead to glottal cramps.	Mouth wide open; lips fully retracted from teeth and gums.
6. "*Infantile scream.*"	Uncertain, but mouth open.

Situation in Which Normally Elicited	*Observed Response of Other Individuals*
4. When disturbed by presence of human; sometimes when two groups meet. Often followed by shaking branches.	May hurry towards the disturbance; may give same call or "waa bark."
5. When threatening superior or animal of another species of which chimpanzee is afraid; may look round to another chimpanzee for "support."	May hurry towards the calling individual and threaten or charge at the caller.
1. When chimpanzee suddenly hears or sees strange object or sound.	Looks in direction of the calling individual and may then peer around.
2. Infant searching for nipple, slipping, trying to reach mother, and so forth. Older ape when begging; when ignored after it has presented for grooming, and so forth.	Mother "cradles" or retrieves infant. Older chimpanzee may be given piece of food, or groomed.
3. When there is no response to the "hoo" whimper; when juveniles lose their mothers; after threat or attack.	May then respond to the begging behavior and so forth; or make reassurance gesture.
4. When juveniles are lost; during "temper tantrums" in juveniles; during "frustration" in older individuals.	Mother hurries towards sound; may reassure child in tantrum.
1. When a subordinate male "bobs" to a mature male.	
2. When chimpanzee approaches to show submissive behavior or to greet a superior.	
3. After threat or attack, while the subordinate is making submissive gestures. Female may make these sounds during copulation.	The dominant individual may touch or embrace the subordinate.
4. Similar to above.	Same as above.
5. When a chimpanzee is fleeing or being attacked; after the attack while making submissive gestures; in many types of aggressive interaction.	Same as above. The mother of a screaming individual, or a "friend" may approach to threaten or attack the aggressor.
6. When infant (during first few weeks of its life) gets a sudden fright, falls from the mother, and so forth.	Mother hastily cradles and embraces. May look down with "pout face" or "grin."

Rowell and Hinde (1962) stressed the fact that in the rhesus monkey there is the possibility of an almost infinite range of intermediates between the main sounds. Rowell (1962) showed that nine major calls given by the rhesus in agonistic situations actually constitute one system linked by a continuous series of intermediates. After analyzing some chimpanzee calls recorded by Reynolds in the field, Marler (1965) has suggested that in this species too there is a similar gradation of sound between some of the calls. I have not yet analyzed tape recordings of chimpanzee calls and therefore, as a temporary measure to clarify the present discussion, I have merely listed the more readily distinguishable sounds of the chimpanzee in Table 2–2.

It should be mentioned that not only is it possible to distinguish between males and females, and adults and youngsters on the basis of their "voices," but it is also possible to recognize each individual from a distance when he utters "panting hoots." Although there is undoubtedly some degree of individuality in most chimpanzee vocalizations, it may be significant that this is particularly obvious in calls that are mainly connected with long-range communication. Such individual differentiation may help in reuniting two animals after a separation.

It is relevant to this discussion that in one species of shrike (genus *Laniarius*), which inhabits thick forest, the two birds of a pair tend to develop a particular and individualistic repertoire of duet patterns, which may well aid their mutual recognition and help them to maintain contact in conditions where visual display is ineffective (Thorpe 1963).

Thus, although a certain amount of individuality is undoubtedly present in the calls and songs of most types of mammals and birds, it is possible that this characteristic may have particular adaptive significance for an animal that is not in constant, or nearly constant, visual contact with its partner or other members of the group.

Facial Expressions

The expression of a chimpanzee's face alters continually with relation to each new behavioral context, and, according to van Hooff (1962), is one of the most important parts of the body concerned with visual communication. This is certainly true with regard to the subtler aspects; for instance, when a subordinate is begging from (or merely sitting beside) a superior, the subordinate is constantly watching the face of the other, presumably for slight changes of expression that may reveal a change in "mood."

The normal attentive facial expression of the wild chimpanzee is shown by the mother in Fig. 2–1. Other expressions range from a slight pouting of the lips (when, for instance, a young animal is denied food) to a wide open mouth with lips drawn back from teeth and gums (as when an individual

Fig. 2–1. Three-month-old infant with "pout face." *(Photograph by Baron Hugo van Lawick; © National Geographic Society)*

screams at the top of its voice after an attack). Other expressions are listed in Table 2–2. There are, of course, expressions intermediate between all the main types. Often these intermediate forms can be observed in a complete sequence – when, for instance, a lost juvenile initially pouts its lips, then commences to whimper (Fig. 2–2), and finally "cries" loudly (Fig. 2–3).

Each chimpanzee can easily be recognized individually by its face structure. In addition, some animals showed expressions that were never

Fig. 2-2. Juvenile with "whimper face" after losing her mother. *(Photograph by Baron Hugo van Lawick; © National Geographic Society)*

seen in others. Thus one male frequently pressed his lips together, drew back the corners of his mouth, and "mock smiled." This was in no way related to the "play face," which has been referred to in the literature as "smiling" (Yerkes and Yerkes 1929), but occurred in a variety of contexts, none of which was associated with play or good temper. Some animals typically droop their lower lips when they are relaxed; others never do so. One mature male, when "determined" to get more bananas, progressively pushed out his lower lip farther and farther as he inspected box after box.

Autonomic Behavior

Chimpanzees show pronounced hair erection in a number of situations. These include arriving at a food source, meeting another group, suddenly noticing chimpanzees in the distance, hearing a strange sound, finding themselves in an unfamiliar situation, threatening or attacking an opponent, or during courtship prior to copulation. Sometimes only the hair of the back and arms may be raised; at other times every hair on the body stands on end.

In some of these situations (feeding, greeting, and courtship) male

Fig. 2–3.
The same juvenile shows a low intensity "cry face" as her distress increases. (*Photograph by Baron Hugo van Lawick;* © *National Geographic Society*)

chimpanzees may also show penile erections. In the field I once observed a very slight swelling in the sex skin of an anoestrus female in response to prolonged inspection by a male.

Mature males, particularly during social excitement, sometimes emitted a strong smell, slightly similar to very pungent human sweat. Although I was often close to individuals when they were sweating in the sun, the odor was not detectable on such occasions. Schaller (1963) detected a similar smell, which he thought emanated from silver-back males; he also concluded that normal sweating alone could not account for the odor.

It should also be mentioned that when chimpanzees are frightened, as during a severe attack or prolonged chase, they may defecate repeatedly, and the excrement is usually diarrheic.

FLIGHT AND AVOIDANCE

When a chimpanzee is frightened, either by a superior or by some other animal or object in its environment, it shows a variety of flight or avoidance patterns in addition to the submissive behavior patterns. The form used depends on the situation and the individual or individuals concerned.

Flight

Flight is defined as a rapid progression away from an alarming stimulus. When the latter is another chimpanzee, flight is characteristically accompanied by loud screaming with the mouth open and the lips fully retracted from the teeth and gums. A chimpanzee fleeing a social superior may either rush along the ground and take refuge in a tree or, if the incident commenced in a tree, it may hurl itself through the branches and then to the ground. This is usually sufficient to shake off the pursuer, although sometimes the latter may persist.

When a chimpanzee flees from some new or alarming object in the environment he does not call out, although there may be some noise as he rushes through undergrowth or dead leaves. When the chimpanzees suddenly came across me (before they were habituated to my presence) they sometimes paused to stare before running off out of sight; at other times they rushed off for a few yards and then stopped to peer at me before vanishing silently into the forest. When I surprised them in a tree, they invariably climbed rapidly and silently to the ground and then ran off. Reynolds and Reynolds (1965) report "panic diarrhea" under such circumstances.

Startled Reaction

When a chimpanzee is alarmed by a sudden noise or movement nearby —from a low-flying bird, large insect, snake, or the like—the immediate response is to duck its head and fling one or both arms across its face or to throw both hands up in the air. A reaction similar to the first is observed in the gorilla (Schaller 1963) and in man, and one similar to the second is observed in the baboon (Hall 1962). After a second the chimpanzee peers around, and if the stimulus is not dangerous he relaxes. Occasionally these reflex actions were followed up by a hitting-away movement with the back of the hand toward the object. This movement, probably defensive in origin, has been incorporated into the repertoire of threat gestures.

"Hiding"

Before the chimpanzees were completely tolerant of my presence they sometimes "hid" when I approached to observe them. Thus an individual who had been feeding in full view might either move behind a tree trunk (where he then continued to feed out of my sight, peering round at me every so often), or reach for a large branch, and pull it down between us. Both Christy and Schaller (Reynolds and Reynolds 1965) reported seeing chimpanzees climb into old nests when they saw humans; they then remained there, occasionally peering over the edge. Reynolds and Reynolds (1965) report that one female constructed a nest for hiding. The orangutan may also hide from humans in a nest (Harrisson 1962).

On some occasions I observed chimpanzees hiding in similar ways from individuals of their own species. One dominant male (Goliath), who had been sitting peering uneasily in all directions suddenly hurried silently away, hair erected, when two other mature males appeared. (These three males were well acquainted.) The two ran after him, and then moved about for a few minutes in the bushes where he had disappeared. After this they commenced to feed. Some minutes later I saw Goliath sitting behind a tree trunk a little way up the opposite slope. He remained there for ten minutes, occasionally peering around and looking at the others below. When one of them moved suddenly or looked up, he at once ducked behind his tree trunk.

On many occasions individuals were seen to move behind tree trunks or other large objects when violent social activities were taking place around them.

Creeping

Occasionally an individual that had been attacked or threatened tried to "creep" from the vicinity of its aggressor. One adolescent male tried to creep from a tree in this manner, proceeding with extreme caution and moving each limb slowly while peering in the direction of his aggressor. He kept his body close to the branch and "grinned" or whimpered every time the aggressor looked toward him. When the dominant male made a sudden movement (which was not made with reference to him) he stopped moving altogether. On another occasion a young female spent 15 minutes trying to creep away from the vicinity of an older female who had been presenting to her for grooming.

"FRUSTRATION" (OR BEHAVIOR IN UNCERTAINTY)

Frustration is used here as a convenient heading under which to group behavior patterns that occur in conflict situations, when the animal is motivated by conflicting tendencies (Hinde and Tinbergen 1958): when fear of another individual prevents the chimpanzee from obtaining a desired objective; when the capabilities of the individual itself are insufficient to achieve a desired goal; or when another individual does not make the appropriate response to a communicatory signal.

An example of the first type of situation is when a dominant individual takes food in the possession of a subordinate—the latter may then be motivated by a strong aggressive tendency that is counteracted by its fear of the other. An example of the second type is when a chimpanzee is unable to get at bananas that it can see or smell inside a box. An example of the third type is when an infant, unable to make a large jump from one branch to another, looks toward its mother with a pout face and "hoo whimper" but is then ignored.

The various behavior patterns observed in the situations described consist in large part of actions that appear irrelevant to the situation and out of context (displacement activities), and actions that are directed away from the eliciting stimulus and that may involve redirection to other objects or individuals.

Scratching

The chimpanzee makes deliberate scratching movements, typically on the arm from elbow to wrist, from below the armpit to the elbow, or from thigh to belly over the groin. The more intense the anxiety or conflict situation, the more vigorous the scratching becomes. It typically occurred when the chimpanzees were worried or frightened by my presence or that of a high-ranking chimpanzee. Scratching, as the response to the presence of a human, has also been recorded in gorillas, baboons, and patas monkeys (Schaller 1963; Hall and DeVore, this volume; Hall and Gartlan 1965). It may be compared with head scratching in man under similar circumstances (Schaller 1963).

Yawning

The chimpanzee yawns repeatedly, looking either toward or away from the disturbance. This typically occurred as a result of my presence in individuals not fully habituated, and also in youngsters that were unable to take bananas in the presence of their superiors. Yawning in uneasy or aggressive animals occurs in gorillas, gibbons, baboons, rhesus monkeys, patas monkeys, and, rarely in vervet monkeys (Schaller 1963; Carpenter 1940; Hall 1962; Hinde and Rowell 1962; Hall and Gartlan 1965). Yawning in man sometimes occurs in mild conflict situations (Schaller 1963).

Grooming

The chimpanzee makes rapid and normally ineffective grooming movements either on itself or on a companion. This behavior is often seen when a chimpanzee is waiting for a desirable food (such as meat or bananas). Similar behavior has been observed in chacma and olive baboons and in hamadryas baboons (Hall and DeVore, this volume; Kummer, personal communication).

Masturbation

Males of all ages sometimes held and toyed with their erect penes in situations when they were apparently frustrated, and one female juvenile frequently made thrusting movements of the pelvis while rubbing her clitoris against a branch or the ground, usually when she was frustrated. A juvenile male occasionally made the same pelvic thrusting movements while standing quadrupedally and looking back between his arms and legs as his scrotum bounced against his penis; this always occurred when he was frustrated.

Rocking

The chimpanzee rocks very slightly, either from side to side or backward and forward (a matter of individual preference). At times the movements are barely perceptible and the individual may only move its head. The violence of rocking may continue to increase gradually, even though there is no corresponding change in the external stimulus eliciting the behavior. The most vigorous form of rocking occurs when the individual stands bipedally and sways from foot to foot, grasping a branch in one or both hands and waving this also from side to side. Rocking has been seen in a number of contexts, the most common being when individuals were frustrated in feeding situations or when they were uneasy in my presence. This rocking movement occurs commonly in captive chimpanzees, normally in situations when they are upset (Berkson, Mason, and Saxon 1963).

Shaking and Swaying of Branches

A branch may be swayed as a chimpanzee rocks its own body. Sometimes a whole sapling is swayed from side to side, with the chimpanzee either in the sapling or standing on the ground. At other times the branch-shaking is not accompanied by rocking, but the chimpanzee, while glaring at the eliciting stimulus, seizes hold of a nearby branch and shakes it violently. This behavior typically occurred as a reaction to my presence when the chimpanzees were uneasy. Shaking of branches has been observed in a number of other primates such as gorillas, red spider monkeys, baboons, and rhesus monkeys (Schaller 1963; Carpenter 1935; Hall 1962; Hall and DeVore 1965; Hinde and Rowell 1962).

Charging, Slapping, Stamping, Dragging, Throwing, and Drumming

In frustration situations slapping, stamping, dragging, throwing, and drumming[4] may accompany the chimpanzee charge away from the eliciting stimulus. As he charges he may slap on the ground with his hands, stamp with his feet, seize and drag branches after him along the ground, and hurl rocks or branches. Finally, he may leap up at a tree and drum on the trunk or buttresses with his feet—usually the two feet pound down one after the other in quick succession making a double beat; there is then a slight pause before the next double beat. From one to three double beats are normal.[5] Two males, out of the eleven that visited the feeding area, beat their chests during such displays in a manner similar to that described for gorillas (Schaller 1963).

The context in which these actions were usually observed was when mature males were unable to open banana boxes or when adolescent males were unable to obtain food because mature males were close by. When one adolescent male was unable to get bananas he frequently whimpered, hurried some 100 yards away from the group, broke into a run, drummed on a tree, swayed branches, and stamped and slapped on the ground—meanwhile uttering "panting hoots" and screams; he then returned, apparently relaxed, to his group. An individual often seemed to be more relaxed after performing one or more of these display patterns.

Temper Tantrums

Temper tantrums are a characteristic performance of the infant and young juvenile chimpanzee. The animal, screaming loudly, either leaps into the air with its arms above its head or hurls itself to the ground, writhing about and often hitting itself against surrounding objects. The first temper tantrum observed in one infant occurred when he was eleven months old; he looked around and was unable to see his mother. With a loud scream he flung himself to the ground and beat at it with his hands. His mother at once rushed to gather him up. Mothers of older infants and juveniles, although they sometimes reached out to touch their offspring during a tantrum, frequently ignored the behavior. Yerkes (1943), when describing tantrums, comments that he often saw a youngster "in the midst of a tantrum glance

[4] Charging displays, which involved all or some of these patterns, occurred invariably when mature or adolescent males approached a desirable food source, when mature (and occasionally adolescent) males joined other chimpanzees, and at the onset of, or during, heavy rain.

[5] Frequently, when a number of chimpanzees traveling together come upon some favored "drumming tree" along the track (a tree with wide buttresses) each male in turn drums in this manner. This results in a whole series of one to three double beats with irregular intervals between each. Adult females have never been observed to take part in such drumming rituals, but an infant and a juvenile female were each seen to do so once.

furtively at its mother or the caretaker as if to discover whether its action was attracting attention." In captivity, individuals are less prone to indulge in temper tantrums as they grow older, and this was also true of wild chimpanzees. Adolescents on five occasions went into tantrums after being mildly attacked by mature individuals or after making submissive gestures and not at once being reassured. I saw only three mature animals display temper tantrums; these were subordinate males who had been attacked by superiors and who were obviously too afraid to retaliate. Temper tantrums in connection with weaning have been described for infant baboons and langurs (DeVore 1963; Jay 1963a).

Redirection of Aggression or Transferred Threat

A chimpanzee that has been attacked by a higher ranking individual or that has been unable to obtain food owing to the presence of a social superior may chase after, threaten, or actually attack an individual subordinate to itself. When mature males attacked infants or juveniles it was always in this context. This type of behavior occurs frequently in rhesus macaques and baboons (Altmann 1962; Hall and DeVore 1965) and, less often, in other monkeys such as vervets or langurs (Hall and Gartlan 1965; Jay 1965b).

AGGRESSIVE BEHAVIOR

In a given group of animals aggressive behavior falls into three main categories: aggressive interactions with (1) animals of another species; (2) other groups within the species; and (3) individuals within the group itself. The chimpanzees of the Gombe Stream Reserve normally ignore or avoid the larger mammals of the area (such as buffalo, adult bushbuck, and bushpig), and often ignore or tolerate the close proximity of the other primates. However, aggressive interactions between chimpanzees and baboons (Papio anubis) over food are fairly frequent. In addition, these chimpanzees are predators of some consequence in the area, and from time to time hunt and kill various types of monkey and the young of the bushbuck and bushpig.[6] The chimpanzees themselves, however, are not apparently preyed upon.

Since the chimpanzee population of the Gombe Stream Reserve comprises one loosely organized social group, I have no data on interactions between different groups of chimpanzees. When individuals or associations

[6] Red Colobus Monkey (Colobus badius grauri), nine times; young bushbuck (Tragelaphus scriptus), six times; young bushpig (Potamochoerus koiropotamus sp.), four times; female Redtail monkey (Cercopithecus ascanius Schmidti) with new-born or fetal infant, once; adolescent baboon, once. On other occasions it was not possible to identify the kill. During one year, the total number of known kills (not including birds) was 20; these were all made by one or another of the known individuals.

of individuals met, there was seldom any aggressive behavior; at such a time, if a member of one association did threaten or attack a member of the other, this could usually be traced to a previous aggressive incident between the two. Reynolds and Reynolds (1965) describe occasions in the Budongo Forest area when loud chimpanzee calling and drumming continued for many hours at a time. They suggest that this may have been associated with relatively unfamiliar groups gathering at a food source; however, they were unable to verify this or to determine whether aggressive behavior had occurred.

Frequency of aggressive interactions between the individuals of a primate group vary from species to species. However, in all for which data are available most squabbles are settled by threat gestures and submissive behavior, rather than by actual attack (Jay 1968b). At least in part, this is because individuals living in constant association know each other, and their relative social positions are determined by habit. In the chimpanzee, however, the picture is slightly complicated because the individuals comprising the community as a whole are constantly splitting up and reuniting in different associations. This means that all mature chimpanzees, whatever their status in relation to the group as a whole, may from time to time be the highest ranking individuals of a temporary association. A further complication of the pattern is that the social position of some chimpanzees depends, in part, on the presence or absence of certain other individuals. One mature male (J.B.) frequently associated with the top-ranking male (Mike). When the two were together J.B. normally dominated the other males with them, but when Mike was not with him, J.B. was often dominated by one of the other high-ranking males. There were other close associations of this type, in which the presence or absence of the higher ranking of the two influenced the status of the other. These fluctuations in status are presumably due to the higher ranking partner's willingness to go to the aid of the other on occasion.

Usually, however, even within this constantly changing pattern order is maintained by means of threatening gestures and displays rather than by physical conflict. It is in this context that the branch-waving display of the male chimpanzee as he joins other individuals may be adaptive, since it reasserts social status by bluff, rather than by actual fighting. The greeting behavior that occurs when individuals meet after a separation can also be considered within this framework.

Threat

In aggressive contexts, the chimpanzee has a repertoire of gestures, postures, and calls that appear to elicit submissive behavior in the individual they are directed toward. They are not normally followed by physical attack. Since behavior of this sort invariably involves a combination of such gestures, postures, and sounds, and since it was seldom possible to estimate the precise significance of each individual component in the complex, I have merely listed each one, together with the behavioral contexts in which it was most usually observed. Most of these gestures were also directed toward humans or baboons during aggressive interactions.

GLARING, which is described in Table 2–2, was frequently associated with other threat patterns; a fixed stare is a form of threat in other primate species and also in man (Andrew 1963b; Schaller 1963).

HEAD TIPPING is a slight upward and backward jerk (or dorsalward tipping) of the head, and is invariably accompanied by the "soft bark." It is a low-intensity threat that was never seen to result in attack, and was often made by a feeding individual when a subordinate approached too closely. Andrew (1963a) comments that a sudden noisy expiration of air (used as a threat) is common throughout the mammals, and he suggests that a tipping back of the head (as when a dog howls) may lower the epiglottis so that the main current of air can pass through the mouth. The gesture of head tipping in chimpanzees may have originated in this way.

ARM RAISING occurs when either the forearm or the entire arm is raised with a rapid movement; the palm of the hand is normally orientated toward the threatened individual, and the fingers are slightly flexed. The gesture is often accompanied by head tipping and the "soft bark" or the "waa bark." The context in which this occurs is similar to that described for head tipping.

HITTING AWAY is a hitting movement with the back of the hand directed toward the threatened animal, and is often accompanied by the "soft bark." It may occur when a subordinate approaches too closely to a male who is feeding, for example. The same movement may also be a defensive reaction when a chimpanzee is startled by a large insect, a snake, or the like.

FLAPPING is a downward slapping movement of the hand in the direction of the threatened individual. It can be accompanied by "grinning" and "squeak calls," screaming, or the "waa bark." This gesture frequently occurred during female squabbles, and often led to "slapping." It may also occur when a chimpanzee, after being attacked or threatened, has sought reassurance contact with a third individual more dominant than the aggressor; after such contact the attacked animal may "flap" in the direction of the attacker.

BRANCHING describes the chimpanzee's taking hold of a branch or twig and shaking it from side to side or backward and forward. It is normally directed toward animals of other species. However, Fig. 2–4 shows a juvenile "branching" an individual older than herself, but subordinate to her mother.

Fig. 2–4. Juvenile female "branching" adolescent female. The juvenile's adolescent male sibling is near her, and their mother is also close by. *(Photograph by Baron Hugo van Lawick; © National Geographic Society)*

STAMPING AND SLAPPING have been described in the section on frustration. This sometimes occurred during an aggressive encounter, typically when a chimpanzee (particularly a female) chased a fleeing individual. The aggressor often gave panting hoots and calls, or screamed loudly.

THROWING of sticks, stones, or vegetation, apparently at random, frequently formed part of the frustration and charging displays of mature and adolescent males in the Gombe Stream area. In addition, I observed eight different individuals throwing articles toward an objective with what appeared to be definite aim, within a context indicating that the action was aggressive. An earlier paper on throwing (Goodall 1964) suggested that the chimpanzees threw anything that was at hand. This may be true, but subsequent data reveal that 51 percent of the objects I saw thrown by the chimpanzees during two years were large enough to intimidate baboons, and certainly humans! Nevertheless, the rarity of aimed throwing as compared with random throwing, together with the fact that I observed this behavior in only 8 out of 17 males and that only 5 out of 44 objects thrown with apparent aim hit their targets, suggests that this behavior is not highly developed. It can more appropriately be described as threat or intimidation, rather than actual attack.

BIPEDAL ARM WAVING AND RUNNING at another individual is a form of threat behavior. The chimpanzee may stand upright facing the threatened individual, and then raise one or both arms rapidly in the air while uttering the "waa bark" or screaming. At other times the chimpanzee runs bipedally toward the threatened animal, waving its arms in the air. This behavior was most usually directed against baboons, both at the feeding area and elsewhere, when they approached chimpanzees too closely.

THE BIPEDAL SWAGGER is typically a male posture, and occurs only rarely in females. The chimpanzee stands upright and sways rhythmically from foot to foot, his shoulders slightly hunched and his arms held out and away from the body, usually to the side. He may swagger in one spot or he may move forward in this manner. This posture occurs most commonly as a courtship display, but it also occurs when one male threatens another of similar social status. I observed the swagger 31 times in the latter context, and twice it was followed by actual attack. Twice during aggressive interactions between two females one swaggered at the other, and one female swaggered at a mature male.

THE SITTING HUNCH has been observed only in male chimpanzees. The displaying animal hunches his shoulders and raises his arms in front of him or to the side while he sits. This is the typical courtship display of the adolescent male. I observed only one male showing this posture as a threat in a nonsexual context, and that was during an aggressive interaction with a young female.

THE QUADRUPEDAL HUNCH is a stance in which the individual stands with back rounded and head bent and slightly pulled back between the shoulders. The chimpanzee may then move forward slowly or in a rapid charge. This was usually a high-intensity threat to another individual of a similar social status, and was sometimes followed by attack.

Attack

During one year I recorded 284 instances of attack; the frequency of attack behavior in the different age and sex classes is shown in Table 2–3. Of these, 66 percent were almost certainly due to the abnormal situation at the artificial feeding area where high-ranking individuals did not always have access to a desirable food. Therefore, although these attacks throw an interesting light on the mechanics of behavior and the social status of various individuals, their frequency cannot be regarded as typical. Of the remaining 34 percent of attacks, 15 were redirected aggression, 5 occurred during branch-waving displays, and the others were observed in a variety of contexts; the cause of many attacks was not apparent to the human observer.

Table 2–3 and the accompanying histogram show the result of a pre-liminary analysis of some of the data on attack behavior. The analysis was carried out in the following manner:

1. The number of attacks made by each age-sex class on each age-sex class was tabulated. These results, together with the total number of attacks made by each class on each class are presented in a table.

2. In order to obtain a more accurate estimate of the relative frequency of attack by each age-sex class (that is, on all other classes) the following calculations were made.

a. The total number of attacks made by each individual was divided by the number of occasions on which that particular animal was seen (see Table 2–1).

b. The resultant figure was multiplied by 1000 (extrapolated) to give the number of attacks made by that individual per 1000 observations.

c. The range of the figures calculated in *b* for all individuals in each age-sex class is given.

d. The median for the figures calculated in *b* for all individuals in each age-sex class is given.

e. Procedures *a, b, c,* and *d* were then repeated in order to determine the relative frequency with which each age-sex class was attacked.

This method of analysis does not tell us whether the observed fre-quency of attacks initiated by a certain age-sex class and directed (by that class) toward another class is a consequence of the relative frequency with which each of the classes was observed; it does not, in fact, tell us whether the initiator of the behavior exercises selection.

Table 2–3 shows that mature males initiated attack about four and one-half times more frequently (per 1000 observations) than did adolescent males, and five times more frequently than mature females. It also shows that mature females were attacked more frequently (per 1000 observations) than was any other age-sex class, and that infants and juveniles were at-tacked significantly less often.

The range for the number of attacks (per 1000 observations) within each age-sex class indicates that some individuals are extremely aggressive as compared with some others. In particular, the old female (Flo) attacked approximately twice as frequently as all the other females combined. It may be significant that all the attacks recorded for juveniles were initiated by Flo's female offspring. The highest score for individual male attacks was that of the top-ranking male, Mike; the lowest was for David, a high-ranking but extremely nonaggressive individual. Other male scores all ranged be-tween 41 and 87 attacks (per 1000 observations). Within the adolescent male class, the three individuals estimated as the eldest attacked far more frequently than the others.

The range given for the number of attacks on animals in the different age-sex classes suggests that some individuals were frequently involved as

TABLE 2–3 Distribution and frequency of attack behavior in the various age-sex classes

One Attacked	One Attacking						Total No.	Median (2)	Range (2)
	♂	Adol. ♂	♀	Adol. ♀	Juvenile	Infant			
Male	45	2	0	0	0	0	47	10.0	0–38
Adult male	52	11	9	0	0	0	72	31.0	8–95
Female	82	15	6	0	0	0	103	50.5	33–64
Adult female	6	2	3	0	5	0	16	17.5	31–66
Juvenile	14	15	3	2	5	0	39	3.5	39–46
Infant	4	0	1	0	7	0	12	5.5	15–26
Total attacks	203	45	22	2	17	0	289		
Median (1)	59	14.5	3	3	12.5	0			
Range (1)	16–120	0–34	0–30	0–6	0–25	0			

Median and Range (1).   Median number of attacks made by each age/sex class per 1000 observations, and the range for individuals of that class.

Median and Range (2).   Median number of attacks made on each age/sex class per 1000 observations, and the range for individuals of that class.

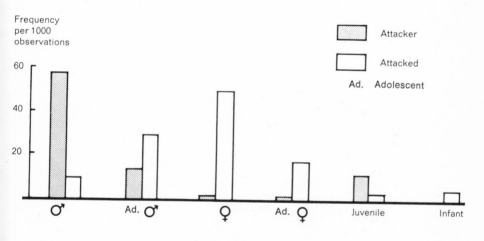

Histogram taken from Table 2–3 to show median frequency of attacking and being attacked in the different age/sex classes.

targets in aggressive incidents, whereas others normally avoided attack. Thus the nonaggressive male, David, was attacked less often than any other male except the dominant, Mike. The three youngest mature females (Mandy, Melissa, and Circe) each were attacked about twice as frequently as the three old mothers (Flo, Olly, and Marina).

Figs. 2–5, 2–6 and 2–7. High-ranking mature male subjects an old female to low intensity attack on arriving at the feeding area. Her two-year-old infant is clinging on the ventral position. After briefly pounding on her back and then kicking her he immediately embraces her with one arm. *(Photographs by Baron Hugo van Lawick;* © *National Geographic Society)*

Attacks were invariably of short duration, seldom lasting more than a few seconds; the longest attack lasted about one and one-half minutes. After a fight the aggressor often charged off, sometimes dragging branches behind him. The various methods of attack are listed below; most attacks comprise several of these components.

THE ATTACKING CHARGE. An attack is frequently preceded by a charge that is normally quadrupedal, silent, and fast. The hair of the attacking individual is usually fully erect at this stage, although on a few occasions it was sleek. In the five attacks by males during branch-dragging displays their charges were accompanied by hooting calls, and the victims (all females) were subjected to only a very mild form of attack (Figs. 2–5, 2–6, and 2–7).

STAMPING ON THE BACK. The attacker seizes the victim by the hair and endeavors to leap onto its back. If this is achieved he then stamps on the victim with both feet. I have seen between one and three double foot-beats during such stamping, and the sound carries at least 50 yards. Occasionally, the male gives a final kick before moving off. Mature females were not seen to attack in this manner, but a juvenile female did so on several occasions.

LIFTING, SLAMMING, AND DRAGGING.   When the victim is smaller than the aggressor it may be lifted bodily from the ground by its hair or limbs and slammed down again (Fig. 2–8). This may be repeated several times. The attacked individual may also be dragged along the ground by its hair or by one limb. Only mature males were seen to fight in this way.

BITING.   During an attack the aggressor often puts his mouth to the body of his victim as though biting; normally, however, the only visible aftereffect is some saliva on the hair of the other. On only four occasions was serious biting observed: (1) once a male bit a female and left deep puncture

Fig. 2–8.   Mature male attacking adolescent female. Grasping her by hair and one leg he lifts her from the ground. The female is screaming loudly. (*Photograph by Baron Hugo van Lawick;* © *National Geographic Society*)

marks in her arm; (2) another male bit an adolescent male in his foot, again leaving deep wounds; (3) an adolescent female bit the clitoris of a juvenile during a squabble; and (4) one mature male, during redirected aggression, dragged a juvenile female along with her foot in his mouth. On other occasions, however, chimpanzees were seen with puncture marks and gashes that might have been inflicted by biting. Yerkes (1943) frequently observed captive chimpanzees using their teeth during fighting, but records only one instance of a jaw-grip maintained.

HAIR-PULLING.   The attacked animal may lose fairly large handfuls of hair during a fight. When females or youngsters squabble they sometimes grab at and pull on each other's hair (rather in the manner of human children), particularly hair on the head and shoulder region. In this context, also, handfuls of hair may be pulled out.

SLAPPING.   One form of attack is a downward movement of the arm in which the palm of the hand slaps the body of the other animal. It occurs frequently when a female attacks another chimpanzee; both males and females may slap humans and baboons during aggressive interaction.

SCRATCHING.   The chimpanzee may scratch by slightly flexing its fingers and drawing its nails rapidly across the skin of the victim. Only females were seen to scratch in this way. When the victim was human, such a scratch broke the skin; no visible marks were left on the skin of scratched chimpanzees.

Only 10 percent of the 284 attacks were classified as "violent," and even attacks that appeared punishing to me often resulted in no discernible injury apart from the occasional wrenching out of hair. Other attacks consisted merely of a brief pounding, hitting, or rolling (by hitting or kicking) of the individual, after which the aggressor often touched or embraced the other immediately (Fig. 2–7).

Most often the victims, screaming loudly, tried to escape from their aggressors. At other times they crouched during an attack, presenting their backs to the attacker and protecting their faces, limbs, and abdomens. This was particularly true of mothers with infants clinging ventrally (see Figs. 2–5, 2–6, and 2–7).   In only 18 instances, when the fight was between two individuals of similar size and social status, did the attacked individual fight back fiercely.

SUBMISSIVE BEHAVIOR

This heading is used as a rather loose term to describe various non-aggressive postures and gestures that are directed toward a dominant chimpanzee in the following behavioral contexts: (1) When the subordinate has

Fig. 2–9. Adolescent male presents with the "extreme crouch" after being attacked by a dominant mature male. The latter reaches out to touch him with a reassurance gesture. *(Photograph by Baron Hugo van Lawick; © National Geographic Society)*

been threatened or attacked by the dominant individual (referred to as "appeasement" gestures); (2) when the subordinate approaches or is approached by, or passes or is passed by a high-ranking individual which has not, apparently, shown aggression toward the gesturing animal; and (3) when the subordinate, after being attacked or threatened, approaches a third individual of higher social status than the aggressor and makes gestures similar to appeasement gestures.

The various postures, gestures, and calls that are observed in submissive behavior of these types are described, together with the contexts in which they were normally seen.

PRESENTING. Presenting, in a nonsexual context, occurs when a subordinate individual turns its rump towards a higher ranking one. It is a posture common to many primates, such as rhesus macaques, baboons, and langurs (Altmann 1962; Hall 1962; Jay 1965*b*), in a variety of nonsexual situations. It was one of the most frequently observed elements in submissive behavior patterns among the Gombe Stream chimpanzees.

The posture of the presenting individual varies in relation to the situation. After a particularly severe attack the subordinate (screaming loudly with open mouth and lips retracted from teeth and gums) may back toward the aggressor from a distance of eight feet or more, looking over its shoulder. It then adopts the "extreme crouch," with all its limbs completely flexed and its head almost on the ground (Fig. 2–9). After a less severe attack, or when the dominance status of the two individuals involved in the attack is more nearly equal, the presenting posture is less extreme. The animal may stand with half-flexed limbs, looking back over its shoulder and "grinning" while screaming or giving "squeak calls," or showing the "silent grin," or whimpering.

After a low-intensity threat, the subordinate may flex its limbs only slightly while presenting; it may grin and give "squeak calls" or it may whimper. When a subordinate comes close to a higher ranking individual (usually a male) it may present with slightly flexed limbs (Fig. 2–10), or it may simply turn its rump toward the other with no obvious change in posture or facial expression (Fig. 2–12). A juvenile once turned her hind quarters toward her adolescent sibling in this manner after he had mildly threatened her.

I have roughly analyzed 324 instances of presenting that occurred in the three types of behavioral contexts described earlier. Of this number, 32 percent occurred when the subordinate had been attacked or threatened (Category 1); 54 percent occurred when subordinates came into close proximity to high-ranking individuals (Category 2); and the remaining 14 percent occurred when subordinates, after being attacked, presented to a third individual (Category 3). Table 2–4 shows the percentage of the total number of presentings that fell into these three categories made by individuals of the various age-sex classes.

These data were further analyzed, and Table 2–5 shows the actual distribution of the presentings observed during one year. It will be noted that males seldom presented but were frequently presented to, and that females

TABLE 2–4 **Percentage of total number of presentings in each age/sex class elicited by three different behavioral contexts**

	Percentages of Presentings Category			Total Number of Presentings
Age/Sex Class	1	2	3	
Male	25	25	50	4
Adolescent male	68	22	10	48
Female	16	80	4	136
Adolescent female	56	43	1	40
Juvenile	55	43	2	35
Infant	10	90	0	61
Total				324

Categories (1) Subordinate had been attacked or threatened.
(2) Subordinate came into proximity to higher ranking individual.
(3) Subordinate, after attack, presented to individual of higher rank than its aggressor.

Fig. 2–10. Adolescent male reaches out to touch the genital area of a juvenile female who has paused to present to him as she passes by. Her limbs are half-flexed but her facial expression is relaxed as she looks back at him. *(Photograph by Baron Hugo van Lawick;* © *National Geographic Society)*

were seen to present more often than any other age-sex class (per 1000 observations). High-ranking males were presented to by individuals of all other age-sex classes (including lower-ranking mature males), but no male was seen to present to a female or adolescent. On two occasions a juvenile was presented to; on both occasions this occurred after the same juvenile female had threatened individuals older than herself, but subordinate to her mother who was close by. The data also show that there is a wide range for the individuals of the different age-sex classes, both in the frequency of presenting and in being presented to. Thus, one female (Melissa) presented twice as frequently as another young mature female, and all the young females presented more often than the older ones. The top-ranking male, Mike, was presented to most frequently. Among the adolescent males the range for the number of times presenting was directed toward them (per 1000 observations) varied from 44 to 50 times for the three oldest to 12 to 25 for the others.

BOWING, BOBBING, AND CROUCHING are shown by a subordinate when it is facing the dominant individual; these actions involve various degrees of limb flexion. On the few occasions when I saw an individual bob slowly, he stood briefly and bipedally, but with his body only slightly raised from the horizontal position. He dropped back onto four limbs, and as a continuation of the movement flexed his elbows until his chest was close to the ground.

Fig. 2–11. Young mature female with start of ano-genital swelling presents, looking back over her shoulder, to an old mature male. (*Photograph by Baron Hugo van Lawick;* © *National Geographic Society*)

He then jerked himself back to the bipedal position prior to repeating the entire sequence several times; his knees were slightly flexed throughout. Usually these movements follow each other rapidly, and often the elbows are flexed and straightened while the hands remain on the ground. Any number (up to ten) bobs is normal; the behavior is usually accompanied by "bobbing pants" or occasionally by "panting shrieks." This behavior is most commonly seen in adolescent males: occasionally low-ranking males bob, and I saw two females do so. It is a typical response of a subordinate animal when a high-ranking male moves into its vicinity (Fig. 2–13).

Bowing occurs when the subordinate goes up to a higher ranking individual and flexes its elbows to a much greater extent than it flexes its knees. It is usually associated with soft panting. It is seen frequently when a subordinate (particularly, a female) approaches a high-ranking male who has previously shown signs of aggressive behavior, or after he has performed a branch waving display.

Crouching is an extreme form of bowing: the subordinate flexes all limb joints so that its body is close to the ground. The animal normally pants, or it may "grin" and give squeak calls. Crouching appears in a context similar to bowing, but may also occur as appeasement behavior after an individual has been threatened or attacked.

A lowering of the body toward the ground in a submissive context occurs in other species of primate. Gorillas and baboons when frightened during aggressive interactions may crouch prone to the ground (Schaller

TABLE 2-5   Distribution and frequency of nonsexual presenting in the different age/sex classes

| One Presented to | One Presenting | | | | | | Total No. | Median (2) | Range (2) |
	♂	Adol. ♂	♀	Adol. ♀	Juve-nile	Infant			
Male	4	44	90	20	24	40	222	71.0	16–170
Adolescent male	0	2	42	13	6	21	84	34.0	12–50
Female	0	2	3	6	4	0	15	3.5	0–15
Adolescent female	0	0	0	1	0	0	1	0.0	0–1
Juvenile	0	0	1	0	1	0	2	1.0	0–2
Infant	0	0	0	0	0	0	0	—	—
Total No.	4	48	136	40	35	61	324		
Median (1)	3	13	39	80	39	102			
Range (1)	0–4	9–38	10–220	98–132	37	55–150			

Median and Range (1).   Median number of presentings made by each age/sex class per 1000 observations, and the range for individuals of that class.

Median and Range (2).   Median number of presentings made to each age/sex class per 1000 observations, and the range for individuals of that class.

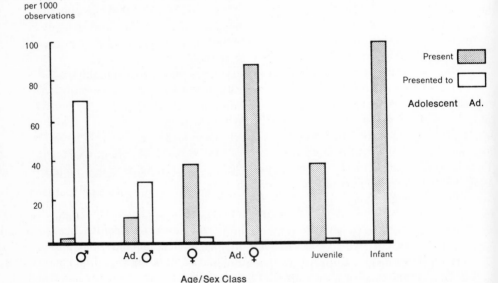

Histogram taken from Table 2–5 to show median frequency of presenting and being presented to in the different age/sex classes.

Fig. 2–12. Adolescent female pauses to present as she passes a mature female; the latter peers at her genital area before briefly touching it. (*Photograph by Baron Hugo van Lawick;* © *National Geographic Society*)

1963; Hall and DeVore, this volume). In man, too, bowing has long been a submissive gesture, and until a few years ago even civilized peoples prostrated themselves when appealing for mercy.

SUBMISSIVE KISSING.   Kissing, a pressing of the lips or teeth to the body of another individual, occurs in both submissive and dominant individuals. As a submissive gesture it frequently accompanies bowing or crouching, in which case the kiss is typically in the groin of the other individual. Sometimes when a chimpanzee moved close to a high-ranking individual (or vice versa), it touched the latter's face with its mouth. On rare occasions a female who had been threatened or startled by a sudden movement of a mature male kissed his hand when he reached out to touch her.

Behavior of this sort, described as "mouth to mouth touching," may occur in both dominant and submissive baboons (Hall and DeVore 1965; personal observation), and again until a few years ago man showed submissive kissing when he pressed his lips to the feet of his master or king.

HAND AND ARM MOVEMENTS.   Various hand and arm movements are seen in a submissive context. The most frequently observed is when the

Fig. 2–13. Low-ranking mature male hurrying toward a high-ranking male (who shows hair erection) in order to "bob" as the latter passes him. *(Photograph by Baron Hugo van Lawick; © National Geographic Society)*

Fig. 2–14. Mature female lays her hand on the back of a higher ranking female as the latter passes her. *(Photograph by Baron Hugo van Lawick; © National Geographic Society)*

subordinate reached out to touch a higher ranking chimpanzee; virtually any part of the body may be touched, but the head, back, or rump are the most usual. The subordinate usually pants as it touches the other, but may "grin" or make "squeak calls." This behavior is seen most commonly when a subordinate is passing or being passed by a high-ranking individual (Fig. 2–14). It occasionally occurred as an appeasement gesture when a chimpanzee had been threatened or mildly attacked.

Touching, in a similar behavioral context, was observed in the Gombe Stream baboons; during nonsexual mounting the mounted individual sometimes reached back and touched the other on the leg with one hand. In addition, when baboons presented in a nonsexual context, they frequently reached back one leg and touched the dominant animal on the thigh or even on the chest with one foot.

Another gesture observed in the chimpanzees is that of holding the hand toward a higher ranking individual without actually touching. The wrist and fingers are extended, and the hand may be held palm upward or, occasionally, downward. (The gesturing individual may also whimper, "grin," make "squeak calls," or scream.) This gesture, which was seen most often in females, occurred in a variety of contexts. When a mature male made a sudden movement a nearby female sometimes held her hand toward him. One female (Melissa) invariably held her hand out toward the highest ranking male in the group several times before moving off to get a hidden banana that she had noticed.[7] When a chimpanzee had been attacked or threatened, it sometimes approached and held its hand toward an individual dominant to its aggressor.

Another gesture falling under this heading is "wrist bending": the wrist is flexed, and the back of the hand or wrist may then be held toward the lips of a dominant chimpanzee. In the typical submissive context this was seen only rarely, when females or juveniles approached mature males. However, a similar gesture was observed when juvenile, adolescent, or mature individuals reached toward infants (under one year of age), since they invariably did so with the backs of their hands, keeping the fingers bent away. In some instances this gesture seemed to be made with reference to the mother of the infant concerned, but further data are required before this can be substantiated.

Another movement in this series is "bending away." This occurs when the subordinate individual flexes its elbow and wrist, at the same time drawing its arm close to its body and leaning slightly away from the higher ranking animal (Fig. 2–15). This may be accompanied by soft panting, or, occasionally, by low-intensity grinning and squeak calls. Bending away was usually seen when mature males passed close to youngsters, particularly if

[7] Sometimes we hid bananas in trees so that the young chimpanzees could find them while the others fed from the boxes.

Fig. 2–15.
Adolescent male "bend-
ing away" as a mature
male passes him. (*Photo-
graph by Baron Hugo
van Lawick;* © *Na-
tional Geographic So-
ciety*)

the male showed, or had previously shown, some signs of aggression or
frustration.

SUBMISSIVE MOUNTING.   The subordinate holds the other chimpanzee
around the waist, or lays its hands and lower arms along the back of the
other from the rear. Normally it leans forward so that its chest is close to the
other's back, and then gives thrusting movements of its pelvis. Both males
and females have been seen to mount in a submissive context, usually after
they had been charged or attacked. When the mounted animal was a male,
the subordinate occasionally reached up with one foot to touch or grasp the
other's scrotum.

REASSURANCE BEHAVIOR

In many circumstances one chimpanzee appears to be calmed or "reassured" as a result of touching or being touched by another.[8] The three
main contexts in which such behavior is typically observed can be described

---

[8] This may be related, sometimes, to the fact that the skin, at least in man, is a highly
complex and versatile organ of communication. It has been suggested that the "sympathetic
innervation, perhaps that of the sweat glands and capillaries, is also conductive to the viscera
and to other organ systems" (Frank 1958). Certainly the rhythmic patting and caressing of a
human baby has a soothing effect (Frank 1958), and this may be compared with the calming
effects of physical contact described in this section for chimpanzees.

under the following headings: (1) reassurance gestures made to a subordinate; (2) calming behavior directed toward a high-ranking chimpanzee; and (3) reassurance contact with individuals of any rank.

Reassurance Gestures Made to a Subordinate

A dominant individual may respond to submissive behavior directed toward it by a subordinate in any one of the following ways:

TOUCHING.   The dominant animal reaches out to touch the subordinate with one hand or, on occasions, with one foot. The latter may be thus touched anywhere on its body, depending mainly on its position relative to the dominant individual at the time. Touching the genital area or rump is the typical response to submissive presenting; occasionally this involves an "inspection" of the vulva of the female. When the subordinate bows or crouches, it is normally touched on the head, shoulders, or back; when it holds out its hand, the hand is commonly touched or even held by the dominant animal (Figs. 2–16 and 2–17). Occasionally the higher ranking chimpanzee will touch the other with his toes; for example, when a female, in passing, presented to a reclining male he idly reached to touch her genital area in this manner.

PATTING.   The dominant chimpanzee may reach out and make patting movements on the body, head, or face of the other with the palmar surface of its hand or fingers. These movements are slight and very rapid, and may be compared with those made by a human mother when patting her baby. Patting behavior typically occurred when the subordinate was extremely agitated (for example, when it had been attacked by the dominant chimpanzee and had then approached with a submissive gesture). Sometimes the subordinate was screaming loudly and on such occasions the dominant continued to pat for 30 seconds or more or until the screaming had died down. Figure 2–18 shows a nervous adolescent female being reassured in this way.

Mature and adolescent chimpanzees, particularly males, frequently responded to the approach of a small infant in this manner (Fig. 2–19).

EMBRACING.   The higher ranking animal may embrace the subordinate with one or both arms from the front, the side, or the rear. This type of reassurance sometimes occurred after a mature male had attacked a female during an arrival branch-dragging display (Fig. 2–7). It was also the typical response shown by a mother when her infant was hurt or frightened.

REASSURANCE MOUNTING.   The position is similar to that adopted during submissive mounting, but it is the subordinate animal that is mounted.

Fig. 2–16. The female ("Melissa") holds her hand toward a high-ranking mature male as she approaches when her infant is two days old. *(Photograph by Baron Hugo van Lawick; © National Geographic Society)*

Fig. 2–17. The mature male responds by reaching out to touch her hand. *(Photograph by Baron Hugo van Lawick; © National Geographic Society)*

This was the response to screaming and presenting on 21 occasions. Both mature and adolescent males sometimes placed their hands around infants and made thrusting movements over their backs when the infants screamed and crouched in submission.

REASSURANCE KISSING.   In a reassurance context, kissing sometimes occurred during an embrace, and was occasionally seen as a response to a subordinate's reaching out its hand, at which time the dominant individual held and kissed the hand. Occasionally a mature male pressed his lips to the face of a female when she approached and slightly crouched in front of him, or after he himself had been kissed submissively. When the dominant individual responded to a submissive kiss with a similar gesture, this sometimes resulted in mouth-to-mouth kissing.

RUMP TURNING.   The dominant animal turns its rump toward the subordinate, and may even back toward the latter. This was not a common response to submissive behavior, but it was seen 16 times. On each occasion the behavior was apparently elicited by submissive postures or

Fig. 2–18.   Adolescent female after being mildly threatened by top-ranking male ("Mike") turned away screaming and presented her rump. Mike reached out and touched her genital area; she turned back toward him, still uttering fear squeaks. As the final stage in this chain of events Mike is now touching her under the chin. After this the female was quiet. *(Photograph by Baron Hugo van Lawick;* © *National Geographic Society)*

Fig. 2-19. Eleven-month-old infant approaches mature male who pats the infant under the chin. *(Photograph by Baron Hugo van Lawick; © National Geographic Society)*

gestures on the part of the subordinate and sometimes resulted in submissive mounting. Seven times when rump turning was shown in response to presenting this led to the animals' pressing their bottoms together (Fig. 2-20); each of these incidents involved the same adolescent male. Three times this animal turned his rump when a mature female held out her hand toward him; she then touched his scrotum and he made thrusting movements on her hand.

Submissive behavior did not always elicit reassurance gestures; sometimes the dominant individual ignored the subordinate altogether. On such occasions the latter either moved away or (particularly if it was a juvenile or young adolescent) maintained its submissive posture while continuing to scream or whimper until it was finally touched. Indeed, some chimpanzees went into temper tantrums if reassurance was withheld. This apparent "need" for reassurance contact sometimes resulted in an obvious conflict situation. Thus, after a particularly savage attack, an adolescent male often showed a tendency to approach the aggressor that was strongly counteracted by a tendency to flee; he therefore moved forward in a series of circles or zig-zags as he alternately approached and turned away from the dominant male concerned.

The underlying motivational state of the dominant individual that induces reassurance gestures of the type described here is unclear, but the effect on the subordinate is apparent in many instances. The two following examples will serve to illustrate this.

An adolescent male (Pepe) approached a mature male (Goliath) that was feeding on a huge pile of bananas. Pepe did not immediately take a

Fig. 2–20. Adolescent male turns his rump towards a female in response to her submissive presenting. *(Photograph by Baron Hugo van Lawick;* © *National Geographic Society)*

fruit, but instead crouched, whimpering and screaming, while looking toward Goliath and occasionally reaching out one hand toward him. After a few moments Goliath reached out and began to pat the youngster on his face and head. After this, Pepe reached slowly toward a banana, but at the last moment jerked back and began to scream again. Goliath once more reached out and patted him for several seconds. Finally the adolescent stopped screaming, and, watching the big male carefully, gathered up a few fruits and moved away with them. Similar behavior was seen on many occasions.

The second example concerns the female Melissa. We have seen that she invariably held out her hand to a dominant male when she wanted to go and take a hidden banana; not until she had been touched or patted once or several times did she actually move away to take the food.

This type of behavior occurs in other primates. Dominant olive baboons sometimes reach out to touch a presenting subordinate (personal observation; DeVore, personal communication), and male hamadryas baboons may, perhaps, show a similar gesture (Kummer, personal communication). Langurs often reach out and place their hands on the body of a subordinate at the end of an aggressive interaction (Jay 1965a; Jay, personal communication). I observed a young red colobus approaching a large mature male that was feeding, and sit before him with slightly bowed head. The big male reached out and touched its head, after which the youngster moved past him and commenced to feed nearby. Lip-smacking in the rhesus monkey may have the effect of "reducing fear" in a subordinate, and a dominant monkey sometimes lip-smacks when a subordinate shows frightened grinning (Hinde and Rowell 1962). Man, too, may respond to an abject subordinate with a smile, a touch, a pat on the shoulder or, in some circumstances, an embrace.

CALMING BEHAVIOR.   On several occasions a high-ranking male, after he had attacked some individual or performed violent charging displays, was touched, patted, or groomed by other males. During such bouts of physical contact the male in question gradually "calmed down" (that is, he appeared to relax and slowly lost his hair and penile erections).

Reassurance Contact with Individuals of Any Rank

A chimpanzee may seek to initiate physical contact with another chimpanzee (often, but not necessarily, a higher ranking one) in a number of behavioral contexts when it is afraid, agitated, or intensely stimulated by social activity or the sight of food. The gestures most frequently involved in

Fig. 2–21.   A high-ranking male who is trying to get at the bananas inside this box pauses and starts to utter "panting hoots." A subordinate joins in, at the same time reaching out to make a few grooming movements on the neck of the other. (*Photograph by Baron Hugo van Lawick;* © *National Geographic Society*)

Fig. 2–22.
Adolescent male touches his own scrotum
when he suddenly sees chimpanzees ap-
proaching in the distance. *(Photograph by
Baron Hugo van Lawick;* © *National
Geographic Society)*

the behavior are touching, holding the hand toward another, embracing,
mounting, and grooming movements. In addition, one adolescent male frequently touched or held his own scrotum or penis under such circumstances.
This type of behavior is illustrated by the following examples.

When chimpanzees were suddenly startled (for example, by a strange
noise or sudden movement) they frequently reached out to touch another
individual. When some violent social activity broke out in a group (such as
fighting or branch-dragging displays) individuals not participating often
reached out to touch or make grooming movements on each other; sometimes one mounted or embraced a companion. The same patterns often
occurred when chimpanzees heard or saw others in the distance (Fig. 2–23).
When chimpanzees were confronted with an especially large pile of bananas
or saw us opening a big box of the fruit, the individuals concerned not only
touched, patted, kissed, or embraced each other, but invariably uttered loud
food "barks" at the same time.

One mature male embraced a three-year-old infant three times – twice
after the adult had been attacked by another male and once when he had a
sudden fright from seeing his own reflection in some glass. Even contact
with the infant had a marked effect in calming him.

Fig. 2–23.   Two males hooting in response to calls of chimpanzee. The mature male
is reaching out to touch the higher ranking male. *(Photograph by Baron
Hugo van Lawick;* © *National Geographic Society)*

When a chimpanzee hurried to touch a high-ranking individual after
being attacked or threatened by a less dominant one, it often looked around
after the contact and made threatening sounds or gestures in the direction
of its aggressor. Similarly, when an infant was suddenly frightened it often
ran to its mother and suckled her, briefly held her nipple in its mouth, or
simply reached out to touch her; he then looked around at the alarming
stimulus (van Lawick-Goodall 1968*b*).

This type of contact-seeking behavior also occurs in other primates.
On one occasion a gorilla was seen to reach its hand toward another when
it suddenly saw a human observer; the second responded by taking the
hand (Osborn 1963). An old baboon at the Gombe Stream Reserve was seen
to reach out and touch another individual while threatening an observer
(Miss Koning, personal communication). Hall (Hall and DeVore, this
volume) observed behavior in the chacma baboon that may be comparable;
when a pair of animals was threatening another baboon, one sometimes
briefly mounted the other, placing its hands lightly on its companion's back
while standing slightly to the side and pointing its muzzle in the direction of
the threatened individual. Man also shows a number of similar gestures in
such situations. Frank (1958) notes that "a person who is strongly reacting
emotionally, as in acute fear or pain, or grief, may be able to recover his

physiological equilibrium through close tactile contacts with another sympathetic person." A small child may hold its mother's hand or skirt in the presence of strangers; similarly an adult may reach out and touch a companion or hold his hand when suddenly frightened or emotionally upset. In a different type of situation (probably comparable to a chimpanzee seeing a lot of food) two men may embrace or clap each other on the back when they suddenly hear good news.

EXPRESSIVE GESTURES OBSERVED IN A SEXUAL CONTEXT

There are various gestures and postures that occur during sexual interactions between male and receptive female chimpanzees. (Many of these have already been described, as they occur also during frustration and aggressive behavior.) Of the 213 copulations or attempted copulations I observed in one year, males took the initiative 176 times, either by approaching the female with both hair and penis erect or by giving a "courtship display."[9]

Four courtship gestures have already been described: the bipedal swagger, the sitting hunch, branching, and glaring. Two other displays are (1) "tree leaping," when the male concerned (if he is in a tree) may execute a series of leaps and rhythmic swings through the branches, his body usually in an upright position while he faces in the general direction of the female; and (2) "beckoning," when the male, in a bipedal posture, raises one arm level with his head or higher and then makes a swift "sweeping toward himself" movement, his hand making an arc in the air.[10]

Females responded to 82 percent of the approaches or courtship displays of mature males, and to 77 percent of those of adolescent males by presenting for copulation (by remaining crouched where they were while the male approached or by running toward him and presenting). On eight occasions the males lightly touched females on their rumps before the latter presented. The percentage of success of the various types of displays of mature and adolescent males, that is, the percentage of those resulting in copulation, is shown in Table 2-6.

On 20 occasions, females ran away screaming when males approached to initiate copulation; on 10 of these occasions the males pursued until the females stopped and presented, on the other 10 occasions the males "gave up" and after shaking branches in the direction of the females moved away.

---

[9] This term is used simply as a means of describing postures and gestures commonly directed toward a receptive female prior to copulation; it is not intended to imply that these were observed only in a sexual context.

[10] This gesture is not unlike an exaggerated form of that made by females when they gather their infants into the ventral position.

TABLE 2–6 Relative frequency of various courtship displays and of solicitation prior to copulation (or attempted copulation) by mature and adolescent males with receptive females

Display	Percent of All Mature Male Displays	Percent of Success*	Percent of All Adolescent Male Displays	Percent of Success
Male approach	43	95	35	96
Bipedal swagger	20	78	6	100
"Branching"	8	94	24	61
Sitting hunch	1	100	15	73
Tree leaping	9	89	1	100
Beckoning	6	100	1	100
Glare	1	100	–	–
Initiated by female	12	83	19	100
Total Percentage	100		100	
Total Success		90		84

* Percent of success represents occasions when the male or female responded to the behavior.

On 12 occasions, females ignored courtship displays; the males concerned either persisted until the females presented, or they moved off.

On 37 occasions females solicited males. Typically, the female approached to within six feet, flattened herself in front of the male with her limbs flexed, and looked back at him over her shoulder. Five times the five soliciting females were ignored; four of them walked away, but the fifth persisted until the male mounted and copulated with her.

The male copulatory position is not a stereotyped one; normally he mounts, places one hand on the female's back, and adopts a "squatting" position, his buttocks scarcely more than an inch from the ground and his body slightly inclined forward. At other times, however, males either placed both hands on the ground or both on the back of the female, or held on to a branch overhead with both hands. Occasionally copulation took place while the female stood quadrupedally with only very slightly flexed limbs; the male either stood behind her and leaned directly forward, his chest on her back and his arms encircling her body, or he stood bipedally behind her and copulated while holding onto an overhead branch.

During some copulations slow lip-smacking was observed in the male; three males invariably gave "copulatory pants" at the culmination of the sexual act (which was normally completed after 5 to 10 seconds). During intercourse, females sometimes looked ahead, sometimes looked around at the male, and often gave short high-pitched squeak calls with "grinning." After sexual contact females remained still, moved away calmly, or rushed off screaming. Copulation was sometimes followed by brief grooming by either partner.

One behavior pattern, "inspection" of the genital area, was frequently observed when a nonreceptive female presented to a male or stopped near him. Sometimes he merely put his nose near her vaginal opening and appeared to sniff; at other times he poked his finger into the opening carefully, and then sniffed the finger; occasionally he used both hands to part the lips of the vulva, and then poked, peered, and sniffed. Such behavior was often repeated two or three times; one male inspected 18 times in 10 minutes while the female reclined beside him.

Inspection seldom occurred when females had large sexual swellings. That the behavior is, in fact, related to the estrous cycle of the female is suggested by the marked increase in the number of inspections usually directed toward any one female at the first signs of sexual swelling. Inspection also occurred frequently in the period immediately following detumescence.

Fig. 2–24.   Adolescent male "inspecting" the genital area of a female who has just arrived in his group. *(Photograph by Baron Hugo van Lawick;* © *National Geographic Society)*

Fig. 2-25. Mature female grooming adolescent male in greeting. Her two-year-old infant has jumped onto the adolescent's "lap" and is closely embracing him. *(Photograph by Baron Hugo van Lawick;* © *National Geographic Society)*

Mature, adolescent, and juvenile females sometimes inspected other females, particularly during greeting behavior; it was a common pattern on the part of infants of both sexes, especially when the female concerned had a sexual swelling.

GREETING BEHAVIOR

Greeting behavior may be defined as the nonaggressive interactions (or, at least, with only a very slight element of aggression) that occur between individuals meeting after a separation. There are many postures and gestures in the repertoire of the behavior. They consist of bobbing, bowing, and crouching, touching, kissing, embracing, grooming, presenting, mounting, inspecting of the genital area, and, occasionally, hand-holding. In addition, males (rarely, females) may precede their greeting with some form of ritualized aggressive pattern, such as the bipedal swagger, the quadrupedal and sitting hunch, stamping, and so forth. Often, but not always, males show erection of the hair and penis prior to and during greeting. As two individuals approach each other they may utter soft or loud panting sounds, particularly the subordinate as it bows, crouches, or bobs. Sometimes both the dominant and the subordinate individuals may "grin"; the latter may also make "squeak-calls."

During a greeting, one or several of the postures and gestures listed may be displayed by both the individuals concerned; at other times the higher ranking individual may ignore the greeting behavior of the other. The 686 greetings that were observed during a seven-month period have been

TABLE 2–7 Distribution and frequency of greetings gestures in the different age/sex classes

One to Which the Gesture Is Made	One Making Greeting Gesture					Total No.
	♂	Adol. ♂	♀	Adol. ♀	Juvenile and Infant	
Male	139	50	176	16	32	413
Adolescent male	209	98	46	15	23	391
Female	12	11	81	22	16	142
Adolescent female	11	13	17	6	4	51
Juvenile and infant	27	13	8	14	12	74
Total No.	398	185	328	73	87	1071

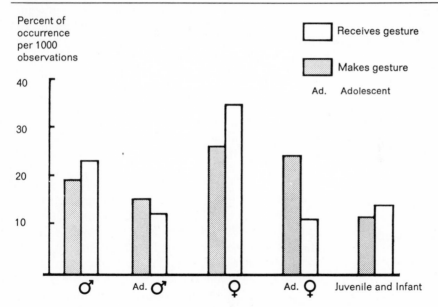

Histogram showing the relative frequency of making and receiving greeting gestures in the different age/sex classes. This is expressed as the mean percentage of occasions that each class made greeting gestures per 1000 observations.

broken down into the 1065 actual gestures and postures of which they were comprised. Table 2–7 shows the observed frequency with which these gestures were directed toward each age-sex class by each class, and the frequency with which they were received by each of these classes. As the median frequency of greetings gestures in each of the classes has not yet been worked out, the histogram below the table shows only the mean percentage of occasions when each age-sex class initiated or was the recipient of greetings gestures (per 1000 observations). This does suggest that mature females show greeting gestures more often than do the other classes, and that greeting behavior is directed toward them more often than toward other classes. It also suggests that the two adolescent classes (particularly the females) are the most frequently ignored during greeting.

The particular gestures and postures displayed during a greeting depend partly on the age-sex class of the animals concerned, partly on their individuality, and partly on their mood at the time and the length of time for which they have been separated. Thus the more exaggerated gestures (such as embracing) are more likely to occur between individuals that are strongly attracted to each other (particularly if they have not seen each other for several days), than between two individuals that seldom interact socially in other ways. Swaggering and charging prior to greeting is seen more frequently when two associations of chimpanzees meet than when one animal quietly approaches a resting group.

Table 2–8 shows that although some of the gestures and postures (such as kissing, grooming, touching, and embracing) are displayed by all

TABLE 2–8 Frequency of different greeting gestures and postures in the age/sex classes, expressed as a percentage of total number of gestures recorded in the greeting behavior of each class

	Percentage of Occurrence				
Gesture	♂	Adol. ♂	♀	Adol. ♀	Juv. and Infant
Kiss	10.0	7	15	8	14
Groom	18.0	20	10	18	8
Touch	24.0	16	21	15	10
Embrace	12.0	9	8	9	8
Present	0.5	3	27	43	30
Inspect	19.0	19	2	1	21
Mount	5.0	8	—	6	1
Bob	0.5	14	—	—	—
Bow and crouch	1.0	4	7	—	—
Hold out hand	—	—	10	—	—
Hold hand	10.0	—	—	—	—
Total Percentage	100	100	100	100	100
Total No. Gestures	398	328	185	73	87

the age-sex classes, others are more or less limited to one or a few classes. Among adults and adolescents, presenting is almost entirely limited to females, whereas inspection is far more common in males. This suggests that there may be a sexual element in greeting behavior between males and females.

I have not yet analyzed the frequency with which each type of gesture was directed by individuals of each of the classes to each of the other classes, or vice versa, but Table 2–9 is an attempt to group greeting behavior into three main categories: (1) gestures made by subordinate chimpanzees to higher ranking ones; (2) those made by dominant individuals to their social inferiors; and (3) those seen when two individuals of similar social status greet each other.[11]

Table 2–9 shows that those gestures and postures that may appear in both submissive and reassurance contexts (such as kissing, touching, grooming, and embracing) were observed in all three types of greeting; those made only in submissive contexts were never directed by higher ranking individuals toward their subordinates, nor did they occur between animals of similar rank. This suggests that gestures and postures of submission and reassurance play a major role in greeting behavior, while the pregreeting displays, such as the bipedal swagger, add an aggressive element.

TABLE 2–9   Types of greetings gestures between individuals of similar social status, expressed as percentage of total number of observations of each type of gesture when it occurred in the three different types of greeting

Gesture	Subordinate Greets Dominant (%)	Dominant Greets Subordinate (%)	Similar Status (%)	Total No. Gestures
Bob	100	—	—	27
Crouch and bow	100	—	—	28
Present	100	—	—	87
Reach hand to	100	—	—	10
Hold hand	—	100	—	10
Mount	—	100	—	15
Kiss	60	30	10	82
Touch	41	51	8	189
Groom	30	60	10	120
Embrace	40	46	14	95

Greeting behavior may be considered adaptive in relation to the loose social structure of the chimpanzee community. When two individuals meet

[11] Sometimes it was difficult to determine the social status of one individual in relation to the other at the time of greeting (for example, when an older chimpanzee greeted an infant the status of the latter was sometimes influenced by the proximity of its mother). Greetings involving two such individuals have been omitted from this table.

after a lapse of time that may have been anything from a few hours to a few weeks, some clear-cut gestures may be important for the re-establishment of a status recognition between them. When two humans greet, the basic principle involved is similar — the acknowledgment of a similar or different social standing of the person greeted. The deep bowing of the Japanese or the nodding of the head in Western civilization may have been derived from a submissive act (comparable to bowing or bobbing in the chimpanzee) that was originally intended to convey the meaning that no aggressiveness was intended (Schaller 1963). Similarly, reaching to shake an outstretched hand may have derived from some gesture of reassurance to an individual of lower rank (such as hand touching or holding in the chimpanzee). In fact, many of the greeting gestures described for chimpanzees have striking parallels in the greeting behavior of man.

SOME EXPRESSIVE MOVEMENTS THAT OCCUR IN OTHER CONTEXTS

There are gestures and postures that can be observed in behavioral contexts not mentioned in the previous pages. Some of these will be briefly outlined below.

In Social Grooming

The chimpanzee has a number of gestures and postures that apparently serve to elicit grooming behavior in another individual. In the most usual of these a chimpanzee approaches another and stands or sits, usually with head slightly bowed, so that some portion of its anatomy is placed in close proximity to the chosen partner. Most frequently presented in this manner are the head, back, and rump — those parts of the body that a chimpanzee cannot easily groom itself. Usually, such solicitation elicited a grooming response. Sometimes one animal simply approached and started to groom another chimpanzee; if the latter did not reciprocate after a few moments, the initiator frequently stopped grooming and adopted a soliciting attitude, or scratched some part of its body that was facing the partner. This often elicited grooming behavior and the two then groomed mutually.

There were many occasions on which an animal soliciting grooming was ignored; its reaction then varied, depending on its social status relative to that of the other chimpanzee and on its individuality. Thus when females were ignored by males they usually either moved away, started to groom the male, or groomed themselves. One female (Melissa), however, often rocked and whimpered if she was ignored, and sometimes reached out to poke the male concerned. This behavior invariably elicited a grooming response from the male. One male, on five occasions when his partner stopped grooming him, moved a few feet away and commenced to "branch" the other vigorously, at the same time showing full hair erection and glaring; each time, the partner got up immediately and began to groom again.

In Play Behavior

There are various gestures and postures associated with playful behavior.[12] These include the "play face," "laughing," and the "play walk." In play walking, the chimpanzee walks with a rounded back, head slightly bent down and pulled back between the shoulders, and takes small, rather "stilted" steps. Often there is a pronounced side-to-side movement as he moves forward, rather like the seaman's roll.

Play between youngsters was often initiated when one approached the other with the play walk or a gamboling-type run. When mature males initiated play with adolescents or juveniles they normally approached with the play walk and then reached out to prod or tickle the youngster. On the seven occasions on which mature individuals were seen to initiate play with other adults they always did so by "finger wrestling" (that is, they reached to the other's hand or foot and began to push, pull, and squeeze the fingers or toes).

In Food Sharing

Chimpanzees frequently show begging behavior when one individual (often, but not necessarily, a higher ranking one) is in possession of a food that is in short supply (such as meat or, at the feeding area, bananas). A begging individual may reach out to touch the food or the lips of the possessor of the food, or he may hold out his hand toward him (palm up), sometimes uttering small whimpers.

The response to such gestures varied according to the individuals involved and the amount of food. Often the possessor pulled the food away from the begging individual or threatened him. Sometimes, however, individuals detached pieces of their food and handed them to begging subordinates. Almost always when chimpanzees held their hands to the mouth of the possessor, the latter eventually responded by pushing out a half-chewed lump of food that was promptly chewed by the one that had begged. It was not uncommon to see a high-ranking chimpanzee reaching to the lips of a feeding subordinate.

The type of food-sharing described above, in which a dominant individual not only allows a subordinate to take a piece of food, but also may

[12] I have described as social play behavior that consists of nonaggressive interactions between two or more individuals, including one or a combination of the following: tickling, wrestling, mock-biting, sparring (hitting and pushing), and chasing.

actually "give" a piece, is not reported for other nonhuman primates in the wild.

In Mother-Infant Communication

There are a number of movements, gestures, and sounds made by infants that appear to act as signals to their mothers, and others made by the mothers that serve as signals to the child. Some of these do not ordinarily occur during interactions between adults, and are discussed in detail elsewhere (van Lawick-Goodall 1968*b*). Some of the more obvious of these communication patterns are the following: The mother may lightly touch her infant when she is ready to move off; the infant then normally responds by clinging to her. She may stand with bent knees, look back over her shoulder, and make a beckoning movement; such a signal is directed toward an older infant, which normally climbs onto her back as she moves off. She may, as she approaches thick undergrowth, touch the child when it is riding in the dorsal position; the youngster then slides round into the ventral position. When an infant first starts to make short exploratory trips away from the mother, either she or the child may seek to re-establish physical contact by reaching toward the other, usually showing a pout face and utter-

Fig. 2–26. Juvenile puts her hands to the mouth of her adolescent male sibling, begging for the chewed food from his mouth. *(Photograph by Baron Hugo van Lawick; © National Geographic Society)*

ing a "hoo whimper" at the same time. When an infant being carried by its mother screams, the mother at once responds by embracing it tightly; if the infant is some distance away and screams, she rushes over to rescue it.

Gradually, during its second year the infant begins to show the expressive movements that are typical of interactions between the older chimpanzees of its community.

DISCUSSION

The preceding pages have done little more than provide an outline of the complex communicatory system of a free-ranging chimpanzee community; a great deal of analysis remains to be carried out on my existing data, and many more observations are needed from the field. However, even this incomplete picture suggests that tactile, gestural, and postural communication plays an important role in interindividual relationships. Thus it may be of value to discuss, tentatively, the origin and evolution of some of these communicatory patterns.

The importance of the role played by tactile experience in the development of the human infant, and subsequently in the development of human communications, has been discussed fully by Frank (1958). It is certain that tactile experience is of at least equal importance to the development of the nonhuman primate infant and of subsequent nonhuman primate communication patterns. It has, in fact, been suggested that the infantile clinging response may be a major factor in the development of filial attachments in nonhuman primates. Thus it has been shown that in infant rhesus monkeys reared with an artificial cloth mother this response was one of the main determinants in the formation of filial attachment to the surrogate; it was further demonstrated that the opportunity to cling to the cloth mother, particularly when the subjects were frightened or upset, was an incentive for performing simple tests (Harlow 1960; Harlow and Zimmermann 1958). Certainly the clinging response, as well as a wide variety of tactile signals, plays an important role both in mother-infant coordination and in the formation of mother-infant social ties in the wild chimpanzee (van Lawick-Goodall 1968b).

It has been suggested that some infantile responses, particularly clinging, may persist after the primate has become independent of its mother and may then be directed toward other individuals in the group. Thus they may form the basis for later aspects of sociability and social organization (Mason 1964). This theory is strengthened by the fact that, in stressful situations, chimpanzees reared with cloth mothers tended to cling to familiar inanimate objects, whereas under similar circumstances mother-reared chimpanzees ignored such objects, but clung to other chimpanzees or to humans. In its natural habitat the growing primate is "usually surrounded from infancy by individuals whose appearance and reactions to him are similar to those of the mother, thus providing seemingly optimal conditions for the generalization or extension of filial attachments" (Mason 1964).

Observations of chimpanzee behavior in the field led me to form the same hypothesis as that discussed above with regard to the derivation of some of the expressive movements described (particularly those pertaining to contact-seeking behavior). Thus a mature chimpanzee may embrace, reach out to touch, or mount another animal under similar circumstances and in more or less the same manner as a frightened or apprehensive infant runs to embrace and be embraced by its mother, reaches out to touch or grasp her hair, or stands upright behind her grasping her rump and standing ready to climb on if the situation warrants it. Figure 2–27 shows a juvenile laying one hand on her mother's back as she hears chimpanzees calling in the distance, and Figure 2–28 shows a mature female laying her hand, in exactly the same manner, on the back of a young male in response to the same stimulus. It is possible also that some forms of "kissing" may have been derived from the infantile response of suckling, or holding the mother's nipple in its mouth when alarmed or hurt.

Many other expressive movements have not evolved from such primitive social responses. Let us consider, for example, presenting, bobbing, and bowing. It is suggested, tentatively, that these patterns may have been derived from conflict situations. Bowing and bobbing may represent a desire to approach a dominant individual counteracted by a desire to flee. The flexing of the elbows may be an intentional movement of flight and does, in fact, leave the individual in a good position to jerk away if necessary. (On many occasions when adolescent males hurried to bob to dominant males during greeting they did actually leap away when the male concerned started a charging display.)

Presenting, in a nonsexual context, may have evolved from a similar conflict situation, and here again, except in the "presenting crouch," the individual is in an excellent position for immediate flight. Zuckerman (1932) suggested that presenting in baboons (in nonsexual contexts) might represent an incipient flight movement in animals that were subordinate.

In this context it should perhaps be mentioned that a mature chimpanzee (particularly, a male) frequently tickles the skin at the base of the white tail tuft of an infant when it approaches, and occasionally strokes its penis or clitoris. When the infant first shows presenting behavior (in the start of its second year) it may present several times in succession to the same adult, particularly if this results in "tuft tickling." This pleasant stimulus, which occurs as a result of presenting, may be related to the high incidence of infants' presenting to males (see Table 2–5). It is also interesting to note that the mature baboon frequently lifts up the hind legs of an infant in "greeting," and "kisses" its bottom (Hall and DeVore, this volume). One might speculate, therefore, that this type of pleasant stimulation of the

Fig. 2–27 (left).
As another group of chimpanzees call in the distance, a juvenile places her hand on her mother. (*Photograph by Baron Hugo van Lawick;* © *National Geographic Society*)

Fig. 2–28 (below).
Mature female places one hand on back of adolescent male as another group of chimpanzees call in the distance. (*Photograph by Baron Hugo van Lawick;* © *National Geographic Society*)

genital area of the infant may have played some part in the high frequency of rump presenting observed in frightened or upset adult chimpanzees and baboons.

Finally, there is one aspect of chimpanzee communicatory behavior that must not be overlooked — the similarity of many of these gestures and postures to some of those made by man. These similarities, which have been pointed out in the relevant sections in the text, do not lie solely in the structure of the movements themselves, but also, more important, in the frequently close correspondence of the behavioral situations in which they occur. It seems almost certain, therefore, that those gestures common to both man and chimpanzee formed part of the behavioral repertoire not only of early man, but also of an earlier primate type ancestral to both ape and man.

chapter three

THE BEHAVIOR OF
THE MOUNTAIN GORILLA[1]

George B. Schaller

Few animals have fired the imagination of man as much as the gorilla. Ever since this large ape was discovered in 1847 in West Africa, tales of the gorilla's ferocity and strength have stirred popular and scientific interest. The following excerpt from Owen (1859) provides an example of this literature:

> Negroes when stealing through the shades of the tropical forest become sometimes aware of the proximity of one of these frightfully formidable apes by the sudden disappearance of one of their companions, who is hoisted up into the tree, uttering, perhaps, a short choking cry. In a few minutes he falls to the ground a strangled corpse.

In spite of this considerable interest little factual information about the habits of the gorilla in nature became known, as Yerkes and Yerkes (1929) clearly illustrated in summarizing the available literature. The lowland gorilla *(Gorilla gorilla gorilla)*, which inhabits the forests of southern Nigeria, Gabon, Cameroun, Rio Muni, and elsewhere in West Africa, still remains unstudied, although such authors as Merfield and Miller (1956) and Sabater Pi (1960) present useful local notes. The mountain gorilla *(Gorilla gorilla beringei)*, separated by 650 miles of forest from the lowland gorilla, was discovered in 1902 in the eastern Congo and western Uganda. It closely resembles the lowland gorilla, the primary difference between the two subspecies being minor physical characters such as the length of the palate and the length of hair (Schultz 1934). Prominent among

[1] The study was financed by the National Science Foundation and the New York Zoological Society, and the latter institution also acted as sponsor. Local sponsors were the Institute of the Parks of the Congo, and Makerere College, Uganda. I am extremely grateful to the University of Chicago Press for permission to quote and to reproduce the tables and figures from my earlier report. Dr. J. T. Emlen kindly read this manuscript critically.

the previous studies of the mountain gorilla are those of Bingham (1932), Donisthorpe (1958), and Kawai and Mizuhara (1959).

This chapter is based on data which were collected between February 1959 and October 1960 on an expedition under the leadership of Dr. John T. Emlen, professor of zoology, University of Wisconsin. The main purposes of the expedition included: (1) a general six-month survey of mountain gorilla range with emphasis on distribution and ecological diversity (Emlen and Schaller February–July 1959), and (2) sustained observations into the life history of the mountain gorilla in a selected area for at least one year (Schaller August 1959–September 1960).

The first phase of the project involved long journeys by car and considerable hiking through various types of forests. We usually asked the local population about the presence or absence of gorillas and whenever possible, checked the information by personal inspection. In addition, we took detailed notes on the vegetation types and food plants that the gorillas used, and obtained comparative data on the ape's relative abundance in various parts of its range.

Most of the observations of gorilla behavior were made on ten groups, comprising nearly 200 animals, in the Virunga Volcanoes of Albert National Park, Congo. My wife and I camped at an altitude of 10,200 feet in the saddle between Mts. Mikeno and Karisimbi, two dormant volcanoes both nearly 15,000 feet high. This main study area, called Kabara, included about 25 square miles of *Hagenia* woodland growing on the precipitous slopes of the two volcanoes.

The purpose of this chapter is to present a brief summary of those aspects of the gorilla's behavior which are of greatest interest as they relate to comparative primate studies. For a detailed account of behavior and for full documentation see Schaller (1963).

STUDY METHODS

Gorillas are usually not difficult to locate, for if a fresh trail is found the trampled vegetation, food remnants, dung, and other spoor aid in leading the tracker to the animals. In most areas, however, gorillas inhabit such dense vegetation that prolonged daily observations are difficult, if not impossible. The habitat with the best visibility was the Hagenia woodland in the Virunga Volcanoes, primarily because a shrub stratum was almost absent.

Two methods were used in observing gorillas. They were viewed from the cover of a tree trunk, with the observer remaining undetected by the animals; or they were approached by the observer slowly, alone, and in full view, to within about 150 feet, with the hope that after repeated contacts they would become habituated to his presence. The former method had only limited value, for the gorillas, with senses comparable to those of man, frequently detected me. The value of the latter method was shown by the fact that after from 10 to 15 prolonged contact periods some groups became so well habituated that their daily routine was little affected. Animals occasionally approached to within 15 feet of me, and once climbed into the same tree in which I was sitting. An excerpt from my field notes illustrates the casual manner in which group VII responded to my close proximity after 78 observation periods.

> 08:30 I advance inadvertently to within 30 feet of a foraging group. After I have settled myself on a low branch 5 feet above ground, a female with an infant on her back spots me, looks intently, and then continues to feed. A juvenile backs into the vegetation upon seeing me, and when the silver-backed male glances over from a distance of 45 feet he emits an annoyed grunt. However, the animals continue to sit and feed leisurely in the sun within 25 to 60 feet of my observation post.

> 09:30 Foraging slows down and ceases entirely as the animals rest in the sun within 40 to 60 feet of me.

> 10:30 Two animals begin to snack, and 10 minutes later the silver-backed male rises, looks at me seemingly startled and roars three times. He begins to feed and the whole group joins him. During the following 25 minutes the group moves some 500 feet and then settles into another rest area.

A group was usually observed daily for from one to four weeks until it moved outside the main study area, perhaps to reappear a few weeks later. I was always able to find at least one of the ten groups which intermittently frequented the forest around camp. In more than 300 encounters and 466 hours of direct observation, six groups became habituated to me. All members of these six groups (II, IV, V, VI, VII, VIII), as well as several gorillas in each of the other four groups and seven lone males, were recognized individually.

In spite of the excellent possibilities for observing gorillas at Kabara, the dense vegetation and the wandering habits of the animals made it difficult at times to see clearly and to obtain continuous observations on one group for more than a few successive hours. Once a group had voluntarily moved out of sight it was rarely pursued for this tended to frighten the animals and increase the chance of attacks. Most observation periods ranged from one to three hours, with a few as long as seven hours.

DISTRIBUTION AND ECOLOGY

Distribution

The range of the mountain gorilla extends over an area of about 35,000 square miles from the equator southward to 4°20′ S. latitude, a distance of some 300 miles, and from longitude 26°30′ E. to the eastern escarpment of the western Rift Valley (29°45′ E.), a distance of some 220 miles. Gorillas are not randomly distributed through the forest, but are concentrated in isolated population units in about 8000 square miles of terrain (Fig. 3–1). The boundaries of some of these units coincide with such natural barriers to the forest-dwelling gorilla as extensive grasslands, cultivation, or broad rivers, but frequently the forest continues unchanged at the boundary of an inhabited area. Stragglers, lone males or small groups, are sometimes found 20 or more miles from the nearest gorilla population. The erratic wanderings of such animals may be a means of colonizing new territory, accounting for some of the isolated population units.

Local concentrations of animals also exist within larger population units, especially in the lowland rain forest of the Congo basin. Populations of gorillas probably drift or are pushed about as local conditions change. The most important disturbance to the region during the past 200 or more years has been the repeated clearing of the forest for cultivation in small patches. Fields are worked by the Africans for three or four years and then abandoned for at least ten years, allowing the forest to regenerate. Gorillas favor the dense secondary growth, where forage in the form of herbs, shrubs, and vines is plentiful, over primary forest, which supports only a sparse ground cover in its shadowy interior. Many gorilla concentrations occur near roads and around villages where disturbance to the forest has been most recent (Emlen and Schaller 1960).

Ecology

The range of the mountain gorilla, although small, has considerable ecological diversity. The animals inhabit three major forest types which are stratified altitudinally from the Congo basin eastward to the summits bordering the Rift Valley.

About three-fourths of the gorillas are found, not in the mountains as their common name implies, but in lowland or equatorial rain forest at an altitude of from 1500 to 5000 feet (Fig. 3–1). A seemingly endless expanse of forest covers the flat to undulating terrain, broken here and there by a road, a village, or some ephemeral native fields. The evergreen trees are from 120 to 180 feet high and the canopy is almost continuous. Lianas entwine the branches and epiphytes are common. Although sunlight filters to the ground in places, it is usually insufficient to support a herbaceous understory. Only where the canopy is broken—along rivers, at the edge of fields, and where a large tree has crashed to the ground—is there the tangle of low vegetation that the gorillas favor. The climate is enervating with high humidity and temperatures that tend to remain above 20° C. throughout the day.

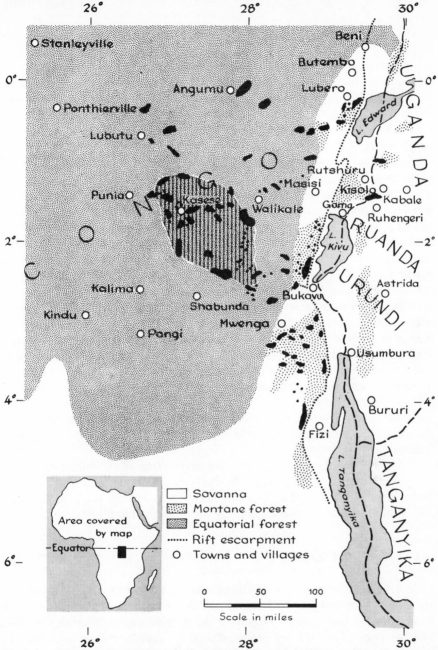

Fig. 3–1. The distribution of the mountain gorilla with respect to vegetation types. Mountain rain forest and bamboo are lumped under montane forest. The black areas indicate the location and approximate shape of isolated gorilla populations or gorilla concentrations in areas of continuous distribution. The small dots represent records of gorillas outside these areas. The hatching marks a central region of continuous but sparse distribution. The Virunga Volcanoes are represented by the black area near the village of Kisolo; the Kayonza Forest lies just north of the Virunga Volcanoes.

Along the eastern edge of the Congo basin, the mountains become rugged, reaching a height of over 10,000 feet. Much of the area between 5000 and 8000 feet is covered with mountain rain forest. This rain forest differs from that in the lowlands primarily in the somewhat smaller size of the trees and in the presence of gymnosperms. Gorillas are usually found in the lush valleys among tree ferns and vines rather than on the more open ridges. The climate in the highlands is temperate with temperatures rarely far above 20° C. and occasionally down to near the freezing point.

Gorillas are also found in bamboo *(Arundinaria alpina)* which occurs as a definite vegetation zone between 8000 and 10,000 feet. This grass forms a somewhat translucent and usually continuous canopy of some 20 to 35 feet high over miles of rolling uplands. Ground cover is sparse or absent except along trails and clearings. Because of its limited distribution, bamboo is of relatively minor importance in the total ecology of the gorilla. The main food of gorillas in this zone consists of bamboo shoots, which are abundant only during the rainy periods of the year. In the absence of shoots the apes have to seek other forage.

In addition to these three major forest types, there are minor local ones. The Hagenia woodland, a subtype of mountain rain forest in which the major part of the study was conducted, is found only in the Virunga Volcanoes. The Hagenia forest resembles a parkland, with gnarled trees widely spaced and averaging only about 60 feet in height, and with a dense herbaceous understory consisting of *Senecio, Lobelia,* and other succulent plants. Above 11,500 feet the forest ceases and the open slopes are covered with scattered groves of giant lobelias and senecios. Gorillas penetrate into this zone to an altitude of 13,500 feet, where temperatures drop to or below freezing nightly.

Though the climate and vegetation in the range of the mountain gorilla vary from tropical to temperate, the habitats are similar in being lush and damp throughout the year. Gorillas show considerable ecological adaptability, but they have remained entirely within the humid forests.

AGE AND SEX CLASSES

The development of one infant was traced in the wild from birth to the age of one and one-half years, and that of others of various ages for from 10 to 12 consecutive months. This enabled me to compare the sizes and behavior of infants and provided a fairly accurate means of estimating age. For further comparison, weight and size data of captive gorillas of approximately known age were obtained and compared to those of wild ones – another method for deriving a crude age scale. The following age and sex criteria are used in this chapter:

Age and Sex Class	Approximate Age (in years)	Definition and Distinguishing Field Marks
Infant	0–3	Any animal carried by a female for prolonged periods and weighing less than about 50 to 60 pounds
Juvenile	3–6	Any small animal not carried by a female and weighing 60 to 120 pounds
Subadult and adult	6+	Any animal larger than a juvenile
Female	6+	Any animal that carries an infant for prolonged periods; any large gorilla (weighing 150 to 250 pounds) with sagging breasts and long nipples
Black-backed male	6–10	Any gorilla weighing 150 to 250 pounds with angular, muscular body, and with few or no gray hairs in its saddle
Silver-backed male	10+	Any very large gorilla (300 to 450 pounds) with prominent sagittal crest and with a gray or silver back

POPULATION AND GROUP DYNAMICS

Population Density

Population estimates are by necessity tentative, for accurate counts of gorillas in their forest environment are difficult to make. Two areas were sampled intensively, and the following figures obtained:

Area	Approximate Number Square Miles	Estimated Gorilla Population	Number Gorillas Per Square Mile
Virunga Volcanoes	155	450	2.9
Kayonza Forest, Uganda	96	150	1.5

Gorillas or their spoor were encountered more often in these two isolated and protected areas than in the other forests sampled, suggesting that the population density for the mountain-gorilla range as a whole does not exceed one animal per square mile. If this estimate is accurate, about

8000 gorillas exist in the 8000 or so square miles of forest inhabited by the subspecies. However, until further work amplifies the population data in more regions, a figure of from 5000 to 15,000 is suggested.

Population Dynamics

We know very little about the factors that govern the dynamics of gorilla populations. For certain information, such as age at sexual maturity and longevity, only a few records from zoological gardens are available, and these may not be entirely applicable to free-living animals. The data I was able to obtain on wild gorillas reflect only the characteristics of the study population during a one-year period.

Age at Sexual Maturity

Three pairs of lowland gorillas have reproduced in zoos (at Columbus, Basel, and Washington), and several females have shown behavior indicating their sexual maturity. Two of the males were about 9 to 9½ years old when they impregnated the females, but a third male was only about 7 years old. The 3 females were about 7, 7¼, and 10 years old at the time of conception; 2 other females first showed cyclic sexual receptivity between the ages of 6 and 7 years. This evidence suggests that females reach sexual maturity between 6 and 7 years of age, and males usually at about 9.

Birth Rates

Infants remain with their mothers for about 3 years, and at least that amount of time elapses between births. Some juveniles about 4 years old occasionally associated with infantless females which were presumably their mothers. These observations suggest that females give birth every 3½ to 4½ years unless the infant dies. Of 27 females in 4 groups whose status was traced for more than 8 months, only 2 females lacked an infant or failed to give birth to one. Both these females seemed elderly and physically below par, one having a scabby skin, the other a blind eye.

Mortality Factors

Injury, predation, and disease were three general causes of mortality observed. Fourteen injuries were noted during the study. Most of these were minor, probably caused by collisions with branches or falls from trees. One female appeared to have a broken jaw, the only bone injury noted. Two of the wounds seemed to be bites; one of the bitten animals was an infant and the wound was so severe that it probably died.

Interactions between gorillas and most other large mammals in the forest, such as the elephant *(Loxodonta africana)* and buffalo *(Syncerus caffer),* appeared to be peaceful. One lone bull buffalo was observed to flee when a female gorilla approached to within 30 feet. Golden monkeys *(Cercopithecus mitis kandti)* were twice seen within 10 feet of gorillas, but the two species did not visibly respond to each other. Although gorillas and chimpanzees *(Pan troglodytes schweinfurthi)* inhabit the same forests in several areas, no interactions were seen. I found no evidence that leopards preyed on gorillas in my study area, but there is one reliable account from Uganda which indicates that they may occasionally do so. Some Bantu tribes persistently snare, spear, shoot, or net gorillas for food and to protect their crops, making man the only major predator on the apes.

Diseases are probably the chief cause of death in the mountain gorilla. Microfilaria have been collected from its blood, and such helminths as *Anaplocephala, Ascaris,* and *Anaglostoma* from its intestine. One adult male, autopsied by a veterinarian, died of gastro enteritis. Several of my study animals showed symptoms resembling those of the common cold. I found nematode ova, similar to those of the human hookworm, in 53.3 percent of 45 dung samples collected. Even minor ailments, such as the nematode infestation, can lower the resistance of the animal and make it susceptible to more serious disorders.

Longevity

At least five lowland gorillas have lived longer than 20 years in captivity. One of these died at the age of 34½ years and looked definitely old at that time. Very old-appearing animals are uncommon in the wild, indicating that longevity there perhaps rarely exceeds from 25 to 30 years.

Population Structure

Only the ten groups in the main study area were known sufficiently well to permit an analysis of population structure. Males, silver-backed and black-backed combined, constituted 18.9 percent of these groups, and females 36.7 percent (Table 3–1, A)—a ratio of approximately 2 females to 1 male. However, if all males, including the lone individuals, are considered, the ratio changes to about 1½ females to 1 male (Table 3–1, B). Although additional lone males undoubtedly existed in the area, their number was not sufficient to raise the total to the expected 1 : 1 ratio. Juveniles were difficult to sex accurately in the wild, but their sex ratio seemed to me roughly equal. It is possible that males have a higher postjuvenile mortality rate than females.

About 45 percent of the population was juvenile or infant. Infants outnumbered juveniles 52 (27.2 percent) to 31 (16.2 percent), or by about 30 percent. The 54 infants which were present at one or another time in the population indicate a yearly birthrate of 17/193 or 90/1000.

TABLE 3–1   Sex and age class composition of the Kabara gorilla population

Sex and Age Class	A Animals in Groups at the Beginning of the Study*		B Total Kabara Population Including All Known Lone Males, Births, and Animals Who Have Left Groups or Joined Them by the End of the Study	
	Number of Animals	Percent	Number of Animals	Percent
Silver-backed male	17	10.0	25	13.1
Black-backed male	15	8.9	18	9.4
Female	62	36.7	65	34.1
Juvenile	29	17.2	31	16.2
Infant	46	27.2	52**	27.2
Total	169	100.0	191	100.0

* Indicates animals present at the time of the first complete count of each group.
** Two infants that died soon after birth are not included.

Infant and Juvenile Mortality Rates

Of 13 infants born between August 1959 and August 1960, one died, one disappeared, and one was so seriously wounded that it probably died—a mortality rate of 23 percent in the first year of life. Given the birth rate and the population figures in Table 3–1, B, it can be deduced that, assuming a 50 : 50 sex ratio at birth, the number of males declines about 47 percent between birth and the age of 6 years. Because it was impossible to differentiate two age classes in adult females, it could not be determined if their mortality rate was of similar magnitude.

Group Dynamics

Group Size

Gorilla groups varied in size from 2 to about 30 animals. C. Cordier, an animal dealer, has trapped or seen groups of 4, 13, 14, 15, 19, and 25 animals in the lowland rain forests near Utu. In the Kayonza Forest, a mountain rain forest, I obtained counts of about 2, 5, 5, 7, 14, and 15 animals, per group. In my main study area the smallest group consisted of 5 animals, the largest of 27 (Table 3–2), but changes occurred in one group which brought its number temporarily up to 30. Although there appeared to be no difference in the size of groups in the various vegetation types, the average size of groups varied from region to region. Around Kabara average group size was 16.9 animals; in the Uganda portion of the Virunga Volcanoes and in the Kayonza Forest it was only from 7 to 8 animals. It is possible that in the absence of predation by man such factors as availability and type of forage influence the size of groups.

Group Composition

The composition of the 10 study groups at the time of the first accurate count of each group is presented in Table 3–2. All groups contained at least 1 silver-backed male, 1 or more females, and a variable number of young. Average group composition was: 1.7 silver-backed males, 1.5 black-backed males, 6.2 females, 2.9 juveniles, and 4.6 infants.

Changes in Group Composition

Changes in composition occurred with relative frequency in some groups, but only rarely in others. Three examples illustrate this.

Group VIII: This group was observed at intervals between November 1959 and May 1960. Its composition of November 19, 1959, was 1 silver-backed male, 2 black-backed males, 8 females, 3 juveniles, and 7 infants—a total of 21. No changes occurred in the composition of this group.

Group VII: I studied group VII in greater detail than any other group from October 1959 to September 1960. Its composition on October 10, 1959, was 1 silver-backed male, 2 black-backed males, 6 females, 4 juveniles, and 5 infants—a total of 18. Between February 1 and 6 an infant was born. Between February 14 and March 16 an unknown female with infant joined the group, the only change of this type noted during the study.

TABLE 3–2 The composition of gorilla groups at Kabara

Group	Silver-backed Male	Black-backed Male	Female	Juvenile	Infant	Total
I	1	0	3	2	2	8
II	1	3	6	5	4	19
III	1	0	2	1	1	5
IV	4	1	10	3	6	24
V	2	2	3	2	2	11
VI	1	1	9	2	7	20
VII	1	2	6	4	5	18
VIII	1	2	8	3	7	21
IX	4	3	9	5	6	27
XI	1	1	6	2	6	16
Total	17	15	62	29	46	169
Percent	10.0	8.9	36.7	17.2	27.2	100

Group IV: Detailed records of the composition of this group were obtained at intervals between March 1959 and August 1960. These records follow:

Composition on September 1, 1959: 4 silver-backed males, 1 black-backed male, 10 females, 3 juveniles, 6 infants—a total of 24.

March 12, 1959: An infant was born.

August 28 to August 30: A peripheral silver-backed male left; then rejoined the group.

September 9: A new silver-back joined the group. The peripheral male of August left again but rejoined it by September 22.

September 18–20: An infant was born.

Between October 2, 1959, and January 11, 1960: The peripheral male and the No. 2 male in the hierarchy left. A new silver-backed male and two females both with infants were added. They probably represented a small group which joined. An infant was born in late December.

April 24, 1960: A silver-backed male who had been with the group at least since August 1959 left.

April 25, 1960: An infant was born but died two days later.

Between May 1 and 15: A new silver-backed male joined the group and remained at least to May 24.

Between May 24 and August 12: The male who joined in early May left.

In the course of 17 months, 7 different silver-backed males were known to have associated with the group, but of the 4 present in August 1959 only the dominant male remained one year later. Figure 3–2 illustrates the coming and going of the males, and also indicates their dominance status and comparative size. Four infants were born, and 2 females with infants were added. At my last encounter, on August 13, 1960, the composition was 3 silver-backed males, 1 black-backed male, 12 females, 3 juveniles, and 10 infants, a total of 29.

In general, groups tended to remain quite stable in composition over many months. One group of 6 animals, which was repeatedly encountered by many different observers in the eastern part of the Virunga Volcanoes, showed no changes in two years except for the birth of one infant. Most changes in my study resulted from arrival and departure of males and the births and deaths of infants.

Subgroups

Subgrouping, or the temporary splitting of one group into two distinct units, was infrequent. Once group IV split into two units in the morning, moving about 600 feet apart, but the subgroups joined again the same eve-

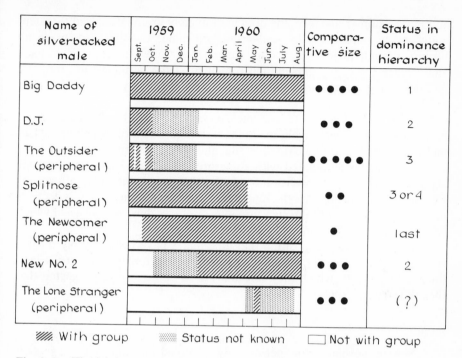

Name of silverbacked male	1959 Sept. Oct. Nov. Dec.	1960 Jan. Feb. Mar. April May June July Aug.	Compara- tive size	Status in dominance hierarchy
Big Daddy			● ● ● ●	1
D.J.			● ● ●	2
The Outsider (peripheral)			● ● ● ● ●	3
Splitnose (peripheral)			● ●	3 or 4
The Newcomer (peripheral)			●	last
New No. 2			● ● ●	2
The Lone Stranger (peripheral)			● ● ●	(?)

▨▨ With group   ▦ Status not known   ▢ Not with group

Fig. 3–2. The joining and parting of silver-backed males in group IV.

ning. Two juveniles and two silver-backed males apparently parted from group V for several days before rejoining, raising the total number of silver-backed males and juveniles in the group to four each (see Table 3–2).

Lone Males

The term "lone male" was applied to animals which were not associated with a group at the time of observation. At least four silver-backed males and three black-backed males were lone animals all or most of the time in the study area. Lone females were not encountered.

The most detailed observations on the behavior of males were made on group IV (Fig. 3–2). Males which tended to remain at the outer edge of the group were designated as "peripheral." Only the dominant male remained with the group the whole year; one male joined and left twice within about 20 days; others remained for several months before departing. Between November and May, one of the lone males, "The Lone Stranger," was seen repeatedly and usually alone. Once I encountered him at the periphery of group VI, but he left it the same day. Six months later he joined group IV, where he remained for at least one week.

The response of groups to lone males was observed three times. Once a silver-backed male merely walked into group IV without eliciting a re-

action. In another instance, the dominant male of group VI stared threaten-
ingly at an approaching lone silver-backed male, and the latter made no
further attempt to join the group. Again, two "lone" males, a silver-back
and a black-back, approached and remained at the periphery of group VI
one day; they were seen in the group the following day and had both left
by the third day.

Lone males fell into several age classes. The three black-backed males
in the study area were small, medium, and large. Most of the silver-backed
males appeared to be in the prime of life, and only one seemed old.

Lone males associated with some groups but not with others. Of the
six groups which were studied intensively, they were seen only with groups
IV and VI, suggesting that the lone males exercised some form of selectiveness. It is possible that through previous contacts they have learned which
group will accept them and which will not.

It is a popular supposition that lone males have been forcibly thrown
from groups by rivals. I observed no strife between males in a group, and
the readiness with which they left and joined some groups suggested that
they did so quite freely.

Intergroup Interactions

Several gorilla groups periodically occupied the same general section
of the forest, and two or more sometimes wandered close to each other.
On at least 12 occasions, one group heard but could not see another group
in the distance yet the groups made no attempt to approach each other.
However, sometimes groups met face to face. I took notes on the sequence
of events at four such meetings, and close association eight more times was
confirmed by evidence from trails and contact with the animals after the
event. Most interactions involved only two groups, but once three groups
(I, II, III) briefly occupied an area of about 300 feet in diameter.

The responses of groups to each other varied considerably. Group VII
and another group approached each other to within 300 feet or less on two
occasions, but made no attempt to join as they foraged and rested in full
view of one another for several hours. Groups VII and XI once sat in adjoining rest areas, and two members of group VII mingled for about two
minutes with group XI. However, the dominant male of group VII charged
several times silently and on all fours at the dominant male of group XI.
The two males then stared at each other, sometimes with brow ridges almost
touching. The two groups parted later in the day. In contrast, group VII and
group III joined one afternoon, occupied a common nest site (Fig. 3-3), and
parted the following morning. Groups VI and VIII once remained near
each other for three days before I saw them mingle briefly. On the first
day the groups were at one time only 35 feet apart, and that evening they
nested near each other, with 30 feet separating the closest animals of the

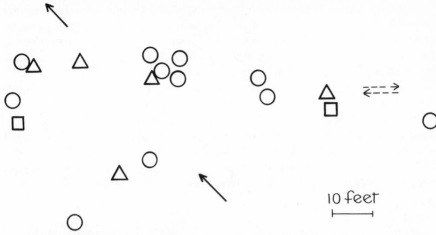

Fig. 3–3. A combined nest site of groups VII and III during the night of October 19–20, 1959. Squares represent nests of silver-backed males, circles of black-backed males or females, and triangles of juveniles. The solid arrow indicates the evening and morning direction of travel of group VII; the dashed arrow of group III.

respective groups. On the second day the dominant male of group VIII approached to within 15 feet of a female of group VI before the latter moved away. The two groups nested 800 feet apart, but finally, on the third day, mingled for about five minutes before parting.

On the whole, interactions between groups were peaceful. Aggressive bluff charges were made by one male as noted above, and I once observed weak aggressiveness, in the form of incipient charges toward intruders from another group by a female, a juvenile, and an infant. Chest-beating displays were not prominent, except once, the animals giving the impression of being only slightly excited.

The most striking aspect of intergroup behavior was the highly variable response of a given group to the presence of others. Group VII, for example, joined one group, merely advanced toward another, and behaved antagonistically toward a third. It is likely that many neighboring groups in a local population such as at Kabara have repeated contacts with each other and that each encounter affects the nature of subsequent meetings.

Groups Ranges and Movements

Home Range

Groups restricted their activities to definite areas or home ranges on the order of from 10 to 15 square miles each. Group VI, for instance, was encountered repeatedly over 8 square miles of forest and group VII over

8½ square miles; however, both groups left the study area for varying periods. Although the range of each group was fairly large, groups occasionally spent many days in a restricted locality. Group VII once remained for 18 days in 1½ square miles of terrain, then switched its center of activity to another part of the forest.

Home ranges were not exclusively occupied by one group (Fig. 3–4). Sometimes the overlap between adjoining ranges was slight, but for several groups it was quite extensive. Six groups intermittently used the same square mile of forest in one area, and three groups periodically occupied a slope by camp, once all at the same time.

The almost complete overlap of some ranges and observations on peaceful interactions between groups indicate that gorillas have no ter-

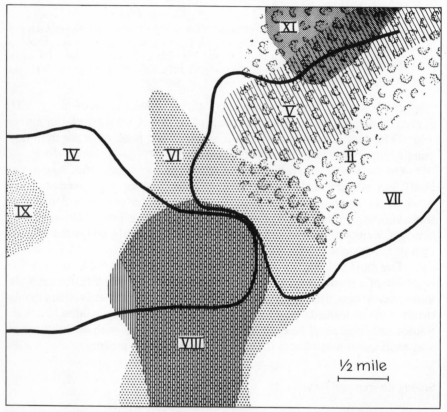

½ mile

Fig. 3–4. A schematic diagram illustrating the extensive overlap of the home ranges of eight gorilla groups in the Kabara area of the Virunga Volcanoes. Two further groups, not indicated on the diagram, were encountered several times in the range of group VII.

ritory in the sense of an exclusive area defended against others of the same species.

Seasonal Movement

Except for bamboo, the vegetation types frequented by gorillas show no conspicuous differences in the abundance of forage throughout the year. No evidence that gorillas move seasonally was obtained, although some groups used the bamboo below Kabara more during the periods of heavy rain, when shoots were present, than during the drier periods of the year.

Patterns of Group Movement

The only generalization about movement which can safely be made is that groups travel continuously within the boundaries of their home range and that they frequent certain parts of this range at irregular intervals. Group VI, for example, made its appearance near our camp at intervals of about 40 days (from 24 to 57 days), and group VIII at intervals of about 60 days (from 52 to 78 days). Groups remained in a limited area anywhere from 1 day to 1 month before moving on to another part of the forest.

Patterns and Distances of Daily Movement

That the direction and distance of travel by a group in the course of a day tends to be unpredictable is illustrated by Fig. 3–5, which shows the peregrinations of group IV over a 17-day period. I paced, or alternately paced and estimated, the complete daily trail of a group from night nest site to night nest site on 114 occasions. The distance of travel in this sample varied from 300 to 6000 feet with an average of 1742 feet. However, the average distance varied somewhat from group to group. Group VI, for instance, averaged 1242 feet (S.D. 660) per day, and group VII averaged 2177 feet (S.D. 353) per day.

Groups tended to move farther in the afternoon than in the morning. After rising from the night nest, usually between 06:00 and 08:00 hours, the animals feed intensively for about 2 hours, moving very slowly. Between 09:00 and 10:00 hours activity slowed down as most gorillas snacked, sunbathed, and slept intermittently until about 14:00 hours. Thus, by noon the average distance of a group from its nest site of the previous night was only about 500 feet, and by midafternoon about 700 feet. Approximately half of the average daily distance was covered between 15:00 and 17:00 hours as the animals walked fairly steadily while feeding. After 17:00 hours activity tended to slow down and finally ceased at dusk, at 18:00 hours, when the gorillas bedded down for the night and remained in their nests during the hours of darkness.

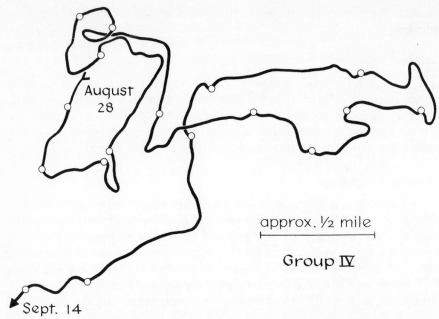

Fig. 3–5. The daily route of travel of group IV on the slopes of Mt. Mikeno between August 28 and September 14, 1959. Each circle represents one night nest site.

Search for food did not account for all the movement observed. A wandering group at Kabara ate only a negligible amount of the available forage in its path. On several occasions, and for unknown reasons, a group moved without stopping some 5000 to 15,000 feet to another part of the forest before settling down again.

SOCIAL BEHAVIOR

Gorilla groups are quite cohesive. The central core of each group is composed of the dominant male and all females and young. The extra males, both black-backed and silver-backed, tend to be peripheral. The diameter of a group at any given moment rarely exceeds 200 feet as every animal remains attentive to the movements of others in the dense forest environment. Except for extra males, single individuals rarely are more than 100 feet from other members of the group. The behavior of the group is coordinated by four means:

1. Postures and gestures are important in communication. A dominant male who stands motionless, facing in a certain direction, indicates that

he is ready to leave and the other members of the group crowd around him; an unexplained sudden run by an animal communicates danger; in order to be groomed a gorilla merely presents a part of its body to another animal.

2. How much facial expressions communicate emotions is difficult to determine, for many occur in conjunction with gestures and vocalizations. Some facial expressions, like the wide-open mouth and exposed teeth of angry animals, probably emphasize and elaborate information communicated by other means.

3. Vocalizations and other sounds are occasionally an important means of communication in dense vegetation, for not until the animal has drawn attention to itself is further communication by means of gestures or facial expressions possible.

4. Physical control appears to be of importance only in females handling their infants. A female pulls her small infant back when it wanders too far from her side, and she regularly shifts it from her chest to her back and vice versa.

The small number of overt social interactions was a most striking aspect of intragroup behavior. The most frequently noted interactions were dominance (at .23 times per hour of observation), mutual grooming (at .28 times), and social play (at .11 times). The relative infrequency of interaction can probably be attributed to the following circumstances. Competition for food and mates provided little basis for strife since forage was abundant and sexual behavior not prominent. The members of a group were alert to the possibility of aggressive encounters, and subordinate animals tended to circumvent issues before they materialized. And, finally, the gorillas gave one the impression of having an independent and self-dependent temperament, appearing stoic, aloof, and reserved in their affective behavior.

Though the members of a group spent most of the day very close to one another and were highly tolerant of each other, persistent aggregations of specific individuals were rare. Once two females in group VII consorted closely for several days. One juvenile tagged behind the sole black-backed male of group IV for more than one month. Various members of the group sometimes sought the vicinity of the dominant male for brief periods. In group IV as many as four silver-backed males occasionally rested within a diameter of 30 feet. Females now and then approached the dominant male to lie by his side, and on nine occasions rested their heads on his saddle or leaned against his body. Juveniles and infants were also attracted to the silver-backed male. At times as many as four youngsters climbed over the reclining male, slid down his rump, and pulled his hair without eliciting a response. Twice within the space of an hour, a large infant grabbed the rump hairs of the dominant male and hitched a ride for from 30 to 50 feet. However, the only persistent associations were between mothers and their young.

Dominance and Leadership

Definite dominance interactions were observed 110 times (Table 3–3). Dominance was most frequently asserted along narrow trails, when one animal claimed the right of way, or in the choice of sitting place, when the dominant animal supplanted the subordinate one. Gorillas showed their dominance with a minimum of actions. Usually an animal low in the rank order simply moved out of the way at the mere approach or brief stare of a high-ranking one. The most frequently noted gesture involving bodily contact was a light tap with the back of the hand of a dominant individual against the body of a subordinate one.

TABLE 3–3 **Dominance interactions between age and sex classes**

Dominant Animal	Subordinate Animal					
	Silver-backed Male	Black-backed Male	Female	Juvenile	Infant	Total
Silver-backed male	13	1	26	9	2	51
Black-backed male	—	—	2	3	—	5
Female	—	4	12	13	11*	40
Juvenile	—	—	—	4	8	12
Infant	—	—	—	—	2	2
(One year or older)						
Total	13	5	40	29	23	110

* Tabulated only if infant was not in direct contact with its mother at the time.

One example from my field notes illustrates a typical dominance interaction:

A juvenile sits under the dry canopy of a leaning tree trunk during a heavy rain storm. A female walks toward the juvenile who rapidly vacates its seat, while she appropriates the dry spot. Shortly thereafter the silver-backed male arrives and pushes the female with the back of his hand on the lower part of her back until she is out in the rain and he is under cover.

Dominance was largely correlated with body size. Silver-backed males were dominant over all black-backed males, females, and young. If more than one silver-backed male was present in a group, they appeared to have a linear hierarchy (Fig. 3–2), seemingly based in part on age, with old and young males occupying subordinate positions. Black-backed males and females were dominant over all juveniles and infants not with their mothers. The relationship between black-backed males and females varied, perhaps being dependent on size and personal differences. As shown in Table 3–3, females were dominant over black-backed males four times and the reverse occurred twice. Females appeared to lack a stable hierarchy among themselves, for situations which resulted in dominance behavior between other group members frequently elicited no response among females. The 12 female-female interactions suggest that females have a changing hierarchy in which mothers with young infants are dominant over those with older infants or with none. Dominance among and between juveniles and infants seemed to be based on size.

The dominant silver-backed male was also the leader of the group. Every independent animal in the group appeared to be aware of the activity of the leader either directly or through the behavior of animals in his vicinity. Cues reflecting a changed pattern of activity were patterned after the leader. Thus, the entire daily routine—the distance of travel, the location of rest stops, and the time and place of nesting—was largely determined by the leader. In times of danger the leader also acted as protector of the group, frequently dropping behind the fleeing animals to face the intruder. Once, when I inadvertently surprised group IV at close range, the leader grabbed a juvenile around the waist and raced from 30 to 40 feet away before releasing it. Variations in the behavior of groups tended to reflect the individual idiosyncrasies of the leaders.

Groups moving rapidly tended to travel in single or double file with the leader at or near the front. One procession in group VII was typical: silver-backed male, female and infant, female and infant, juvenile, juvenile, black-backed male, female and infant, large infant, female and infant, 2 females (one with infant) and 2 juveniles together, female and infant, black-backed male. However, when movement was slow, the leader often traveled in the center or near the end of the group. The number-two silver-backed male in the group IV hierarchy assumed leadership of a subgroup until the two units rejoined.

Mutual Grooming

The 134 observations of instances in which one animal groomed the pelage of another are summarized in Table 3–4. Gorillas did not often groom each other, and grooming was never reciprocal. No female was seen to groom the dominant male, and females groomed each other only five times. Nearly three-fourths of my observations were made on females grooming infants and juveniles, the activity being sometimes intense and prolonged. Juveniles groomed females on nine occasions, sometimes when the latter carried small infants. A 1¼-year-old infant groomed the back of

TABLE 3–4 Number of observations of mutual grooming in the various age and sex classes

Animals Grooming	Animals Groomed					
	Silver-backed Male	Black-backed Male	Female	Juvenile	Infant	Total
Silver-backed male	0	0	0	0	3	3
Black-backed male	0	0	0	0	0	0
Female	0	1	5	13	76	95
Juvenile	1	1	9	10	12	33
Infant	0	0	2	0	1	3
Total	1	2	16	23	92	134

its mother; it was the youngest animal observed to groom either itself or another.

The function of grooming in adult gorillas seems to be primarily utilitarian since they rarely groom each other and when they do, they concentrate on those parts of the body which the animal itself cannot reach with ease. However, the purpose of grooming appears to vary somewhat with age and sex class. At times juveniles seemed to employ grooming as a means of initiating social contact with females and with infants. Generally, animals groomed themselves, with females doing so proportionally twice as often as males, and juveniles another half as many times as females. Infants less than 2 years old were never seen to groom themselves.

Play

I recorded as play any relatively unstereotyped behavior in which an animal was involved in vigorous actions seemingly without definite purpose. Of 156 gorillas involved in 96 observations of play, all but five were juvenile or infant. One female played with an infant by holding it down until it struggled free, the only interaction of this type noted. About half (43 percent) of the animals played alone, and this usually involved swinging and sliding on branches and lianas, batting vegetation, and turning somersaults down slopes. One infant placed a cushion of moss on its head like a cap and walked back and forth on a branch. Infants played alone twice as frequently as juveniles.

In social play infants had their first opportunity to interact closely with other youngsters in the group. Some 81 percent of the play groups consisted of only two animals, but a few contained three or four young. Playing youngsters rarely remained together for more than 15 minutes. Although a juvenile sometimes weighed five times as much as the infant with which it was wrestling vigorously, the disparity in weight never caused quarrels or injury, for the larger animal always contained its strength. Most social play included wrestling and chasing, and such games as follow the leader and king of the mountain.

The following two quotations from my field notes illustrate typical lone and social play sequences:

(1) A 9-month-old bumbles around by the reclining silver-backed male. With a wide overhand motion it swats the male on the nose, but he merely turns his head. The infant then runs downhill and turns a somersault over one shoulder and ends up on its back, kicking its legs in bicycle fashion and waving its arms above the head with great abandon. A 10-month-old infant watches these proceedings while propped against the rump of the male. Suddenly the ten-month-old infant rises, hurries to a sitting juvenile and pulls the hair on its crown with one hand. When this brings no response, the infant yanks at the hair with both hands, but the juvenile remains oblivious. The infant desists, sits briefly, suddenly rolls forward over one shoulder, and with arms and legs flailing like a windmill rolls over and over downhill and disappears in the vegetation.

(2) A juvenile and a 1¾-year-old infant sit about 4 feet apart. Suddenly the juvenile twists around and grabs for the infant, which rushes away hotly pursued by the juvenile. The juvenile catches the infant and covers it with his body, propped on elbows and knees. Twisting and turning, struggling and kicking, more and more of the infant emerges from beneath the juvenile. Freedom gained, the infant grabs an herb stalk at one end and the juvenile snatches the other end. They pull in opposite directions; the juvenile yanks hard and the infant is jerked forward. They then sit facing each other, mouths open, and swing their arms at each other and grapple slowly. Another juvenile comes dashing up, and in passing swipes at the juvenile and all three disappear running into the undergrowth (Schaller 1963).

Female-Young Interactions

The relationship between the mother and her single offspring changes from the infant's complete dependence during the first few months of its life, through a period of gradual lessening of the physical and emotional bonds, to a stage when the youngster becomes integrated into the group.

Female-Infant Interactions

The ties between mothers and their infants remain close for about 3 years. Only females care for infants, feeding, transporting, and protecting them when they are small, and, after they are fairly self-sufficient, providing the social comfort which the young seem to derive from the female's proximity.

FEEDING. I observed only 5 instances of suckling in infants older than 1 year of age, indicating that the young were partly weaned by then. Twice females pushed away their young, 8 months and 1¼ years old respectively, when they attempted to suckle. One infant, 5 months old, suckled on a female who was not its mother. Infants ingested some solid food by the age of 2½ months, and by the age of 7 or 8 months the bulk of their diet probably consisted of forage.

Females rarely exerted direct control over the foraging of their young, who appeared to learn what to eat and what not to eat largely by observing their mothers and by trial and error. Once a female removed a leaf from the hands of her infant and threw it away, the only such instance noted. One infant took some food from the hands of its mother and ate it, and another picked a leaf from its mother's lip. On 5 occasions infants placed plants which adults did not eat into their mouths, but 3 times they spat them out.

TRANSPORT. Over long distances mothers carried only their own infants. Juveniles picked up infants 8 times but never transported them farther than 50 feet before the infant either struggled free or was retrieved by its mother. The method of transport varied with the infant's age. A newborn young appeared to lack the strength to clasp its mother's hair securely, and she supported it to her chest with one or both arms until it was nearly 3 months old. By the age of 1 month an infant was able to hang on by itself long enough to permit the female to use both hands briefly while ascending trees. By the age of 3 months infants began to ride regularly on the broad backs of their mothers. The characteristic position of the infant on the back of a walking female was to lie prostrate with the head near the shoulder region and with the hands grasping the mother's sides, shoulders, or neck.

Although the youngster was carried throughout the period of infancy, the female gave it the opportunity to travel on its own at an early age. One female placed her 3-month-old infant, barely able to crawl, on the ground and walked slowly along while her offspring attempted to follow. By 4 and 5 months of age, when infants assumed the quadrupedal gait of adults, they began to leave their mothers and to seek social contact with others in the group. By 6 and 7 months of age, infants were walking and climbing by themselves. Large infants, from 1 to 1¼ years old, spent much time sitting beside rather than upon their mothers. When infants reached the age of 1½ years and weighed some 25 pounds, females sometimes appeared unwilling to carry them. A frequent sight was a female walking slowly with an infant toddling at her heels. However, at the first sign of danger or the onset of rapid movement all infants up to the age of nearly 3 years rushed to their mothers and climbed aboard.

PROTECTION. The mother guarded the infant from harm during the average day-to-day existence. On two occasions mothers batted away the hands of other females when they attempted to touch a newborn. When younger than about 4 months old, the infant was rarely permitted to stray farther than 10 feet from its mother before being pulled back to her side. By the age of 8 months, young sometimes strayed more than 20 feet from their mothers, and by the age of 1 year they wandered through the resting group, sometimes out of their mothers' sight. One female retraced her steps 8 feet through a tree and pulled her 1½-year-old young across a gap in the branches. Another female showed strong concern for her infant, which had a large wound on its rump; the infant never rode on her back, but was cradled gently in her arm in such a way that no part of the injury was touched.

SOCIAL.   The mother was the only object in the environment to which the infant turned readily at all times even after it had been weaned and was able to travel under its own power. Large infants retained their social ties with their mothers, remaining near them most of the day and all night until they became fully integrated into the group at the age of about 3 years.

Two infants were observed to have strong social bonds with females other than their mothers. One 6-month-old young repeatedly visited an infantless female and stayed with her for an hour or more. Another infant, 1½ years old, spent most of its time from November through May with a female and small infant, returning to its mother only intermittently during the day and apparently at night.

Strong social ties also appeared to bind females to their infants. Sometimes a female walked up to a large infant, sat beside it, and placed her arm over its shoulder, drawing it close. The infant of one female died 2 days after birth and was then carried by her for 4 days before being discarded on the trail.

Female-Juvenile Interactions

Although most females and juveniles interacted only in dominance and grooming, a few juveniles associated closely with certain females, presumably their mothers. One juvenile, about 3 or 3½ years old, was constantly near an infantless female; 1 year later the female still had no infant, but the juvenile associated with her only intermittently. Another juvenile, judged to be 3½ or 4 years old, was frequently seen with an infantless female, but when she gave birth the juvenile no longer stayed near her. Several juveniles approached females with small infants and rested near them. A close examination of dung in nest revealed that juveniles occasionally slept in the nest with a female or with a female and infant. Although females sometimes rebuffed juveniles mildly by swatting at them or by simply walking away from them, the social ties between some animals persisted until juveniles were about 4½ years old.

Mating Behavior

My work on free-living gorillas yielded remarkably few observations on sexual behavior. I witnessed only two copulation sequences and one invitation to copulate. Thus, most of my information is based on captive gorillas, whose behavior is not necessarily comparable to free-living ones. Captive males and females, for example, masturbate and display such eroticisms as fondling each other's genitalia, behavior never seen in the wild. In fact, I saw no instances of play mounting, homosexualism, or deviant sexual behavior in wild gorillas.

Estrous Cycle

The monthly cycle of the female is not readily discernible in nature, for gorillas lack a prominent genital swelling and their menstrual flow of blood is minute. However, captive females exhibit intermittent periods of sexual receptivity, lasting from three to four days, during which they initiate contact with other gorillas or with humans. Reports from several zoos indicate that the average length of this cycle is from about 30 to 31 days.

Copulation

I observed copulation in group IV on September 4 and 23, 1959. Two different silver-backed males were involved, and both were subordinate individuals. One occupied the number-two position in the hierarchy and the other was peripheral. On both occasions the dominant male was resting near the copulating animals and did not prevent them from completing the act. One copulation lasted 15 minutes as the male thrust intermittently while covering the back of the female, who rested on her knees, belly, and elbows. A female initiated the second copulation by mounting the male and thrusting about 20 times. The male then grabbed her by the hips and pulled her into his lap. They copulated first in a squatting position, with the female facing away from the male, and later with the female lying prone and the male squatting at her rump. The animals copulated three times in one hour, resting during the intervals. The male reached orgasm only during the third copulation, after a total of about 300 thrusts.

Copulation in wild gorillas was more direct than in captives, lacking the prolonged preliminary wrestling, chasing, and fondling which I observed in a pair of lowland gorillas at the Columbus zoo. Captive gorillas also copulated in a ventro-ventral position.

Pregnancy, Parturition, and Gestation Period

The abdomen of wild females is ordinarily so distended that pregnancy cannot be detected with certainty. Captives tend to show edema of the ankles, a gain in weight, and a certain irascibility. Although I twice encountered females in the wild within a few hours after they had given birth, I never witnessed the event itself. However, six infants have been born in captivity. The birth itself occurs so rapidly that no trained observer has witnessed it. The native keepers at a stockade in the Congo saw a female mountain gorilla lie down, and within five minutes the head of the infant appeared; the mother pulled the infant out with her hands, but then killed it. Four of the captive females took little or no care of their first-born. However, two of these females gave birth a second time and cared for their infants.

The gestation period of four infant lowland gorillas born in zoos was 251–253, 252, about 266 (256–295), and 289 days, respectively.

Breeding Season

In an effort to determine if gorilla young are born at certain seasons, I estimated the probable month of birth for all 54 infants in my study area. The 27 birth date estimates for 1959 and 1960 are listed below and suggest that gorillas lack a definite breeding season.

Month	1959	1960	Month	1959	1960
January	0	1	July	4	
February	2	1	August	3	
March	2	0	September	4	
April	2	1	October	1	
May	1		November	0	
June	3		December	2	

Aggressive and Submissive Behavior

There was considerable variation in the intensity of aggressive behavior which gorillas exhibited toward each other and toward such intruders as man. In the order of increasing intensity, the responses included: (1) an unwavering but usually brief stare, sometimes with furrowed brow and slightly pursed lips; (2) a jerk of the head or a snap in the direction of the offending animal; (3) an incipient charge, indicated by a light foward lunge of the body, occasionally without moving the feet, but usually accompanied by one or two abrupt steps; (4) a quadrupedal bluff charge over a distance of from 10 to 80 feet; and (5) physical contact in the form of biting or wrestling.

Most intragroup aggressiveness was confined to staring and snapping. I saw incipient charges about 10 times, directed primarily at members of another group or at lone males. One dominant male charged in bluff at the dominant male of another group, the only such instance noted. Occasionally two members of a group ran at and slapped each other in passing, behavior which appeared to be redirected aggressiveness elicited by my presence. I have not witnessed serious aggressive contacts between gorillas. Although females sometimes quarreled, the grappling, screaming, and mock biting never resulted in a discernible injury. However, if a gorilla group is harassed by man, males, and occasionally females, may attack him and cause serious injury. According to several observers, such attacks usually consist of a lunge forward, brief contact during which the gorilla bites, and retreat.

The following excerpt from my field notes describes a typical quarrel between females:

> A female walks leisurely past another one sitting by the trail. The latter slaps her on the back for a reason unknown to observer. She in turn wheels around and runs with open mouth straight at the female who swatted her. This female cowers down with legs and arms tucked under, but with head raised screaming loudly. Her lips are curled up and the teeth and gums show. The two females then grapple briefly and mock-bite each other's shoulder.
>
> As the two fight, two other females run up and join the melee. All four then scream, grapple with each other, and run around with teeth bared. The rest of the group watches; that is, all but the silver-backed male who sits five feet from the nearest combatants and does not even turn his head toward them. After about 15 to 20 seconds three females cease fighting and walk away. Only one female remains in the battle area and emits short screams. Suddenly she takes two steps after one of the retreating females and slaps at her hind leg, whereupon the latter one turns and advances screaming. The former one backs away and collides with the silver-backed male, who gives an annoyed grunt.
>
> The two females meet and wrestle briefly as a third runs up to join the hassle. Finally all part after the whole sequence has lasted about one minute (Schaller 1963).

Gorillas indicated their submissiveness during an aggressive encounter in three ways. Usually they simply averted their eyes by turning their heads to one side. When an animal was within 60 feet of me and presumably nervous, it sometimes shook its head rapidly back and forth, a gesture that appeared to mean, "I intend no harm." Occasionally when a male, excited by my presence, slapped at a female or youngster the latter cowered down on the abdomen, head lowered, and arms tucked under, presenting only the broad back.

INDIVIDUAL BEHAVIOR

Locomotion

Gorillas are primarily quadrupedal and terrestrial, walking with the soles of the feet flat on the ground and with the anterior part of the body supported on the middle phalanges of the fingers. Although gorillas spent much time on their hindlimbs with their hands free for such tasks as grooming and feeding, bipedal locomotion was rare. The animals frequently took a step or two in a bipedal position, but only twice were they seen to walk farther than 20 feet in this manner.

Gorillas ascended trees with ease to sit, feed, rest, and nest, though they did so cautiously. Only young animals swung on lianas and ran along branches. No gorilla was ever seen to brachiate — that is, to swing from branch to branch by means of its arms alone. Proportionally females ascended trees to a height of 10 or more feet twice as often as silver-backed and black-backed males, and juveniles more than 4 times as often. Infants less then 1¼ years old usually hesitated to climb high into trees, but the infant class as a whole climbed with a frequency comparable to that of females.

Feeding Behavior

Gorillas are herbivorous and subsist on a wide variety of vines, leaves, barks, roots, and some fruits. I obtained no evidence that they feed on animal matter. The apes usually foraged leisurely, alternately sitting and walking. Food was collected almost entirely manually, although a leaf was sometimes detached directly by mouth. The gorillas were highly selective feeders, consuming only certain parts of certain plants. They used hands and teeth equally to tear and shred a food item, with the hands prominent in manipulating, holding, and pulling, and the teeth in biting and gnawing. Food was rarely transported and never for more than 25 feet. No animal offered food to another and no gorilla was seen to use a tool to obtain its food.

The following quote from my field notes describes the leisurely feeding behavior of "Junior," a black-backed male:

> Junior sits and peers intently at the vegetation, reaches over, and bends the stalk of a *Senecio tichopterygius* to one side. He stretches far out and with a quick twist decapitates a *Helichrysum*. After stuffing the leafy top into his mouth, he looks around and spots two more plants of the same species which he also eats in similar fashion. He then yanks a *Peucedanum kerstenii*, including the root, from the ground, and with rapid sideways and backwards jerks of the head bites apart the stalk before gnawing out the pith. The sun appears

briefly and Junior rolls onto his back. But soon the sun hides behind a cloud, and Junior changes to his side, holding the sole of his right foot with the right hand. After about 10 motionless minutes he suddenly sits up, reaches far out, slides his hand up the stalk of a *Carduus afromontanus,* thus collecting the leaves in a bouquet which he pushes with petioles first into his mouth. This is followed by a leafy thistle top, prickles and all, and a *Helichrysum.* He then leaves his seat, ambles 10 feet, and returns to his former place, carrying a thistle in one hand and a *Helichrysum* in the other. After eating the plants he sits hunched over for 15 minutes. The rest of the group feeds slightly uphill and Junior suddenly rises and moves toward the other members, plucking and eating a *Helichrysum* on the way. A *Senecio erici-rosenii* has been torn down by another gorilla, and Junior stops and rips off a leafy top. From the stem he bites large splinters until only a two-inch section of pith remains in his hand, which he eats. A strand of *Galium* follows, and just before he moves out of sight, a final *Helichrysum* (Schaller 1963).

One hundred species of food plants were collected in various parts of the gorilla's range. This figure represents a minimum, for some plants were not identified below the generic level and thus are not included in the tabulation.

Number of Species	*Part of Plant Eaten*	*Type of Plant*
4	shoot, base of stem	Grass-sedge
5	whole plant, pith of frond and stem, root	Fern
29	leaf, fruit, stem, bark, pith, flower, root	Herb
22	whole plant, leaf, stem, fruit, bark	Vine
9	leaf, fruit, flower, bark, pith	Shrub
25	leaf, fruit, bark, pith, rotten wood	Tree
6	fruit (maize, peas), pith (banana), root (taro, manioc, carrot)	Cultivated

Gorillas have so well adjusted their food habits to the local vegetation types that I found, for example, no overlap between plants eaten in lowland rain forest and *Hagenia* woodland, although I collected 17 food plants in the former and 29 in the latter. Of the fairly large number of plants eaten in any one area, only a few provided the bulk of the forage. In the young secondary growth of lowland rain forest, the pith of the stem of the cultivated banana and the pith of the herb *Aframomum* were the plants most commonly consumed; around Kabara, 1 vine and 3 herbs (*Galium simense, Peucedanum linderi, Carduus afromontanus, Laportia alatipes*) furnished at least 80 percent of the daily food supply.

Several forage plants were widespread and tolerated considerable altitudinal variation. With these I attempted to determine if "cultural" differences in food habits existed from area to area. The vine *Urera hypselendron* and bamboo were eaten wherever they occurred. On the other hand, gorillas in the isolated Kayonza Forest of Uganda were not seen to eat five plants (*Galiniera coffeoides, Aframomum* sp. [stem], *Palisota* sp., *Marattia fraxinea,* and *Pennisetum purpureum*), although these plants were consumed by gorillas in other forests. Perhaps the animals ate them only sparingly, but my observations suggested qualitative rather than quantitative differences.

Gorillas were twice seen to feed on volcanic soil on the slopes of Mt. Mikeno. Analysis of the soil showed that it was high in sodium and potassium.

Drinking Behavior

I never saw gorillas drink in the wild. In the Virunga Volcanoes, where permanent water was scarce, the animals probably obtained most of their moisture from the succulent forage.

Nesting Behavior

Gorillas built crude platforms, either on the ground or in trees, to which the term "nest" was applied. Nests occasionally were built for resting at any time of the day, commonly between 09:00 and 13:00, and usually for sleeping at dusk. The animal stood or sat and broke and bent nearby branches, vines, and other vegetation from all sides, placing them around and under its body to form a roughly circular structure. There was no particular sequence in the placement of the vegetation, nor was there interlacing, knot-tying, or other involved manipulation. Gorillas never used the same nest two nights in succession.

The following quotation from my field notes describes a typical instance of nest construction:

> Group VII; April 1, 1960. A juvenile sits at the base of a tree and bends four to five handfuls of small herbs toward its left side with the right hand. It then stands on two legs, grabs the top of a mass of *Senecio trichopterygius* heavily overgrown with *Galium,* and pulls it in. It sits and breaks or bends the tips of the herbs to fit in a semicircle around its body before pressing the mass down with both hands. Standing on three legs, it reaches far out and breaks two to three more *Senecio* stalks off at the base and pulls them in. After placing these individually along the edge of the nest, it breaks their protruding tops to fit the rim. It sits, turns around, and sits again. The time required for building was about one minute (Schaller 1963).

Ground nests varied considerably in the care with which they were constructed. In the Hagenia woodland 10 percent of the animals bedded down for the night without constructing a definite nest. Numerous nests consisted merely of two or three handfuls of herbs pushed down to form a partial rim. However, some nests had rims nearly two feet high and central cups a foot deep. There was no definite correlation between the age and sex of the animals and the complexity of the nests they constructed, although juveniles tended to build crude ones. Ground nests seemed to have little or no function. Most were so crude that they could offer but little comfort, or insulation from the bare soil. Nests did not protect the animal from rain and other adverse weather unless they were placed in a sheltered position. When it rained at bedtime, 18.5 percent of the animals built their nests under the leaning boles of trees and other shelters, as compared to 4.7 percent when it did not rain. In general, nest building on the ground appeared to be vestigial behavior pointing to the ape's arboreal ancestry.

The principle of constructing tree nests was the same as that for ground nests, except that they were usually solidly built, and, in the absence of a firm substrate, considerable attention was given to the nest bottom. Nests in trees were functional in that they provided a platform on which the gorilla could recline without danger of falling.

TABLE 3–5   Height of gorilla nests above ground

Location	Hagenia Woodland. Kabara, Virunga Volcanoes 2488 Nests (percent)	Mountain Woodland and Bamboo. Uganda Side, Virunga Volcanoes 106 Nests (percent)	Mountain Rain Forest. Kayonza Forest, Uganda 179 Nests (percent)	Lowland Rain Forest. Near Utu, Congo 110 Nests (percent)
On ground	97.1	45.3	53.5	21.8
2–10 feet above ground	2.3	50.0	35.4	13.6
11–20 feet above ground	0.3	4.7	8.9	26.4
21+ feet above ground	0.3	0	2.2	38.2

Although gorillas nested both on the ground and in trees, the percentage of ground nests varied from area to area and seemed to be partly correlated with the availability of suitable trees. Table 3–5 shows that 97.1 percent of the nests in Hagenia woodland were on the ground as compared to only 21.8 percent in lowland rain forest. In the latter vegetation type, gorillas often nested from 20 to 60 feet above ground, a height rarely reached in the other areas. Analysis of dung in 50 nests above ground in Hagenia woodland showed that no silver-backed males nested in trees and that juveniles nested above ground twice as often as females and black-backed males combined. However, in the eastern portion of the Virunga Volcanoes, I saw nests of large males as high as 8 feet above ground.

Each juvenile and adult gorilla independently built its own nest for the night, although a juvenile occasionally slept with a female, or two juveniles occupied the same nest. Infants usually remained with their mothers, but sometimes a large one slept in its own nest adjacent to that of the female.

Some infants built crude practice nests, both on the ground and in shrubs, but these were rarely used. The earliest age at which I observed nest-building behavior was 8 months.

On the 4 occasions when I watched the behavior of the whole group, the dominant male was the first animal to construct the night nest, and the other members then nested in his vicinity. Most bedding areas were compact (Fig. 3–6a), but occasionally some were split into 2 parts (Fig. 3–6b). Splitting occurred when a few members of the group continued to forage after the others had bedded down. Black-backed males sometimes nested 60 or more feet from the main site. The distance to the nearest neighboring nests for the various age and sex classes was measured in 146 sites at Kabara, with the following results.

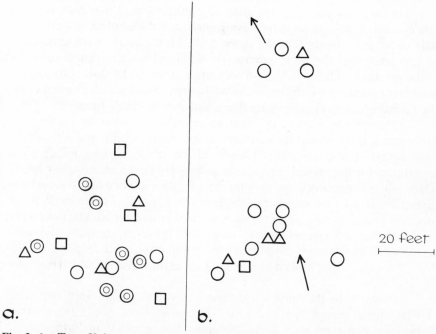

a.   b.

Fig. 3–6.   Two Kabara night nest sites—one compact, the other split. Squares
            represent nests of silver-backed males, circles of black-backed males or
            females, circles within a circle of females with infants (in Fig. 3–6a only),
            and triangles of juveniles. The arrow indicates the direction of travel.
            a. (Left) Group IV   b. (Right) Group VII

The data suggest that most members of the group do not place their nests with reference to a certain sex class or particular animal but that medium-sized gorillas and juveniles may do so.

Total Number Observations	Age and Sex Class	Mean Distance between the Nearest Nests (in feet)
24	Silver-backed male/silver-backed male	34.0
103	Silver-backed male/medium animal (female and black-backed male)	13.4
81	Silver-backed male/juvenile	20.7
144	Medium/medium	5.4
120	Medium/juvenile	3.0
76	Juvenile/juvenile	12.0

Vocalizations

Undisturbed gorillas were generally quite silent. I noted 21 more or less distinct vocalizations in free-living gorillas, but four of these were heard only once and 7 fewer than 10 times each. Most sounds were abrupt and of low pitch and the most intense vocalizations were given in response to the presence of man. Table 3–6 presents a list of the 13 most common and easily recognizable vocalizations, as well as the situations which elicit them, the probable emotion underlying them, and their seeming function.

Although the number of basic vocalizations emitted by gorillas was fairly small, there was great variation in the pitch, pattern, and intensity of each sound. These variations broadened the scope of the gorilla's vocal repertoire, for the animals responded selectively to the sounds they heard. Their reaction depended not only on the condition under which a sound was given, but also on the member of the group that gave it. For example, harsh staccato grunts, signifying annoyance, emitted by a silver-backed male when females quarreled, caused them to subside. If, however, the male gave what appeared to be the same sound when the group was quietly resting and feeding, all members first looked at him and then in the direction which occupied his attention.

Infants were the least vocal members of the group. Soft, tremulous whines, with pursed lips and raised forehead, were noted three times. Screeches of distress were fairly common when infants thought themselves left behind by their mothers. Playing youngsters sometimes chuckled audibly with their mouths wide open and lips drawn back into a smile. Juveniles barked and grunted harshly in response to my presence, vocalizations not heard in free-living infants under similar conditions. Females emitted 12 different vocalizations, the largest number of any age and sex class; a panting ho-ho-ho during the chest-beating display appeared to be peculiar to them alone. Black-backed males were vocally at an awkward age; their voice seemed to be changing from one resembling the female's to that of

TABLE 3–6   The most conspicuous or prevalent vocalizations of free-living gorillas

Description of Vocalization	Probable Emotion	Stimulating Situation	Probable Function
*Soft grunting	Content-ment	Feeding and resting peacefully	Indicates that all is well
*Series of abrupt grunts of low pitch	?	Given primarily by male, but also by females and juveniles, when moving out of rest areas and when group is scattered	Aids in group cohesion — denotes "here I am"
*Series of rapid, high-pitched bo-bo-bo or similar sound	?	Given when group is scattered widely	Probably aids in group cohesion
*Loud, clear, but low-pitched series of hoots ending in a growl	Excitement	Given by silver-backed males as part of chest-beat-ing display in tension-pro-ducing situations: the presence of man, another group, etc.	Generates ex-citement; in-timidation as part of display
Rapid, loud, staccato series of ö-ö-ö-ö, with the first vowel more forceful than the others and separated from them by a brief pause	Excitement (?)	Given by silver-backed male during copulation	?
*Harsh, staccato grunts	Annoyance	Given by silver-backed male if females quarrel; given by quarreling animals	Warning of possible danger
*Short, loud barks	Annoyance	Given by animals when quarreling; in response to presence of man	Warning of possible danger
Single, loud explosive roar	Anger	Given by males in response to presence of man and, once, when swooped at by 2 ravens	Intimidation, warning of danger
*Harsh, fairly short screams	Anger	By quarreling females, in response to presence of man	Warning (sometimes contains ele-ments of fear)

TABLE 3–6   The most conspicuous or prevalent vocalizations of free-living gorillas — *Continued*

Description of Vocalization	*Probable Emotion*	*Stimulating Situation*	*Probable Function*
Soft whine	Distress	Given by youngster when in danger of being abandoned or injured	Communicates distress
*One or two high-pitched screeches	Distress	Given by infant when in danger of being left behind by its mother	Communicates distress (sometimes contains elements of anger)
Intense screaming roar	Fear	Given by male to presence of man	Warning of danger; intimidation (usually contains elements of anger)
Loud, long, high-pitched screams	Fear	Given by females and juveniles to presence of man	Warning of danger

* The most frequently heard vocalizations of undisturbed free-living gorillas.

silver-backed males. Thus young males screamed when angry, but larger ones produced a rather squeaky roar. Only large males emitted the hoot preceding the chest beat, the copulation call, and the full roar (Fig. 3–7b).

The Chest-Beating Display

The chest-beating display is the most striking behavior pattern of the gorilla, and one of the most complex and stereotyped displays among mammals. The display consists of nine more or less distinct acts, most of which may be given individually or in several combinations of two or more, with a tendency for some to precede others. The complete sequence is given infrequently and then only by silver-backed males. The entire display occurs typically as follows:

(1) At the start of the display the gorilla sits or stands as it emits a series of some 2 to 40 clear hoots, at first distinct, the more slurred as their tempo increases.

(2) The hoots may be interrupted as the animal plucks a leaf or branch

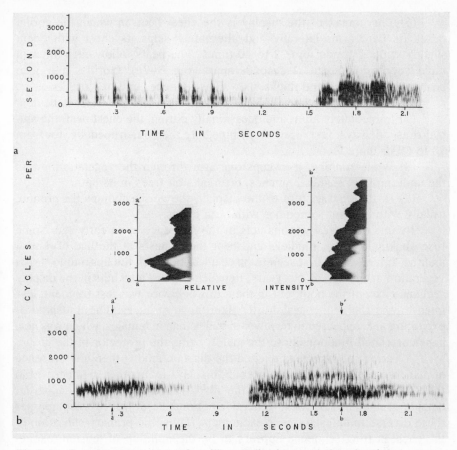

Fig. 3-7. Sound spectrograms of gorilla vocalizations and chest beating.

    a. A female slaps her chest 11 times. The intermittent vertical lines of high frequency represent the impact of the hand against the chest. The sounds between each beat apparently consist of resonance and of background noises. A three-toned roar of low intensity by a silver-backed male is shown on the right of the spectrogram.

    b. Intensely roaring males. The first roar is that of a black-backed male, the second that of a silver-backed male. The section above each roar represents the relative amplitude of the sound at the instant of time a-a′ or b-b′. The instant of time is 1/24 seconds long. The sections show relative intensity throughout the entire spectrum of the sound, including overtones that did not register on the spectrogram.

from the surrounding vegetation and places it between its lips, seemingly a gesture of "symbolic feeding."

    (3) Just before the climax of the display, the gorilla rises on its hind legs and remains bipedal for several seconds.

    (4) As it rises, the animal often grabs a handful of vegetation and throws it upward, sideways, or downward.

(5) The climax of the display is the chest beat, in which the gorilla raises its bent arms laterally and alternately slaps its chest with open, slightly cupped hands some 2 to 20 times. The beats follow each other in rapid succession about .1 seconds apart (Fig. 3–7a). Gorillas may also beat their abdomens and thighs, as well as branches and tree trunks.

(6) A leg is sometimes kicked into the air while the chest is beaten.

(7) Immediately after, and occasionally during, the chest beat, the animal runs sideways, first a few steps bipedally then quadrupedally, for from 10 to 60 or more feet.

(8) While running, it sweeps one arm through the vegetation, swats the undergrowth, shakes branches, or breaks off trees in its path.

(9) In the final gesture of the display, the gorilla thumps the ground, usually with one but sometimes with both palms.

Infants displayed various acts in the sequence at an early age. Some rose shakily on their hindlegs and beat their chests at the age of 4 to 5 months. Infants were first seen to place a leaf between the lips and to throw vegetation at the age of 1½ years. Females showed all 9 acts in the display sequence except the hoot, but in them this behavior was less frequent, less intense, and the sequences shorter than those of males. The threshold of excitation also appeared to be lower in males than in females, which was perhaps correlated functionally to the male's being the protector of the group.

A variety of situations elicited the display, among them the presence of man, the presence of another group, displays by another member within the group, and play; sometimes the display occurred without apparent outside stimulus. The most general emotional term which encompasses these diverse manifestations is excitement. Thus, the primary causation of the chest-beating sequence appears to be the build-up of tension (excitement) above a certain threshold. The display itself serves to make the animal conspicuous, it advertises its presence, and probably functions in intimidation.

According to the concept of causation as developed by ethologists, a potentially dangerous situation arouses the impulses to flee and to attack. The conflicting tendencies generate tension, which finds release in some functionally inappropriate act, like the throwing of objects and the beating of the chest in gorillas. Behavior which is not actually relevant to the situation at hand has been termed a displacement activity. A displacement activity may become stereotyped, incorporated into a definite display, and achieve secondary functions, such as intimidation. In ethological terminology, it has become ritualized.

Several acts in the gorilla's display, especially the chest beating and the "symbolic feeding," appear to represent ritualized displacement activities. The displacement and ritualized elements are especially evident in "symbolic feeding." At the first sight of the observer, males occasionally began to feed very intensively, stuffing large handfuls of food into their mouths. This appeared to be displacement feeding. Sometimes, just before rising to beat his chest, a male pushed two or three handfuls of vegetation into his mouth, but at other times he placed only a single plant, branch, or leaf between his lips. This latter act probably represented ritualization of displacement feeding.

CONCLUSION

Gorillas are rather amiable vegetarians, who, though primarily terrestrial, reveal their arboreal ancestry in their structure and in some aspects of their behavior. The arms and trunk of the gorilla are adapted more for climbing, hanging, and reaching in trees than for walking quadrupedally on the ground, but the huge size and reduced agility of the ape have made a life in trees impracticable. The gorilla has, however, retained the habit of building nests, behavior which appears to be of adaptive value in trees and which on the ground seems to be an anachronism serving little or no function. On the other hand, the elaborate chest-beating display probably evolved after the gorilla assumed its terrestrial mode of life. Although single components of the display such as beating, shaking, and throwing are seen in a variety of arboreal primates, the full sequence of the gorilla's display, which includes rising up on the hindlegs while facing the opponent and running sideways, is unlikely to have adaptive significance in the forest canopy. Not only is the efficacy of the full display in such a situation questionable, but when gorillas attempt to display in trees they easily lose their balance and are in imminent danger of falling.

The social life of gorillas differs in some respects from that of the other primates studied so far. Groups are quite cohesive; they are led by the dominant adult male who by his actions and idiosyncrasies determines to a large extent the daily routine of the group and its response to other groups and to man. In spite of the continuous and usually peaceful contacts between the members of the group, overt interactions between adults are infrequent. For instance, grooming, which is often thought to strengthen the social bonds of primates, was never seen between adult males and females. The composition of many groups tends to remain stable over several months like that of the stable social groups of the monkey species described in this volume. However, some groups are fairly unstable, with certain males leaving freely to lead a lone life and others joining for a time. A further difference in the behavior of groups is apparent when groups, or groups and lone males, meet. Some join readily, others merely approach each other closely, and a few behave antagonistically toward each other. These individual responses are striking but not surprising if it is remembered that, because of the extensive overlap of ranges, each group undoubtedly contacts others fairly often. Many groups are probably well acquainted and may even be related to each other; they have had favorable or unfavorable meetings in the past, and this, together with the fact that gorillas are often highly individualistic in their behavior, may strongly influence the kind of interaction witnessed by the observer. The variable responses shown by gorillas in a given situation emphasize that the behavior of primates is highly adaptable and that generalizations based on a few observations in one area may be quite misleading.

chapter four

BABOON SOCIAL BEHAVIOR

K. R. L. Hall and Irven DeVore

Zuckerman's (1932) field observations of chacma behavior were confined to a few days in the eastern Cape Province, South Africa, by far the bulk of his data coming from detailed observations on *P. hamadryas* in the Regent's Park Zoo, London. No data on behavior are yet available in English on the baboon colony at Sukhumi, Black Sea coast, although this colony has been established for about 30 years, and the Russian film "Threshold of Consciousness" indicates that behavior studies of at least one hamadryas group living in a compound have been carried out. The major captivity studies so far have been those of Kummer (1956, 1957) on hamadryas at the Zürich zoo. This is at present the only species which has been thoroughly studied under restricted conditions as well as in the natural habitat, Kummer and Kurt (1963) having recently completed a year's field work on groups in Ethiopia. Bolwig (1959) combined some detailed observations on the behavior of two young chacma baboons that he kept in captivity with data obtained from watching baboon groups foraging around the camp rubbish heaps in the Kruger National Park, but the scope of his study was restricted by these conditions. No systematic data whatever on the behavior of baboons elsewhere in Africa were available until the DeVore and Washburn studies.

Baboons have been used very little in laboratory studies of learning or "intelligence." As adults they have a reputation for being difficult to handle, and their comparatively large size makes them less convenient than macaques for most laboratory purposes. Where they have been used in problem-solving studies (Watson 1914; Harlow and Settlage 1934; Bolwig 1961), the data from one or two animals on one class of problem have yielded very little information that can be meaningfully related to field data.

Reprinted from *Primate Behavior: Field Studies of Monkeys and Apes,* Irven DeVore, ed. Copyright © 1965 by Holt, Rinehart and Winston, Inc. pp. 53-110.

Most of the behavior data we shall discuss were obtained by one or two observers, who worked at close range to the baboon group. A condition of neutrality between the observer and the animals was first achieved so that they were neither positively attracted toward the observer by expectation of food, nor frightened of him, unless he introduced some marked change into their environment. The presence of two observers permitted each to concentrate on a certain class of individuals in a group or upon a certain aspect of behavior, and film or tape recordings could be made without loss of observational data.

Marking of individuals was not attempted and, in fact, was not necessary for the immediate purpose of the studies, although it would be highly desirable as a basis for long-term investigations. All individuals of two Nairobi Park groups (SR and LT) were recognizable to DeVore, and most of the adults in four other groups in the same area were individually distinguishable. All individuals of one Cape Reserve group (S) were recognizable to Hall and Robert Wingfield, as were also many of the adults of the large C group and the Kariba main group in Southern Rhodesia. By memorizing blemishes and scars observed at close range as well as learning major physical differences such identification could be made.

Although sustained observation of the "natural" behavior of these animals was the main objective of these studies, minor experimental interferences were occasionally introduced to elicit unusual behavior. As a method of supplementing normal field data, we have hardly begun to explore the possibilities of systematic use of field experiments or of exploiting accidental occurrences not planned by the investigator but creating drastic alterations in the environment of a group.

SOCIAL ORGANIZATION

The baboon group is organized around the dominance hierarchy of adult males. The nature of this hierarchy varies between groups according to the constitution of each group. The simplest form of organization is probably that of groups in which one, and only one, adult male is conspicuously dominant. Such groups, numbering from 15 to 35 animals, were the S group in the Cape, both in 1960 and 1961, the N group at all periods of observation, and the LT, AR, and MR groups in Nairobi Park in 1959. A much more complex relationship was that observed in the SR group in Nairobi Park; during the time of study there were 6 adult males and only 7 adult females in this group, which totaled 28 animals. However, the proportion of adult males to adult females was nearer to the average in the SV group, the other group intensively studied in that area, in which there were 5 adult males and 14 adult females out of a total of 40 animals. In 1961 in C group, Cape, when the group totaled 80 animals, 8 identifiable large males took part at various times in threat behavior among themselves, and there were at least 30 adult females in the group. It was not possible during the limited observation period to work out the dominance relations among them. It remains one of the most important research tasks of the future to concentrate intensive observations for a long period on one group of this size or larger in order to work out accurately the complex interactions among the males and to learn how the adult females with their infants and the subadults and juveniles are organized in relation to them.

Pending a long-term study on a large group of the Kenya or southern African baboons, it is not possible to evaluate the apparent differences in social organization of the hamadryas baboons in Ethiopia, reported on in a preliminary account by Kummer and Kurt (1963). According to these observers, a one-male group is the characteristic social unit in the population area they studied. It consists typically of from one to four, rarely as many as nine, females who follow a single adult male, as do their offspring until they reach the age of one or one and one-half years. The juveniles and subadult males, and some adult males without females, live outside these units, yet all these, together with other one-male groups, tend to congregate in very large numbers at the sleeping cliffs. It would appear that tolerance between adult hamadryas males is significantly less than it is in olive or chacma baboons. However, until the full report of the Ethiopian study is available, we cannot judge whether the difference in social organization is one of degree, imposed upon the population by the ecological circumstances, or whether, as the authors suggest, it is so fundamental as to express itself in the same way even in captivity (*cf.* Zuckerman 1932).

Male Dominance

Dominance is a complex conception assessed by observation of the frequency and the quality of several types of behavior in various kinds of situations, with reference both to the other animals within the group and to external events — such as the presence of a predator or some other disturbing stimulus. The "peck-order" concept of a linear kind of social relationship, derived from only one kind of situation — competition for food — is scarcely applicable to wild baboons, where in the normal course of foraging the animals are widely spaced and competition for any item of food is a rare event among adult males. Special food-incentive tests to determine relative dominance between a pair of adult males, used in several laboratory studies, have a limited usefulness in the field. In nature it is difficult to isolate the test pair so that it is unaffected by the presence of other adult males, though these may be temporarily in the background. Occasionally a situation occurs, such as in S group, Cape, in 1961, in a group numbering 38 animals that was sometimes to be found on the roadside about six miles to the north of S, and in the LT and AR groups, in which one and only one of the adult males has a consistent priority of access to any food object thrown down within its view.

Dominance as expressed in the natural behavior of a group can be illustrated by the straightforward system of S group, Cape. At the time of intensive study of this group there were only three adult males in the group, one somewhat larger than the other two. The key functions or behavior patterns most prominently associated with the largest male can be summarized as follows:

1. He mated exclusively with some, but not all, of the females as they came to maximum turgescence in the estrous period, and drove away males 2 or 3 from the female at this time. (Those females with whom he did not form an exclusive mating relation were not fully grown, and even when in full estrus were only occasionally mounted by him.)

2. The act of presenting, as a submissive gesture, was directed far more frequently to this male than to the others.

3. The number of aggressive episodes recorded for him within the group was far greater than that for any other male.

4. Whenever some disturbing situation occurred, as when the group was charged by an eland cow, and when a strange baboon was released in the vicinity of the group, this male went ahead of the rest of the group and threatened or attacked.

5. When there were mothers with black (recently born) infants in the group, they tended to cluster near him and to walk close to him during the day range. His retaliation against attacks on the mothers was immediate.

The quantifiable behavior patterns of this male in comparison with those of the other two adult males are summarized in Table 4–1, for the 1960 observation period, and the same kind of dominance relationship was still in evidence in this group in 1961.

This kind of dominance relationship may be no more "typical" of the chacma baboon than the one for the Kenya SR group is typical of the baboon groups in that region, but it reveals very clearly the general nature of the functions of dominance. The aggressive episodes are noisy, very menacing, and effective in breaking up squabbles among the other animals and in protecting mothers and infants from disturbance or injury. In addition, as was observed once in S group and three times in C group, when there is some cause for disturbance from outside the group, members of the group tend to close up rather than to scatter, the generalized aggressiveness of the a male being directed successively at many different animals within the group and having the effect of bringing them together or at least preventing them from wandering away. For example, this was observed the one time C group encountered N group at the limits of their home ranges, the first time a stuffed serval cat was placed near the sleeping cliffs of C group, and once when the

TABLE 4-1 Comparative data on mating, presenting, and participating in aggressive episodes for the *a* male and males *2* and *3* of S group, 1960

	Male Animal		
	α	2	3
Mating frequency			
female in full estrus	101	37	56
female less than full estrus	8	4	28
Total	109	41	84
Presented to			
by females and juveniles	104	3	6
by other males	34	1	0
Total	138	4	6
Aggressive episodes			
against females	37	8	3
against others in group	15	6	6
Total	52	14	9

observers were approaching S group through a thick mist that periodically lifted. The probable mechanism for this kind of generalized aggressiveness will be discussed below. In spite, or perhaps because, of the very pronounced dominance of the *a* male in S group, aggressive episodes were infrequent and were never observed to result in visible injury to any member of the group. The dominant male of this group was aggressive (as expressed in certain characteristic behavior to be described later) about once in every six and a half hours of daylight, except when some special situation arose.

In C group, Cape, in the 1958–1959 observation period, when it numbered about 53 animals, threat behavior and chasing among the adult males was not recorded, and dominance relations may have been, temporarily, in the form of an established hierarchy. In 1961, however, several sequences were observed in this now much larger group (80 animals) in which two or more of the males threatened each other. One such episode resulted from an attack by one adult female on an adult female with an infant on her back. An adult male immediately chased the aggressor, who ran away from him but continued to try to get at the mother and infant. The first adult male now chased a second adult male who at first ran away; as is typical of these male-male encounters in this group, the animals did not make physical contact with each other. The second male then stood and repeatedly threatened the first male, turning his head rapidly from side to side. The second male was joined by two other adult males in threatening the first, and the four chased about without engaging each other. Several similar episodes occurred, usually while the group was still near the sleeping cliffs in the morning. On one occasion five adult males were charging about after each other, again without engaging with one another. As was also true in Kenya,

Fig. 4–1.
Yawning under tension, a young
adult male, Mark, of the SR
group displays his unworn teeth.
The upper canines are not quite
fully erupted.

all the other animals in the group kept well away from these adult males
when such episodes were taking place.

There was strong behavioral evidence that one of the adult males of C
group (Saddleback) was often dominant over the others, but there was no
way of working out the relations among the others. The probable explanation for these aggressive episodes may be drawn from the detailed close-
range data obtained on the SR group, Kenya; in this group each of the six
adult males was individually recognizable to the observer, and the interactions among them were studied continuously over a period of about three
and one-half months. In contrast to most adult males in the Park, the six
males of this group were all large and of approximately equal size. They
ranged in age and physical fitness, however, from young prime (Mark) and
prime (Kula) through late prime (Dano, Mdomo) to old (Pua) and very old
(Kovu). There were far more aggressive episodes among these adult males
than were observed in any other group, and the dominance pattern that
emerges is one that cannot be reconciled with the model of a strictly linear
hierarchy. It became clear that certain of the adult males constantly associated with each other and tended to support each other in aggressive interactions with other males. Some of these males associated so closely that
they were scarcely ever observed acting independently in such episodes, and
on this basis three of them came to be designated a "central hierarchy."

Fig. 4–2.
An old adult male, Pua, of the SR group displays the broken and eroded teeth characteristic of old age.

In trying to work out the dominance pattern in the SR group, several criteria were used, including: (1) success in achieving food objectives in paired tests given when other males were too far away to interfere directly; (2) frequency of successful dominance assertions in "natural" situations arising within the group, for example, success in gaining and maintaining access to an estrous female or in causing another male to move away from a particular resting place or feeding spot; (3) success of combinations of males against other individual males. The frequency data for criteria (1) and (2) given for the interaction of each of the six males in this group indicate only a fairly consistent subordination for Kovu and Mark and dominance for Dano (Table 4–2).

The data in Table 4–2 indicate the difficulty of arriving at a linear ranking for the males by traditional criteria of dominance. Other criteria, such as instances of mounting between the males, confirm the positions of the males at the top and bottom of the hierarchy but do not clarify the positions of all six. Some of the inconsistencies between sections 2a and 2b (Table 4–2) are due to the fact that the three central hierarchy males (Dano, Pua, Kovu) not only combined against the other males, but also were very uncompetitive among themselves with regard to food. These three males would feed side by side when food was thrown to them. Although Kovu was obviously less assertive in these situations, Dano and Pua were never asser-

TABLE 4–2   Dominance interactions among adult males of SR group; number of interactions with dominance expressed*

2a. Success-failure on experimental food-incentive tests

Success	Dano	Pua	Kula	Mdomo	Mark	Kovu	Total Success	Rank Order Successes	Failures
Dano		5	3	3	6	2	19	1=	2
Pua	5		1	6	2	5	19	1=	5=
Kula	3	4		3	3	1	14	4	1
Mdomo	3	4	3		1	5	16	3	3=
Mark		2				2	4	5	3=
Kovu							0	6	5=
Total Failure	11	15	7	12	12	15			

2b. Dominance-subordination in "natural" situations

Dominant	Dano	Pua	Kula	Mdomo	Mark	Kovu	Total Dominant	Rank Order Dominant Scores	Subordinate Scores
Dano		3	6	1	1	1	12	1	1·
Pua						2	2	5	5
Kula	1	1		3	3	2	10	2	4
Mdomo		2				6	8	3	2=
Mark		1				2	3	4	2=
Kovu							0	6	6
Total Subordinate	1	5	8	4	4	13			

* The numerical tabulations represent the number of observed interactions in which the male in the vertical row on the left was dominant to the male in the horizontal top row.

tive toward each other over food. Since only one male at a time can consort with an estrous female, however, even central hierarchy males sometimes competed with each other over access to a receptive female, and some scores reflect this. The rank order for "dominant scores," then, is the most accurate measure of individual dominance status (as shown in Fig. 4–3), but the most significant aspect of the dominance relations in this group is that the central hierarchy males, who stayed together in the center of the group, ordinarily acted in concert and together controlled access to incentives, determined group movement, and so on. In every instance where Dano, Pua, and Kovu or Dano and Pua combined, they were 100 percent successful against any of the other three males, who very rarely combined.

Thus a male's dominance status was a combination of his individual

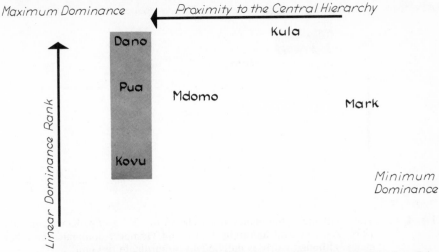

Fig. 4–3. The adult male dominance hierarchy in the SR group, Kenya. A male's position is a combination of his abilities as an individual (his "linear dominance rank"), and his ability to enlist the support of other males in the central hierarchy.

fighting ability ("linear dominance," see Fig. 4–3) and his ability to enlist the support of other males ("proximity to the central hierarchy"). Dano emerges on these criteria as the *a* animal; that is, he was at the top of the central hierarchy. On several occasions, however, it was possible to test dominance between him and Kula when they were well away from other males. Kula was the more dominant and is shown (Fig. 4–3) as the highest ranking *individual*. The fact that Pua and Kovu would support Dano when he was challenged, however, meant that Dano was almost always in control and that Kula could only assert himself briefly.

In the SV group, numbering 40 animals (about the average size of groups in Nairobi Park), there were six adult males and the usual ratio of about twice as many adult females. This group has a special interest for the understanding of the dynamics of the dominance system. It was the only group intensively observed in which one adult male (Humbert) disappeared with a consequent alteration of the dominance balance, and to which an adult male from another, adjacent group (Lone from the AR group) attached itself, with interesting effects on the dominance pattern. All adult males and most adult females in the group were individually recognizable. The dominance pattern was characteristically stable — Curly, a younger and unusually aggressive male, being consistently dominant over the other males so long as the much older Humbert, with whom he allied, was in the group. In fact, Humbert, in spite of the physical disadvantage of worn-down canines, maintained a dominance over Gam (and the other smaller males) through a simple, triangular relationship that existed among the three of

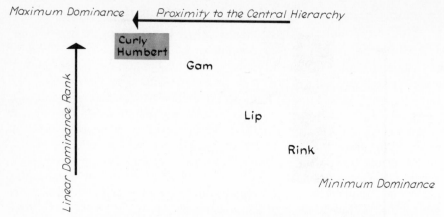

Fig. 4–4a. The adult male dominance hierarchy in the SV group, Kenya, in July 1959. As a central hierarchy Curly and Humbert dominate the other males, although Curly is individually subordinate to Gam.

them (Curly dominant over Humbert; Humbert dominant over Gam; Gam dominant over Curly). The dominance pattern can best be illustrated diagrammatically (Fig. 4–4a) where the position prior to Humbert's disappearance can be compared with that four months later (Fig. 4–4b) when Lone, the male from the AR group, had achieved his position in the SV group.

Before joining the SV group Lone had been very subordinate to the *a* male of the adjoining AR group. When the group was on the move he ranged well ahead of the others; when the group rested the *a* male frequently drove

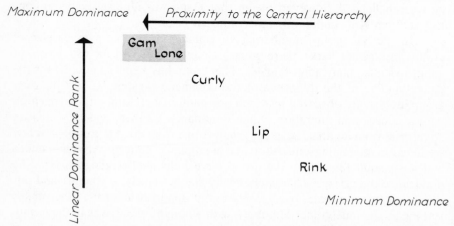

Fig. 4–4b. The SV group hierarchy in November 1959. With the disappearance of Humbert, Gam has asserted his dominant status over Curly, and with a new male, Lone, is beginning to establish a new central hierarchy.

Lone away from resting spots or grooming clusters, and in general asserted his dominance over Lone continually. At first Lone kept 40 or 50 yards away from the SV group; then, after two days, he stayed within 20 yards of them and slept near them at night. During this period he engaged in several vicious fights with Curly and, probably, with Gam. Ten days after Lone was first seen near the SV group in October, he was walking among the SV animals, although Curly continually threatened him, forcing him to the edge of the group. Before Humbert's disappearance, when he and Curly together maintained a central hierarchy, Lone and Gam remained for a time in more peripheral positions, leading the group during progression. When Humbert disappeared in November, Curly dropped significantly below both Gam and Lone, between whom there had been less tension, with the rotating of the dominance relationship shown in Fig. 4–4b. At the end of the observation period (November) Gam and Lone seemed to be forming a new central hierarchy. Lip and Rink were small adult males who avoided fights and who had no observable effect upon the relations among the more dominant males.

During observations of chacma baboons in Northern Rhodesia in 1955, Washburn observed the only other instance we have seen of a baboon's leaving one group and joining another. In a situation analogous to that just described, a male was driven out of a group in which he was at the bottom of a hierarchy of six males. Shifting the same day to a new group, he decisively defeated the one male in it and became its new a male.

In four groups (having two or more adult males) on which intensive observations of individually identifiable adult males have been carried out (S, Cape; SR, SV, and PP, Nairobi Park), the dominance relationships have varied significantly. A dynamic pattern is formed that is dependent not only on such obvious factors as relative size, strength, and age of individuals, and their differences in experience and temperament, but also upon the support which combinations of two or three males may provide for each other. Although a stable and linear hierarchy sometimes exists, it is likely to represent only a temporary stage in the history of any one group, the pattern changing as younger males become full grown and older males disappear. Equally impressive is the evidence of the possibility of shifts in dominance within neighboring groups when an adult male changes from one group to another.

Dynamics of Threat Behavior

The nature of the dominance functions of one a male, or of a group of central hierarchy males, is the clue that leads to an adequate understanding of the major aspects of baboon social organization. In addition, however, the complex relationships stemming from these dominance functions must also be fully studied as they affect the whole group. For example, threat behavior we now know can have at least four forms according to the kinds of social interaction with which it is associated. One of the commonest of these is called, in ethological terms, "redirection of aggression" (Bastock, Morris, and Moynihan 1953), corresponding to "transferred threat" (Altmann 1962). We have chosen the former term to indicate a wide range of behaviors that we believe can be so described. In its simplest form redirected aggression occurs when an animal threatens another, and the threatened animal redirects the aggression to a third party (chasing or threatening him) or, rarely, to an inanimate object (for example, bouncing against a tree or tugging vigorously at a rock). During the tense situation in the SR group, 33 instances of redirected aggression between individuals were recorded. A typical instance would be: Dano chases Mdomo, who chases an adult female. Initiation of a kind of chain reaction of aggressive acts downward through the group structure presumably serves periodically to reinforce the dominance pattern throughout the group.

Redirection of aggression was observed very clearly in a different context in S group, Cape, as a sequel to the release of a tame young male baboon near the group one evening close to the sleeping cliffs. Whenever the *a* male threatened the strange baboon, the latter took refuge behind or very close to the observer, and, apparently as a direct consequence of the *a* male's being prevented from attacking, he began to attack the other animals in the group more or less continuously during the day. His aggressiveness became ten times more frequent than on normal occasions. Aggression, in such a situation, is transferred from the object which elicited it to substitute objects or animals, probably with some kind of equivalence in the transfer. The generalized aggression of the S-group *a* male, however, gave the impression that he was attacking any animal that came near him or that he came upon. Similarly, during the period when Lone was joining the SV group, Curly's aggressiveness toward other group members was far more frequent than normal.

Threat behavior is often seen between combinations of individuals, indicating that an individual may sometimes seek support or "enlist" the threat behavior of another individual. This is done by gestures such as jerking the head rapidly from side to side. It is important, however, not to misinterpret the significance of this kind of social situation and the gestures involved in it. Head-turning and other gestures to be described later are expressive of agitation, and the support—if it comes—may be primarily in response to a conditioned association between the two animals for which the gestures are the signals. The situation where two animals "simultaneously threaten" another is of the same order. What Kummer (1957) described as "protected threat," on the other hand, refers basically to the situation in which a subordinate succeeds in escaping the threat or attack of a more dominant animal by running toward or standing in front of an animal that is still more dominant. The four terms we have used (redirection of aggression, enlistment of threat, simultaneous threat, and protected threat) all refer to social variants due to combinations of behavior within the general dominance pattern. This variety, observed in differently constituted groups, has served to bring out very clearly the dynamic quality of the adult male relationships.

Harassment

The adult sex ratio was atypical in the SR group, as has been mentioned, and this factor seems to account for the occurrence of yet another form of aggressive behavior normally never seen in any other Kenya or southern Africa group, namely the harassing of an adult male when he is in consort with an estrous female. In a harassing sequence, similar to that which could almost invariably be induced in males of any Nairobi Park group by experimental feeding, the harasser would start by slowly pacing around the other male, accompanied by an audible tooth-grinding and yawning, directed straight toward the other in what can be described as a canine display. The harasser would gradually approach closer to the other

Fig. 4–5. An adult male *(left)* displays his canines as he harasses another adult male who is grooming a fully estrous female. SR group.

who, in the consort situation, tended to increase his grooming of the female in an intense and agitated manner while occasionally mildly threatening the harasser. After several canine displays by the harasser, these tending to increase in frequency the closer he approached, a chase, attack, or fight was the usual outcome. In such a situation other adult males would often join either the harasser or the harassed male. When the latter attacked the harasser, the usual conclusion was that a third male would appropriate the estrous female temporarily. Sequences of this sort were never seen in a group where a sufficient quota of adult females was available to the adult males, and never with regard to naturally available food.

Dominance among Females

The dominance hierarchy among the females in the groups, even where all the adults were identifiable, has been much more difficult to determine objectively. This difficulty may in part reflect the tendency of female status to be more variable and perhaps rather more subtly defined than that of the male. Estrous condition is found to alter the status of a female very markedly; it is possible that the female hierarchy is typically unstable, that it is individually based rather than partly organized around coalitions as in the males, and that it is expressed in more-or-less continuous minor bickering with very little real attacking and biting. This description is indicated on the basis of the interactions between the SR females according to social criteria such as being forced from a feeding spot, from a grooming partner, or from the presence of a mother with an infant, together with an assessment on the indicators of female threat shown by the sudden stopping of the female in front of a subordinate and by eyelid-lowering (Fig. 4–6).

While these figures indicate trends of dominance during this observation period (Inama apparently being the most dominant and Notch easily the most subordinate), it is not possible at present to indicate at all accurately the meaning of these differences in the whole social organization of the group. To what extent these differences depend upon the individual character of the female, her age, experience, and so on, and to what extent they may be a function of some fairly consistent relationship with one or more of the adult males (independent of estrous manifestations of the sexual cycle), remains a problem for further research.

In S group, Cape, a similar picture of inconsistency arises from the 1961 data. One of the females (E), however, was consistently subordinate to all the others, including another low-ranking female who was physically smaller than E but far more aggressive. The most significant general finding from the 1960 observations on this group was the extent to which female aggressiveness against other females was correlated with the frequency of mating by the a male with the female who was in full estrus. This

		Subordinate									Total
	Inama	Tita	Ma	Brash♂	Kink	Miss	Pana	Naya	Lank	Notch	
Inama		3	1	1	2	2(1)*		1		1(1)	11
Inama & Tita			2		4(1)			1			7
Tita			2			1		1		5(1)	9
Tita & Kink								2			2
Ma								1		4(3)	5
Brash♂						1		1			2
Kink								1		7(3)	8
Miss			1					2		4(1)	7
Pana								1		6(4)	7
Naya										2	2
Lank											0
Notch				1	1			3			5
Total =	0	3	2	6	7(1)	4(1)	0	11	3	29(13)	

Dominant (vertical label on left)

* times subordinate was in estrus

Fig. 4-6. Dominance interactions among females (and one subadult male, Brash) in the SR group, Kenya. The tabulations represent the number of observed interactions in which the female in the vertical row on the left was dominant over the female in the horizontal top row.

relationship seemed to indicate that the social pattern among all the adult females was considerably affected by the close attention of the markedly dominant *a* male to one of their number, for there is no significant correlation with the mating frequencies of the other males. Although their attacks on the fully estrous female with whom the *a* male was copulating increased markedly at the stage when he was consorting exclusively with her, there

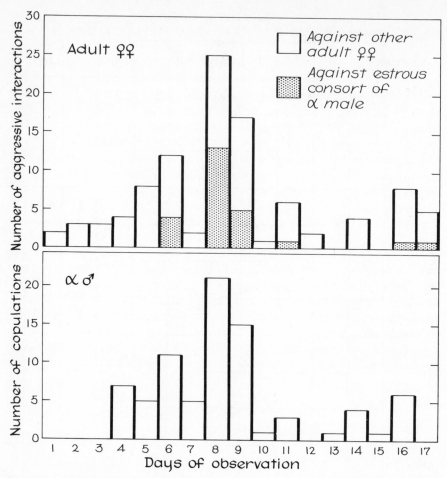

Fig. 4-7. Relationship between frequency of copulation by α male with a fully estrous female, and aggressiveness of adult females toward her and among themselves. S group, Cape.

was also a general increase in aggressiveness among the adult females which indicates an unusual degree of disturbance of the relationship among them (Fig. 4-7).

Although there is evidence of a kind of individual ranking among the females, it is also common for two or more females to "gang up" in threatening and attacking another female. Instances of such simultaneous or enlisted threat were noted on the part of the two subordinate females, E and B. A striking instance of combination was observed in S group when the *a* male was attacking one of the females in a pond and several females threatened him so vigorously that he withdrew from his victim.

Spacing within the Group

As studies of the Japanese macaque have clearly shown, the relative position and distance of the various members of a group from one another reflect the nature of the social relationship between them (Jay 1965a). Mothers carrying their infants tend to be found in the middle of a group when it is on the move; the less dominant males are to the front, at the sides, and to the rear; and the most dominant males are in the middle (see Fig. 4–8). This order of progression was invariable in all groups observed. When some crisis occurs, however, the *a* or central hierarchy males tend to go immediately to the front to meet the threat, a response observed consistently in all groups both in the Cape and in Kenya.

The so-called "sentinel" behavior of chacma baboon groups was found to be entirely consistent with this kind of organization. A subadult or a young adult male, ahead of the group by 200 yards or as much as one-quarter mile, would be observed to give the alarm bark, and one or more of the large adult males would come to the front to see the cause of the disturbance. It was almost always one of the "peripheral" males that continued to watch and to bark, while sitting at the edge of the area where the rest would soon resume feeding.

It is almost impossible to discover the extent to which adults other than dominant males "lead" a group in the sense of determining the route the

Fig. 4–8. The positions of group members during group movement. Dominant adult males accompany females with small infants and a group of older infants in the group's center. A group of young juveniles is shown below the center and older juveniles above. Other adult males and females precede and follow the group's center. Two estrous females (dark hindquarters) are in consort with adult males. Nairobi Park, Kenya.

group will take during the day range. For example, it sometimes appeared from their frontal position that one or more of the peripheral males might be initiating group movement and determining the direction the group was to take. Similarly, in S group, Cape, two of the adult females were often at the front of the group during the day range, and others appeared to follow them even when they changed the direction of movement. In fact, animals in the lead were probably taking a course which was habitual to the group, but might not otherwise be determining the group's behavior. It seems more correct to say that the group as a whole is continually alert to the behavior and location of the dominant males, and that those ahead are mainly anticipating or steering with reference to these males. This is suggested by their behavior whenever even minor disturbances from outside the group occur, and by the fact that the main body of the group occasionally changes direction, forcing the animals who had been in front to make a wide detour before rejoining the group.

The main characteristics of baboon social organization, as revealed in the Kenya and southern Africa studies, are derived from a complex dominance pattern among adult males that usually ensures stability and comparative peacefulness within the group, maximum protection for mothers with infants, and the highest probability that offspring will be fathered by the most dominant males. With all the variations so far apparent in groups of differing constitutions, it still remains to discover accurately the kind of relationship among the many adult males of a large group of 80 or more animals. The nature of the social structure of adult females, and its periodic variations, also remains to be worked out in detail over a longer period than has so far been available.

SEXUAL BEHAVIOR

We have noted in the previous chapter the evidence on seasonal variations in births in different regions of the distribution area, and we have noted also some of the major physiological characteristics that may be assumed to underlie the manifestations of sexual behavior to be discussed in the present section. To recapitulate: the female estrous cycle averages, in captivity, about 35 days; the period of turgescence averages 19 days, deturgescence averages 16 days (Gillman and Gilbert 1946). Sexual swelling (see Fig. 4–5) increases gradually for about ten or twelve days until reaching the maximum, when it remains more or less constant for about eight days, and deturgescence takes place during the next one and one-half to five days. The mean gestation period is about six months. Nursing goes on for from six to eight months (Gilbert and Gillman 1951), and turgescence is usually not observed again until after lactation has ceased (although in some females it has occurred while the infant was still at the breast). These indicators are derived from data on chacma baboons in captivity and suggest that the minimum interval between successive births in the same female may be of the order of from 12 to 18 months. The Kenya field data, however, appear to indicate an even longer interval. The criteria for distinguishing cessation of lactation are difficult to establish in field studies, and these, together with actual observations of resumed sexuality by individual females in wild groups, need to be further studied in long-term field observation. A female begins her estrous cycles at about the age of three and one-half or four years, and the size of perineal swelling thereafter increases for a few years. Less is known about the chronological pattern of events in the male baboon, but puberty in captive hamadryas is reported to vary in onset between four and six years (Zuckerman 1932), this being similar to the age of first copulation in *M. fuscata* — reported as five years.

Much more needs to be discovered about the variants of sexual reproductive physiology of these animals in the wild. It is clear that standard findings derived in the laboratory may diverge importantly from those derived in the widely varying ecological conditions in which the animals live in nature. Nevertheless, it is now possible to describe rather fully the way in which the sexual behavior patterns fit into the over-all pattern of baboon social organization and to analyze in some detail the components of the behavior.

Comparisons of the Kenya and Cape Peninsula data are interesting in part because of the general similarity of the mating pattern in its social context, and in part because copulation has been observed in the Cape groups in all months of the year — with one or more females always in some state of estrus in C and S groups. In Kenya birth seasonality may have corresponding limitations upon estrous occurrences, because there are periods in some of the groups when no females in estrus are found and no mating actively takes place, though no satisfactory statistical data are yet available to determine whether these are seasonal trends. It is generally true that no mature males will attempt to mate with a female unless she is showing distinctive signs of estrous swelling (see below).

Mating in Relation to Dominance

The mating pattern that we find to be typical of both Kenya and southern Africa baboons can be illustrated clearly from our field data. In S group, Cape, the *a* male copulated exclusively with each adult female that came into estrus only when her sexual swelling was most prominent. For two or three days he was extremely alert in watching for the female with whom he was consorting whenever she strayed away from him during the day range. This exclusiveness was not observed to take place in his relations with the two lowest ranking young females when they came into estrus in 1961. Although both these females repeatedly presented to him and groomed him whenever he would tolerate their approach, he rarely mounted and copulated with them at any stage in estrus. Much younger males, two of them estimated at only from about one and one-half to two and one-half years old, repeatedly mounted these females and achieved intromission, (see Fig. 4–9) and males *1, 2,* and *3* copulated with them without increase in tension between them, and with no interference from the *a* male.

Prior to the *a* male's exclusive mating with the fully estrous adult females, other males in the group shared in the mating toward the onset of the female's turgescence. The nature of this relationship is shown by the frequencies of copulation by the *a* male and males *1* and *2* of S group with the same female during five complete day ranges (Fig. 4–10). On August 28 the group spent several hours in the visitors' car park, but, during the five observation hours before and after this, the *a* male copulated with this female four times, male *1* three times, and male *2* not at all. On the third occasion when male *1* copulated with her toward evening, the *a* male, who was feeding 50 yards away from the pair, immediately leapt up and charged after and attacked male *1* who at once rushed away. On August 29, as indicated, male *1* made no attempt to copulate with the female, although she occasionally presented to him, and he remained near the *a* male without threatening him except very briefly on one occasion.

An elaboration of the kind of dominance-mating pattern of S group is

Fig. 4–9. A juvenile of about one-and-a-half years attempts to copulate with a low-ranking estrous female. S group, Cape. *(Photograph by Dr. G. J. Broekhuysen)*

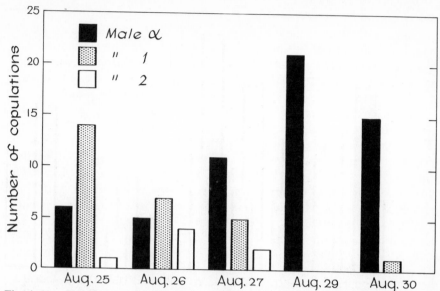

Fig. 4–10. Frequency of copulation by three males with one female (F4) during five full-day observation periods. S group, Cape, 1960.

shown in the Kenya SR group, in which there were six full-grown males of various ages, but of about equal size, and only seven sexually mature females out of the total of twenty-eight animals. At one time only one of these females was in estrus. The tension among the males was very marked and was expressed in almost continuous harassing and fighting for her possession. Consequently, she did not remain in consort with one male for many hours. Later, when there were four females simultaneously in some state of estrus, consort pairs sometimes stayed together for several days. In this group the adult male's dominance is reflected by the number of times he was able to copulate successfully, as is shown by the frequency data for the six adult, one subadult, and juvenile males of this group (Fig. 4–11).

Interruptions of sexual relations were thus a prominent feature of the SR group, contrasting with a pattern of noninterference in other Kenya groups and in S group, Cape. Where several adult males are found in the more usual ratio of 1 : 2 adult females, as in C group, Cape, in 1961, and in most Kenya groups, the tendency has been for consort pairs to form for two or more days. These pairs usually remain at some distance from the rest of the group—as far as 200 yards or more in the Cape. This withdrawal is likely to ensure avoidance of interruptions and harassments. Indeed, except for the SR group, fighting between adult males over estrous females was never observed in any of the Kenya or southern Africa groups. That such fighting might occur in any group, under conditions similar to those in the SR, is indicated by the results of the artificial feeding experiments men-

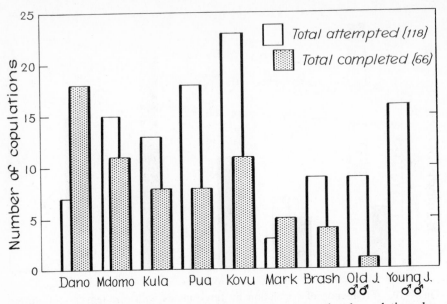

Fig. 4-11. A comparison of each male's attempted and completed copulations during 55 days of observation. SR group, Kenya, 1959.

tioned above. In any Nairobi Park group of more than two adult males, the complete harassment pattern (seen during mating activity in the SR) could be elicited by continuously offering a desired food item to the males.

Despite the prominence of harassment and fighting in the SR, the overall mating pattern was the same as that in other groups: juvenile, subadult, and less dominant adult males copulating with a female in the initial stages of her swelling; dominant males forming consort pairs with her and copulating during her period of maximum turgescence. This pattern is clear in Table 4-3, which includes all mating activity recorded for the S and SR groups. Despite the fact that in the SR group no females were in estrus some of the time and only one was in estrus most of the time, not a single copulation was attempted by the dominant males except when the swelling of a female's sexual skin had reached the maximum. Table 4-3 also illustrates how a male's position in the central hierarchy gives him an advantage over other adult males. in mating activities. Kovu, an old male whose teeth were worn level with his gums and who was *individually* the least dominant adult male in the group, was nevertheless second only to the most dominant male in copulations completed at the time of maximum swelling in the female. Since ovulation occurs during the period of maximum swelling, it is likely that he was therefore one of the most effective breeders in the group.

The mating pattern in a large group with the usual adult sex ratio is still not entirely clear, as indicated in our discussion of the dominance relations

TABLE 4–3 Frequency of copulation or attempted copulation by adult males with females in a state of full or partial estrus

S Group, Cape

Adult Male	Completed Copulations Female in:		With Full Estrus Female (percent)
	Full Estrus	Part Estrus	
α	101	8	93
1	37	4	90
2	56	28	67
Others:			
Young adults	20	11	65

SR Group, Kenya

Adult Male	Completed Copulations Female in:		Attempted Copulations Female in:		With Full Estrus Female (percent)
	Full Estrus	Part Estrus	Full Estrus	Part Estrus	
Dano	18	0	12	—	100
Kovu	11	0	23	—	100
Pua	8	0	18	—	100
Kula	8	0	13	—	100
Mdomo	8	3	13	2	87
Mark	0	5	—	3	0
Others:					
Brash (subadult)	1	3	3	6	33
Old juveniles	—	—	—	9	0
Young juveniles	—	—	1	15	6

in such a group. Although consort pairs are formed and maintained without evident interruption, we do not know whether a particular cluster of females tends to remain in the social milieu of one particular male for any period beyond the consort relationship. It is thus not known whether after her infant is born the female follows the male with whom she consorted for protection, and whether she maintains social contact with him during her gestation period by mutual grooming. Only prolonged study of a large group can determine the consistency or variability of the social and mating patterns and establish whether these differ significantly for the baboons of Kenya and southern Africa from the kind of social system reported by Kummer and Kurt (1963) for hamadryas. They state that in the one-male unit groups typical of their population the adult male makes strenuous efforts to prevent his females from straying into another group when several of these groups come together at the sleeping cliffs. The male does this by going after the female and biting her in the back of the neck, and the screaming female then follows him closely. To quote Kummer and Kurt, "In the one-male-group of

Papio hamadryas the neck-bite obviously was selectively developed as the male's instrument to herd his group together in the crowd of the party." Certainly the neck-bite was never seen in such a context in the Kenya baboons and was only occasionally seen in S group, Cape, when the *a* male was trying to force an unwilling estrous female, with whom he had already been mating, to take up a position of readiness to receive him. In this latter situation the biting was gentle, the female did not scream, and the biting was accompanied by gestures of touching her on the side or flank with one hand. In other words, it was not a "herding" gesture. No herding behavior of any kind was observed except briefly, in SR, when a less dominant male in consort was being hard pressed by another. Kovu, for example, was seen four times to put an arm across the back of the estrous female he was with and try to pull her away from the area of the other males.

Mating Behavior Details

It is possible that between the baboons of Kenya and those of southern Africa there are more clear-cut differences in mating behavior than in any of the other behavior categories. In both areas copulation as well as grooming and other social behavior occurred much more frequently early in the day, while the group was still in the vicinity of its sleeping place. In S group, Cape, for example, copulation was observed almost twice as often, in relation to the number of observation hours, during the first two or three hours after dawn as at any other time of day. The frequency again increases slightly toward evening. Although we have only negative evidence, the fact that the peak of sexual activity occurs in the early morning in both study areas indicates that it is unlikely that copulation occurs very often at night.

The first important difference that seems to exist in comparing data from the two areas is in the performance of the males in copulation. The quantitative data obtained for the three adult males of S group, Cape, suggest that the typical pattern is a series of copulations with the estrous female similar to that described for the rhesus by Carpenter (1942). For example, the *a* male of this group copulated with one fully estrous female 17 times between 0800 and 1400 hours on one day. There was then a three-hour interval when no copulation was observed between them, and four more copulations occurred between 1700 and 1800 hours as the group was nearing its sleeping place. Copulations in such a series occurred at about two-minute intervals, each consisting of between five and ten thrusts and lasting ten to fifteen seconds from time of mounting to time of dismounting. It was obvious that ejaculation did not always occur, and it is thus likely that it occurred only at the culmination of a particular series. However, the copulation ending with ejaculation could usually not be distinguished from any other copulation in the series. Because they may be of some comparative

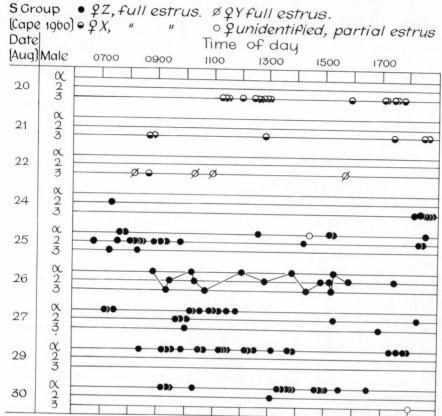

Fig. 4–12a.   Complete day-range observations of mating activity in the S group, Cape, to show copulatory series. Each dot represents one mounting by a male (see text).

significance, the time relations of such series for a sample of complete day-range observations are given for the adult males of S group, Cape (Fig. 4–12*a*) in which they are contrasted with data from Kenya (Fig. 4–12*b*). It is evident that the pattern is very different.

In contrast to the copulation series observed in the Cape, males in Nairobi Park typically ejaculated at the conclusion of only one mounting sequence. Such a copulation is brief, lasting 8 to 20 seconds, and averages only 6.2 pelvic thrusts (based on 54 observations). Rather than 2-minute intervals, the interval between copulations in Kenya was seldom shorter than 30 minutes and was usually longer. When tension was high among the males in the SR group, a male might make repeated, unsuccessful attempts to copulate, but this situation was clearly distinct from the lack of tension associated with copulations in other groups. In Nairobi Park ejaculation by the male is accompanied by a pause of from 2 to 4 seconds during which

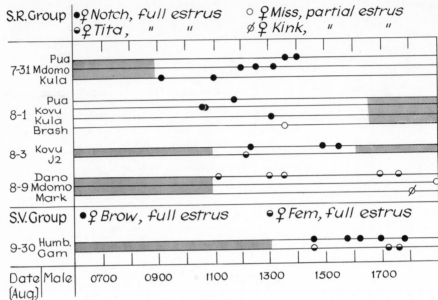

Fig. 4–12b. Mating activity in two Nairobi Park groups to indicate contrast in mounting sequences. Each dot represents a mounting with ejaculation. Shaded portions show times of day when group was not under observation.

the male's whole body remains rigidly fixed in the mounting position. From numerous observations of the "rigid pause" at close range, where ejaculation could be seen, it was usually possible to ascertain whether ejaculation had occurred in mounting sequences which took place at longer distances from the observer.

The actual initiation of mating behavior is likely to vary between partners and between stages of the consort relationship. In Kenya the female initiates sexual activity during the first two-thirds of her cycle by going from one male to another. Although she does not solicit any but adult males, she allows mounting by juvenile and subadult males at this period. Bolwig (1959) describes the chacma female as always initiating mating by coming up to the adult male and stopping in front of him, where she presents to him. It seems likely that Bolwig was observing females during the early part of their swelling, although this is not clear from his account. The data on S group, Cape, show that the initiating act depends on the state of estrous swelling of the female, as well as on the status of the male who is currently mating with her (Table 4–4).

From Table 4–4 it can be seen that prior to the establishment of a consort relationship the female takes the initiative, that is, when the non-dominant males are mating with her. But when a consort relationship is established with the dominant male, either partner might take the initiative.

TABLE 4–4   Copulation-initiating behavior by and to the males of S group, Cape

	Initiative by Male	Initiative by Female
Other males	7	18
α male	26	27
Total	33	45

The act of copulation closely resembles that of other monkeys, the male mounting the female as she stands in front of him, holding her back or side fur in his hands, and gripping her in the ankle region with his feet so that he is clear of the ground. The chacma male (only) sometimes grimaces, his lips being drawn away to expose his teeth, and occasionally the female turns her head to the side in the direction of the male (as does the rhesus female). As the male dismounts, the chacma female gives a distinctive muffled growling call in which her cheeks puff out and her mouth is almost closed. This occurred in 92 percent of the copulations observed at close range in South Africa and Rhodesia (Washburn, personal communication), but vocalizations by either male or female were very infrequent in Kenya (see Table 4–5, item 10), and males were not seen grimacing. After the male has dismounted, the female characteristically runs or walks a few yards away from him, this reaction occurring even when copulation has taken place on a narrow ledge or the top of a bush from which she will leap down. Running away from the male occurred in about 33 percent of the copulations observed in Kenya and in about 75 percent of those observed in the Cape. In S group, Cape, frequency data show that this flight has been far more frequent from the a male than from the other males, and its significance is not yet clear. In Kenya the female sometimes ran away from the male until she came to another male to whom she indicated her readiness to mate, but this is probably typical only of the early phase of estrus. No such running from male to male has been seen in the southern Africa groups. Grooming between partners often occurs before copulation and sometimes directly after copulation, initiated by either the male or the female. In this the baboons of both regions are similar.

Although the dominance pattern associated with mating seems essentially the same for both areas of study, there are apparently behavior differences in the details of copulation that require further intensive investigation. The copulation call of the female chacma is so distinct and so constant an accompaniment of the male's dismounting that if it had occurred it would have been noted in the close-range observations of the Kenya groups. But an even more prominent difference in behavior between the two areas is the copulation *series* of the chacma, as well as the indication, from the behavior samples obtained in the Cape at regular intervals throughout the year 1958–1959, that copulations occur in any month. It is possible, then, that regional differences in sexual reproductive activity are accompanied by differences in the form and frequency of its behavioral expression.

SOCIALIZATION

In dealing with the development of social behavior from birth to maturity, we are probably more dependent upon the continuous checking of field with laboratory data, and vice versa, than in any other aspect of nonhuman primate behavior. The interactions of social learning, of play and exploration, of sensori-motor coordinations, and of reflex systems are such that they can be described only tentatively from field data. It must be left to experiment to try to sort out the complex of factors that determine the final behavior of the mature animal within the group. It is probably a correct assumption that the developing primate learns its functions in the group and its skills in coping with its physical environment in ways that differ importantly according to its age and growth stage. It is likely, however, that learning is necessary for all specific meaningful connections and modifications to be made in the built-in forms of the behavior repertoire. Thus, an infant baboon does not have to learn the basic threat-behavior gestures and vocalizations of its species, but it must continually learn to inhibit, modify, and refine the ways in which its repertoire is used to accord with varying social conditions. Exactly how these learning processes proceed in the natural group can, at present be inferred chiefly from laboratory studies of habituation, conditioning, and observational learning, and a more complete account awaits the result of naturalistic experimental studies of free-ranging animals carried out over several years.

The stages in baboon development from birth to more or less complete independence from the mother will be outlined here under eight arbitrary chronological headings with which various progressive changes in behavior and social relationships within the group are associated. A provisional "typical" picture is thus sketched, but it must be noted that there is much variation between individuals. Furthermore, the data are not yet sufficiently based upon a long-term follow-up of identified individuals through all the stages. We present, rather, a synthetic account derived from observations of the young animals of nine groups in Nairobi Park and of three groups in the Cape Peninsula, together with data compounded from miscellaneous observations of other groups in Kenya and southern Africa. The longest series of observations of an identified individual was made in Kenya in a group observed continuously from June through mid-September, with intermittent data up to late December of the same year (DeVore 1963). The scheme thus derived is shown below where average behavior correlates are set against each of eight stages in age growth. Following this is some elaboration on the behavior stages and their social significance, together

Fig. 4–13.
A mother grooms her infant two or three days after its birth. SR group.

with an assessment, largely in the nature of hypotheses for further study, of the main factors affecting normal behavioral development within the group (other than those of nutrition, sickness, injury, and so on).

A composite account of the chronological stages of normal development and associated social behavior in baboons from birth to adulthood follows.

1. *Newborn, birth to one month. Infant:* locomotion confined mainly to body surface of mother; grasping her ventral surface and suckling.
   Mother: shortly after birth, supporting infant's head with one hand, licks blood from rock on which she has sat and from her fingers, moves closer to other baboons; continues to support infant's head on day ranges and supports its rump on upper surface of her feet when she is sitting; grooms infant frequently; tends to come close to a dominant male and remain in his vicinity; little vocal communication between mother and infant because of direct physical contact (Fig. 4–13).
   Others: soon are attracted to approach close to mother and infant, lip-smacking, touching infant with hands and mouth, and grooming mother; this may include the *a* male; "attractiveness" of mother-infant manifest.
2. *Infant-one, first to fourth month. Infant:* begins to ride briefly on mother's back at about five weeks, though changes its position back to her belly quite readily; infant's developing locomotor and sensory coordination

allows occasional movement away from mother; signals from mother to infant now appear; occasionally touches with hands food plants which the mother is eating, or puts to mouth without chewing.

Mother: allows infant to move some feet away, but sometimes takes hold of it and draws it to her for grooming; tends to remain close to a dominant male.

Others: fully adult and even young adult males are completely tolerant of the infant as it crawls near or over them; they may touch the infant with one hand, and the dominant male may grunt at high frequency while looking at the infant; immediate retaliation by a dominant male against attacks on the mother.

3. *Transition, fourth to sixth month* (period of color change from black to light brown). *Infant:* begins to sit up, "jockey style," on mother's back, though when mother runs or climbs, lies face down and clutches with hands and feet, as at stage 2; begins to eat some solid foods whereas previously only touched or chewed at them; occasionally picks up objects such as sticks, and stones, carrying them sometimes in the hand or mouth; an important stage in exploring immediate physical environment, moving as much as 20 yards or more from the mother, and in beginning social play with peers; much play movement, such as climbing and jumping and tussling.

Mother: grooms the infant regularly, but not so consistently as earlier because of its wandering.

Others: peers and older young participate in play; adults considerably reduce their attentions to the infant, although both females and males may sometimes carry the infant, as though in play, and adult males maintain their protective function.

4. *Infant-two, six months to one year. Infant:* progressive increase in play relations with other young, and increased exploration of physical environment; by about ten months, nurses only rarely, though tends to follow mother closely when she is feeding, and feeds often from the same plants and uses the same feeding movements; probably an important stage in learning feeding habits and discriminations.

Mother: remains tolerant of infant's close presence and of its wandering away from her.

Others: increasing play relations of other young with infant; adult females occasionally threaten or even attack the infant, holding it down or giving it a "token bite"; screeching of the infant may elicit a counterattack by the mother, or by an adult male; adult males remain completely tolerant and protective of the infant.

5. *Weaning, eleventh to fifteenth month. Infant:* tends to be rejected by mother both from taking the nipple and from riding on her back; the infant may continue to press its attentions upon the mother, following her and sometimes grooming her, while the rest of the time it is continually active in playing and exploring.

Mother: females vary considerably in the stage and in intensity of rejection; in general the independent activity and irritability of the mother increases with her sexual swelling; the temporary "privileged" status of the mother, due to the presence of the infant, is now lost.

6. *Young juvenile, second year of life.* The increasing independence of the young animal is partly shown by the fact that now, when danger threatens the group, it tends to flee to the protection of the adult males and not to the mother or other adult females; the juvenile spends most of the day with its peers, feeding with its age group or with other young animals, and playing with these early and late at the sleeping place.

7. *Older juvenile, third and fourth year.* Accurate age estimates beyond the second year were not possible, but the category "older juvenile" conveniently describes the transition from the young, dependent juvenile to the independent young adult. In the early portion of this stage the individual is still oriented to play in a peer group, but play has become rougher and dominance interactions more intense. Adult males no longer tolerate play by older juveniles and in fact threaten them when the juveniles play too roughly with infants and young juveniles. During this period older juvenile females leave the juvenile play group and join the grooming clusters of adult females. By the end of this stage the female's sexual cycles have begun, she is almost fully grown, and her behavior is that of an adult.

8. *Subadult male, fourth year to complete physical maturity* (eighth year?). Although the male probably produces viable sperm by about the fourth year, his period of growth is much longer than that of the female. No records are available, but full skeletal and muscular growth, complete eruption of the canine teeth, and development of the shoulder mantle (olive baboon) may take eight years or even longer. By the fifth year the subadult male is larger than adult females and establishes dominance over them. He is still much smaller than the fully adult males and tends to avoid them. Temperamentally the subadult male is pugnacious within the group and "daring" in his behavior generally. Although in some species of monkeys males of this age may leave the group, no subadult male baboon left his natal group during our studies. Nevertheless, males of this age are more behaviorally peripheral to other group members than is true of either sex at any other stage of development.

In considering this outline scheme it is necessary both to elaborate on it and to attempt some generalization from it. At stage 1, we have no observations of actual birth, but one female of C group, Cape, was seen at 1735 hours sitting on a rock at the sleeping cliffs and holding a wet infant to her belly. Her posterior was bloody, and she was in full view of the observer until dusk at 1850 hours. During this period none of the other animals in the group actually approached her. On the contrary, on several occasions she tried to move close to some of them, and these sometimes moved away from her. For example, juveniles walked close by her and gazed at the infant and the mother (lipsmacked toward her), but they passed on without stopping. Several times she walked on three legs, holding the infant's head with one hand, toward juveniles or females with other infants, none of whom made any attempt to touch her or the infant, and most of whom moved away from her. When she took her infant and sat near the one adult male among this cluster of animals, he made no response to her or the infant. The next day she tended to be to the rear of the group during its foraging, and was seen with a large male and another female with a black infant at one time as much as 400 yards from the rest of the group. In Kenya healthy mothers with infant-ones were typically toward the center of a group, close to one or more adult males, but on several occasions an adult male was seen to drop behind the group and walk with a mother who was having difficulty keeping up with the rest of the group.

It is not yet possible to state precisely what it is in the combination of mother with infant-one that begins so strongly to attract the attentions of the other animals in the group. Certainly the temporary lack of such attention soon after the birth is not an unusual occurrence in baboons, whereas not a single instance has been recorded from either area where a mother and infant-one did not soon begin to attract these attentions. Possibly it is the beginnings of movement in the infant as it crawls about the mother's body that sets off the reactions of friendly behavior, or it may be such minimal movement combined with perception of the infant face with its relatively large, pink ears, and contrasting facial and body coloration. The clearest answer to such a question can best be obtained from experimental study.

Throughout the stages we note the protective function afforded to the mother-infant combination by the adult males. This protection seems to extend, at least in individual cases, well beyond the color-change transition period. The adult males vary quite markedly in this respect, but it is usual for older males to approach fairly frequently and to touch the infant. On several occasions adult males in Kenya and southern Africa have carried infants on their bellies–for twenty minutes on one occasion. The *a* male of S group, Cape, was notably attentive and tolerant of infants and even juveniles. He once carried a black infant ventrally up the sleeping cliffs in the morning, often picked up or touched other infants, and was also seen to sit with one juvenile against his chest for several minutes on a day of strong cold wind. In the SV group, Kenya, an infant-two female with a disease of face and scalp became the constant companion of the B male (Humbert), remaining with him all day, feeding close to him, grooming him, and sleeping beside him at night. She even stayed beside him while he was in consort with an estrous female, and took food in front of him (DeVore 1963).

Precisely what it is that stimulates the aggressive retaliatory behavior of an adult male in relation to mother and infant was often not clear in rapidly occurring events. Sometimes an infant shrieked, apparently because it was separated from its mother, and an adult male looked around slowly and even approached it. A similar sounding shriek, on other occasions, elicited an immediate aggressive charge by the màle, as did also the shrieking of a mother with an infant when she was set upon by another female. It is thus doubtful if there is any one kind of stimulation having a "releaser" effect for this behavior by the male, and it is probably elicited by combinations of stimuli the social significance of which is learned. The prominence of this protective and tolerant behavior by adult males toward infants and juveniles is such that, in general, it prevents injuries from fighting within the group and also prevents predation. What is not yet known from field studies is the association of adolescents and young adults with the adult males. Although some young adult or subadult males are somewhat peripheral, or complementary, to a group, others appear to be assimilated into close relationship with dominant males (see Social Organization above).

In describing the gradual independence of the infant from its mother as it comes to explore the physical environment and to establish play relations with other young, we are dealing with a behavior phenomenon that is characteristic of other young primates and mammals. However, since this is the first time this exploratory play behavior has been systematically observed in wild baboons, it is important to try to evaluate its probable significance. It can be said at once that play-climbing, play-fighting, play-manipulating, running, somersaulting, and so on are strongly in evidence among the young of all stages after stage 1, and continue in evidence even to the subadult stage. It is the first behavior to be seen at the sleeping places before dawn and the last behavior to cease at night. The developmental trend might be generalized as a progressive focussing of the play-manipulatory repertoire of movements and gestures upon particular objectives, early efforts being much more random. It is usually assumed that the frequent social play of young monkeys is a very important kind of trial-and-error experience that may determine adult dominance and the pattern of adult sexuality, but we have to rely upon the experimental evidence (Harlow 1962a; Mason 1965) to demonstrate what can happen in macaques when the social norm is drastically changed.

EXPRESSIVE AND COMMUNICATIVE BEHAVIOR

A straightforward description of the repertoire of expressive movements, postures, gestures, and vocalizations observed in our wild baboon groups would not be realistic or meaningful unless it was accompanied by comments on the social situations in which they occurred and on the variations in individual performance that seem to reflect the particular status of the baboon within its group. Descriptions of the repertoire of baboons are to be found in Zuckerman (1932), Kummer (1957), and Bolwig (1959), and here we shall attempt chiefly to coordinate our own observations and to relate them to the kind of scheme exemplified by the studies on captive rhesus monkeys of Hinde and Rowell (1962), Rowell and Hinde (1962). It will already be apparent from the previous sections that individual patterns of communication vary according to sex, age, and status. With regard to the vocalizations of these animals, it is notable that many hours of the day are spent in almost complete silence. Probably the commonest kind of sound is the grunt. A specialized investigation remains to be carried out of the significance of the many variations in grunting, and in the other types of calls, distinctive to the human listener at least, that are heard during the routine of social behavior. It is not unlikely that the major system that mediates interindividual behavior for baboons is one of visual cues from facial expressions, intention movements, and attitudes, while auditory, tactual, and olfactory cues are of descending order of importance. Except where specifically stated, no difference was observed in either study area in the quality, frequency, or situations of occurrence of movements and gestures. Tables 4–5, 4–6, and 4–7 list the vocalizations and elements of visual and tactile expression so far distinguished in the two study regions, with a summary of the situations in which they occur.

Accurate and detailed comparisons of the vocal repertoire of rhesus macaques and baboons cannot be undertaken until the vocalizations of both species have been tape recorded and analyzed, but it is worth noting briefly what we believe to be similarities. References below are to Rowell and Hinde (1962) and Altmann (1962).[1]

Baboon	Rhesus
1. Two-phase bark	No apparent equivalent
2. Grunting	Comparable to Altmann No. 48
3. Roaring	Probably same as Rowell and Hinde "roar"
4. Grating roar	No apparent equivalent
5. Screeching	Identical to Rowell and Hinde "screech"; Altmann No. 36
6. Yakking	Probably same as Rowell and Hinde "gecker"
7. Chirplike clicking	Probably identical (Rowell and Hinde)
8. Shrill bark	Probably identical (Rowell and Hinde; Altmann No. 55)
9. Ick-ooer	Probably identical (Rowell and Hinde "rising clicks"; Altmann No. 24)
10. Muffled growl	No apparent equivalent
11. Chattering	No apparent equivalent
12. Dog-like bark	Perhaps same (Rowell and Hinde "bark")
13. Grunting	Probably same, including "chorus" given by rhesus in sleeping trees on Cayo Santiago
14. High-frequency grunting	No apparent equivalent

[1] Through the courtesy of Dr. Stuart Altmann, DeVore was able to visit the rhesus colony on Cayo Santiago in 1958; and through the courtesy of Dr. Carl Koford, Hall was able to visit the colony in 1963 after the tabulation of baboon vocalizations had been completed.

TABLE 4–5 Baboon vocalizations

Description	Class of Animal Vocalizing	Area and Situation in Which the Vocalization Was Observed	
		South Africa	Kenya
		A. ATTACK-THREAT	
1. *Two-phase bark* (very far-carrying; audible for 1/2+ mile in mountainous districts; often repeated at about 1 per 2–5 sec. according to intensity of arousal).	Adult males only.	a. Extragroup danger situations; reaction to humans and large predators.	Same.
		b. Intragroup aggression; occurring in adult males' threat among themselves, and in attacks on, e.g., females	Apparently *not* in evidence in such in-group situations; see roaring.
2. *Grunting* (soft, sometimes two-phase uh-huh).	Adult males only.	Occasionally adult male grunts in this way as a prelude to mild threat gestures toward another animal.	Same, except that it usually begins as single grunts, becoming two-phase as the male begins threat gestures. May blend rapidly into (3).
3. *Roaring* (loud, two-phase grunting; a crescendo of grunting with a two-phase character; uh-huh [intense expression of two-phase grunting]).	Adult males only.	Not recorded.	Occurred during all fights between adult males. This accounts for 90 percent.
4. *Grating roar* (resonant but of low intensity).	Adult males only.	Not recorded.	Heard only 5 times. Given by dominant male after fight in the group. The male sits with chin pointing up and brings his head down in a series of 3 or 4 jerks, expelling the vocalization with each jerk.

TABLE 4-5 Baboon vocalizations—Continued

Description	Class of Animal Vocalizing	Area and Situation in Which the Vocalization Was Observed	
		South Africa	*Kenya*
		B. ESCAPE-FEAR	
5. *Screeching* (repeated, high-pitched; sometimes modifies to a repetitive churring noise).	Any animal	High intensity vocal reaction, accompanying other behavior indications of escape-fear, as when fleeing from adult males.	Same.
6. *Yakking* (Rowell and Hinde 1962, "Gecker"; single sharp yak sound, repeated perhaps 2 or 3 times).	Any adult or near adult.	Accompanying fear grimace, when animal withdraws from a threat.	Same.
7. *Chirplike clicking* (Rowell and Hinde 1962)	Infant and juvenile.	Probably juvenile equivalent of yakking; occurs when juvenile is frustrated, as when separated from mother.	Same.
8. *Shrill bark* (Rowell and Hinde 1962; single sharp explosive sound).	Any animal, except possibly adult males.	Has occurred on seeing a snake, or on discovering a scorpion (in the Cape), and on *suddenly* seeing eland or a hyena appear; a startle reaction that usually elicits an instantaneous avoidance response in other baboons near the barker.	Same; has occurred in response to sudden appearance of bush-buck, and mongoose, as well as to large predators, such as lions.
9. *Ick-ooer* (a repeated sharp ick, followed by longer drawn *ooering* sound, thus a two-phase vocalization).	Infants primarily; juveniles occasionally.	Accompanied by twitching of the head, shoulders and arms, and by grinning; apparently a frustration response that may occur in situations similar to (7).	Same.

C. SEXUAL

10. *Muffled growl* (mouth almost closed; cheeks puff out with inspiration-expiration, audible at 50–100 yards).	Estrous females only.	Occurred in 92 percent of copulations, coinciding with the male's dismounting; also occurred occasionally during defecation.	An indistinct call, apparently of a similar kind but with no blowing out of the cheeks, was heard in 10 percent of copulations; uncertain whether from the male or the female.

D. FRIENDLY

11. *Chattering* (nasal, very rapid, short-phased gruntlike series of sounds).	Juveniles	During play.	Probably does not occur.
12. *Doglike bark* (higher-pitched, less staccato and sudden than shrill bark and somewhat quavering).	Probably any adult or young adult.	Seems only to occur when, for example, one animal, or a party of animals, has been temporarily separated from the rest and is about to rejoin them.	Same.
13. *Grunting* (varies in frequency from slow, about 1 per 2 seconds to very rapid, sometimes as a "chorus" through group).	Any animal, except possibly infants.	The most frequent baboon sound, heard at intervals when feeding close together, as well as at night.	Same. The chorus was regularly heard when group was congregated near or in sleeping trees at nightfall, as well as during day; infants and juveniles also gave ooer sound toward end of grunting chorus. Also occurs when the animals are becoming calm after being frightened.
14. *High-frequency grunting* (slightly higher-pitched; possibly an intensity variation of 13).	Adult males only.	When close to, and possibly about to touch, an infant-one.	Probably does not occur.

Nine out of the fourteen most conspicuous vocalizations occur in both study regions in similar situations. Two of the other four (items 11 and 14) probably do not occur in Kenya; item 4 probably not in the Cape. Two more (items 3 and 10) are so infrequent in one region, so common in similar situations in the other, that they may constitute important area differences. One (item 1) seems to differ between the regions in the situations of its occurrence. While sampling differences in the kinds of observations made in the two regions may account for some of this variation, the vocalizations in question are sufficiently distinctive to indicate some qualitative as well as quantitative variations in the vocal repertoires.

TABLE 4–6   Visual communication and expression

Description	Class of Animal	Directed at Another Baboon	Areas Where Observed	Comments
		ATTACK-THREAT		
Head and face				
Tooth-grinding	Adult males only	Yes	Kenya	In threat displays of two or more animals at close range to each other
Yawning (always accompanied by ear-flattening, eyebrow-raising, etc.)	Adult males	Yes	Kenya	In harassment sequences
Staring (directed)	Any adult or young adult	Yes	Both regions	
Eyebrow-raising	Any adult or young adult	Yes	Both regions	
Ear-flattening	Any adult or young adult	Yes	Both regions	Occurs also in friendly behavior
Jerking of head down and forward	Adult and young males only	Yes	Both regions	
Body and limbs				
Mantle hair raised	Adult males only	Yes	Both regions	
Rearing on hindlegs	Any adult or young adult	Yes	Both regions	Can prelude attack or escape
Shoulders forward	Any adult or young adult	Yes	Both regions	

Behavior	Performer		Region	Comments
Slapping ground with hand	Any adult or young adult	Yes	Both regions	
Rotating movement of hands on ground	Any adult or young adult	Yes	Both regions	
Shaking of rocks, branches	Any adult or young adult	Yes	Both regions	
Hitting-away reaction with hand	Any adult or young adult	No	South Africa	Response to noxious small object, especially scorpions
Charging run	Any adult or young adult	Yes	Both regions	Attack
ESCAPE-FEAR-UNCERTAINTY				
Head and face				
Staring (undirected, eyes wide open)	Any animal	Yes	Both regions	
Grin	Any subordinate animal	Yes	Both regions	
Looking away	Any subordinate animal	Yes	Both regions	
Sideways jerking glances	Any adult or young adult	Yes	Both regions	
Yawning	Most frequent in adult males	No	Both regions	Occurs in situations of uncertainty

TABL 4-6 Visual communication and expression — Continued

Description	Class of Animal	Directed at Another Baboon	Areas Where Observed	Comments
Body, limbs, tail				
	ESCAPE-FEAR-UNCERTAINTY			
Tail erect	Any subordinate animal	No	Both regions	Usually adult females, particularly mothers with infants
Body prone to ground, animal rigid if fear is extreme	Any animal	No	Both regions	Most intense fear response; very rare in adult males
Twitching of head, arms, shoulders	Infant, juvenile	No	Both regions	Separation from mother, insecurity
Throwing aside of arms	Any animal	No	South Africa	Response to noxious, small object, especially snake
Scratching	Any animal	No	Both regions	Occurs in situations of uncertainty
Shoulder-shrugging	Adult or young adult male	No	South Africa	Response subsequent to startle-reaction
Muzzle-wiping with hand	Adult or young adult male	No	Both regions	Response subsequent to startle-reaction

FRIENDLY

Head and face			
Lipsmacking	Mostly adults and young adults	Yes	Both regions
Ear-flattening	Mostly adults and young adults	Yes	Both regions
Body and limbs			
Presenting	Any animal	Yes	Both regions
Stand on hindlegs in front of another	Uncertain; mostly adults or young adults	Yes	Both regions

SEXUAL

Head and face				
Turning head back	Adult females in estrus	No	Both regions	Only occasional
Grin	Adult males	No	South Africa	Occasional during copulation

Tension Differences in Normal Posture or Locomotion

The stance of a baboon, independently of any specific gesture, may indicate differences in tension and of individual status. The dominant male macaque walks deliberately and stiffly, with tail elevated. The dominant male baboon tends to walk very directly and "confidently" through different parts of a feeding area or when moving across country. It is too subjective a task to try to judge whether the proximal section of his tail tends to be more upright than that of subordinate animals. Intentions are declared by changes of pace, by halting, and by looking around, and it is such changes in ordinary locomotion that may be inferred, from the behavior of others in the group, to be the main cue to which they respond when moving away from the sleeping place or altering direction during the day range. No vocalization from a dominant animal normally accompanies or precedes such movements. The general tension in the attitude of a watchful animal is clearly to be seen when it is sitting, standing, or walking about, and alertness is sometimes shown by its climbing to a high point or by its standing on hind legs.

Relaxation is equally clear in lying or sitting attitudes. In relaxed sitting the head is lowered, the back is arched, the hands rest loosely near the knees. When the animal is lying down, a variety of doglike postures are seen, including lying on the back, on the stomach, or on the side.

All such postural differences between a tense and a relaxed animal are likely to be in a general way communicative to one or more animals in the group, and thus may affect their behavior, as in determining the distance they keep from the animal and whether they look toward it or ignore its presence.

Attack and Threat

The intensity, duration, and complexity of attack and threat behavior varies greatly with the social situation and its stresses. A dominant male may sufficiently indicate his intention or mood by simply standing up, raising his muzzle, and staring, or by remaining seated and, while staring, slapping once with one hand in the direction of another animal. Or he may go immediately into an attacking charge which culminates in his seizing his victim in his jaws, biting it, holding it, and rubbing it on the ground with his hands. In the latter case the chacma male will repeatedly utter a loud two-phase bark, while the olive male roars in the same situation (Table 4–5). More complex expressions of attack and threat occur when the participants are of more equal status, as in the "harassment" sequences between adult males, or as when several females or young males engage in a quarrel. Quantification of attack-threat episodes, both where no physical contact was made with a victim and where such contact as grappling, beating, and

TABLE 4–7 Tactile communication and expression

Description	Class of Animal	Areas Where Observed	Comments
Head and face		ATTACK-THREAT	
Biting	Any animal; more frequent in males	Both regions	End result of attack; victim sometimes lifted from ground; usual bite areas are nape of neck; occurs in play sometimes as a gape; in Kenya females usually bite at the base of the tail
Body and limbs			
Grappling with hands	Adult males	Both regions	Attack
Rubbing against ground	Adult males	Both regions	Attack
Slapping	Adult males	Kenya	By the attacker, when a threatened adult male is deliberately looking away from the threat. (rare)
		ESCAPE-FEAR	
Mounting to the side	Any adult, juveniles	Both regions	
		FRIENDLY	
Head and face			
Mouth-to-mouth touching	Any animal	South Africa	
Genital-stomach nuzzling	Mostly by adult females and young adults of both sexes, to infants	Both regions	Rare in Kenya

TABLE 4–7   Tactile communication and expression – Continued

Description	Class of Animal	Areas Where Observed	Comments
Head and face		FRIENDLY	
Putting nose to perineal region	Adult males, usually to estrous females	Both regions	Rare; usually sexual
Body and limbs			
Posterior grasping	Adult males	Kenya	Response of male being harassed
Mounting	Dominant animals	Both regions	Animals tense during mounting
Touching on side	Adult males	South Africa	
Tweaking (hands between hindlegs of other)	Adults and young	South Africa	
Grooming	Adults and young	Both regions	
		SEXUAL	
Head and face			
Biting gently at nape of neck	Adult males to estrous females	South Africa	Rare
Body and limbs			
Pushing with hands	Adult males to estrous females	South Africa	Rare

biting occurred, showed in S group, Cape, that aggression by adult males was of the contact kind twice as often as aggression by adult females. In *all* the episodes observed, no-contact threat was a far more frequent culmination than physical attack. Certain components of no-contact threat were relatively more frequent in the females, but all such frequency data are representative only of the dominance structure and of adult male-adult female relationships within the particular group, and no generalization can at present be attempted.

Expressions of attack-threat, very similar to those described for *P. hamadryas* by Kummer (1957) and for *Macaca mulatta* by Hinde and Rowell (1962), are listed here.

CHARGE OR ATTACKING RUN.   This is most clearly seen in the behavior of a dominant male, who may start the attack by leaping to his feet, raising his muzzle, and immediately launching himself after his objective without any preliminary threat gestures. He barks loudly and repeatedly and his mane fur tends to stand up around him, enlarging his appearance. If he catches his victim, he grapples with it and seizes it in his jaws, usually near the nape of the neck, or he may beat it with his hands as it lies prone on the ground. A female thus attacked has been seen to be briefly lifted off the ground in the jaws of the male. Such attacks have usually been of very short duration, and, even when they sound and appear extremely vicious, have only once been seen to result in visible physical injury to the victim. Often the victim evades the attacker by jumping or climbing down a steep cliff, hiding under a boulder, or running up into a tree, in which case the attacker soon gives up. Although such attacks come chiefly from adult males, one instance was recorded in Kenya of two adult females attacking and viciously biting another female and persisting in so doing for twenty minutes. At the end of this attack the animal was badly mauled and bleeding from numerous wounds.

THREAT.   Facial expressions observed in threatening animals consist of "staring," sometimes accompanied by a quick jerking of the head down and then up, in the direction of the opponent, flattening of the ears against the head, and a pronounced raising of the eyebrows with a rapid blinking of the pale eyelids. This reaction, as Van Hooff (1962) points out, is seen in all the macaques, mangabeys, and baboons, and is particularly obvious because the upper eyelids and the skin above them are white.

Grinding of the teeth, the function or significance of which is not clear, has been observed in adult males of Kenya groups when closely threatening each other, as in harassment sequences. Yawning has likewise occurred frequently as two adult males come close to one another. In yawning the large canine teeth of the adult males are clearly displayed and, as Zuckerman (1932) has suggested for baboons, and Hinde and Rowell (1962) for macaques, this may have a secondary intimidating effect. An animal may also frequently yawn, without necessarily directing its gaze toward the source of disturbance, in situations where a conflict of tendencies is highly likely and hence "anxiety" (see below) is present, but this can easily be distinguished from the threat yawn which is always accompanied by other threat gestures, such as ear-flattening, eyebrow-raising, and so on.

Threat gestures and movements of the limbs that may either accompany or follow facial expressions include a variety of intention movements made from sitting, standing, and sometimes crouching positions. They include rapid down-pressing movements of the arms, with accompanying jerking forward and back of head and body, and rapid forward and backward scraping movements with the hands against the ground. Sometimes these are suggestive of ambivalence in the animal, that is, uncertainty as to whether to threaten more closely or to go away. They occur often in young animals or subadults that are disturbed by the presence of an intruder or predator. All such reactions are usually without vocalization, but in the Kenya males a low grunting which grew to a roaring crescendo of rapid two-phase grunts always accompanied movements forward, such as a short run, lunge, and slapping of the ground in the direction of the opponent. Low grunting sometimes occurred in the Cape males in similar circumstances. A grating roar (Table 4-5, item 4) has been heard in the Kenya males on only five occasions, and always after fighting in the group.

Baboons sometimes shake tree branches violently, or pull vigorously backward and forward against rocks, when, for example, they are excited by the too close approach of human beings or large predators. This kind of reaction has rarely been observed in encounters between individual baboons, and it is possible that it is a redirection of aggression onto substitute objects (Bastock, Morris, and Moynihan 1953) rather than a part of interindividual threat behavior.

Although in the main the basic attack-threat repertoire of baboons resembles closely that of macaques, some differences of detail or of social significance may emerge from intensive comparative study. It is not clear, for example, that the behavior described as "backing threat" and "showing hindquarters" by Hinde and Rowell have had their counterparts in our baboon groups.

Escape-Fear

As with attack-threat a range of behavior is included under this heading, from the obvious running away to slight gestures of mouth or limbs.

RUNNING AWAY. When, for example, a female is chased by a dominant male, she typically runs away screeching continuously (Table 4-5, item 5), her tail may go to a vertically upright position, and she may defecate and urinate as she goes. If caught, or even if she escapes down a cliff and continues to be threatened by the male above her, she tends to crouch to the ground, and her teeth are bared as she repeatedly utters a long-drawn, high-pitched, repetitive "churring" (Table 4–5, item 5). Occasionally, when running away from a charging male, a female has succeeded in temporarily arresting his aggression by quickly halting and presenting in front of him.

FEAR GRIMACE.   This, the "grimace" of Altmann and the "frightened grin" of Hinde and Rowell, is characterized by retraction of the lips so that the teeth are exposed but not separated. The ears are usually flattened against the head, and the eyes may have a staring appearance. In chacma baboons this occurs quite frequently as a gesture by a subordinate animal when passing close to a dominant one, and it also occurs when an animal is offered food but is afraid to approach nearer to get it. The fear grimace is rarely seen in adult males. In Kenya only the old male Kovu, when very hard pressed, grimaced. But any unexpected attack on a female, juvenile, or infant almost invariably caused them to grimace. A more intense form of this expression occurs with the mouth open and accompanied by a sharp short "yakking" vocalization (Table 4–5, item 6) usually as the animal begins to withdraw and is looking back over its shoulder (Fig. 4–14).

STARTLE REACTIONS.   These have been most clearly shown in field experiments with S group, Cape, and in field observations of this and C group when reacting to snakes. The action of an individual baboon on seeing a snake in a bush or on the ground has been sometimes to leap instantly away from the spot with all four limbs clear of the ground. It then stands, turns its head cautiously to look at the snake, and walks slowly away. When the whole of C group walked close by a coiled mole-snake about 6 feet long, each animal looked at the snake, but only some of the infants riding on their mother's backs were seen to have their tails go vertically upright. No vocalization was heard. On one occasion, a young female, on seeing a snake, uttered a sharp shrill bark (Table 4–5, item 8) and looked toward the rest of the group, none of whom appeared to respond to her.

In 1961 the animals of S group were discovered to have a complete aversion to live scorpions when these were experimentally given to them. The animals typically showed their aversion by immediately leaping back from the scorpion, with arms spread apart at fullest extent, and usually with the "yak" bark of alarm. Following this the animal would sometimes draw one hand down over its muzzle — apparently a "nervous" reaction with no particular significance. Some of the animals followed their startle behavior by hitting away several times with one hand the scorpion or the bag or box which contained it.

The startle reaction to the very sudden appearance of another kind of animal, such as a group of eland in the Cape and a hyena in Southern Rho-

Fig. 4–14. A female bares her teeth and "yaks" in fear. SR group.

desia, has consisted of one very sharp bark (Table 4–5, item 8), almost explosive in quality, by the baboon that first saw the intruder. Any baboons nearby will instantaneously dash away or up into trees. This is a reaction quite distinctive from the far-carrying, repeated barking of baboons when a lion or other large predator is in sight, or when a human intrudes on a group unaccustomed to human presence. Startle reactions of baboons vary enormously from situation to situation. If the observer comes suddenly upon an unhabituated group, it is likely to rush away silently, and only later to begin barking. Under such circumstances it is likely to disperse rather than keep together.

In many situations affecting behavior of baboons toward humans or other animals that arouse some fear of hostility, or both, as well as in some situations of interaction between baboons, attack-threat and escape-fear expressions may both occur. In addition, behavior expressive of tension or uncertainty may be observed. Yawning has been seen on several occasions in individual baboons in both regions when it was neither preceded nor fol-

Fig. 4–15. An adult male yawns while he sits expectantly beside another male eating a young gazelle. SR group.

lowed by any expression of attack-threat (Fig. 4–15). In these cases it may be primarily a displacement activity and secondarily a part of threat display. It occurs occasionally in adult females, but less commonly.

Scratching the arms or back, wiping the muzzle with the hand, shrugging the shoulders, and fiddling with food objects have all been observed under similar circumstances, the actions tending to be desultory, jerky, and of short duration. When a male in the SR group is being sharply harassed, his normal grooming of the estrous female often increases in rapidity until he is brushing at her side in a frantic, ineffectual way. A rapid copulation was frequently observed when general chasing and fighting had broken out among the males of this group. On a few occasions mounting was observed and apparently true copulation by adult males of the Cape groups when disturbed by experimental objects. Similar "out of context" behavior is reported by Carpenter for the howler monkey (1934) and for *M. mulatta* (1942). It seems likely that in baboons and probably other monkeys, agitation or uncertainty leading to anxiety may express itself through any behavior system that happens to be or to have recently been activated, while the cues as to which system will receive the expression may also come from the presence of the relevant stimuli in the environment. The situations in which certain behavior of infants and young juveniles occurs — such as being ignored by the mother or separated from her—indicate a high degree of frustration or insecurity. In these situations the infant gives a repeated ick-ick-ooer (Table 4–5, item 9) accompanied by spasmodic or twitching movements of the head, shoulders, and arms, and by grimacing.

Agitated Behavior of Unclear Function

Occasionally during quarrels in S and C groups, particularly in the latter, where adult females or subadult males were primarily involved, one animal would appear to mount another, with hands resting lightly on its back, but tending to stand with hindlegs a little to the side. Both animals pointed with their muzzles directly toward one or more other individuals whom they were threatening. The movements of both were jerky, their attitudes tense. Similar behavior has also been observed in Kenya groups in quarrels among adult males and, occasionally, juveniles. It is not clear at present whether this is an action that forms a part of "enlisted threat" and, if it does, whether it derives from mounting.

Friendly Behavior

As in the other categories of behavior, friendly behavior patterns may occur in varying social contexts. Thus, presenting, which is usually a gesture of submission, is often accompanied by nervous, even fearful, behavior on the part of the presenting animal, whereas mounting, conversely, in baboons is usually an indicator of relative dominance. However, all the forms of behavior to be discussed here seem to have as their chief function the establishing or maintaining of peaceful social relationships within the group.

GROOMING.   Grooming in baboons is a very prominent form of social behavior, occurring in the wild most frequently early in the day, during a midday quiet period, and when the group approaches its sleeping place at the end of the day. Initiation of grooming may be made by an adult male, for example, by going up to a female and lying down beside her. The grooming response of the female is usually immediate. A female who is anestrous or partly in estrus may approach an adult male and present to him, upon which he may briefly pick over the perineal fur with his hands and put his nose close up to her. Initiative may come from the animal who intends to groom, as when a female, lipsmacking, approaches a mother with an infant and begins to groom the mother. The groomer usually lipsmacks while grooming. The movements of grooming are very similar to those observed in other monkeys, namely a brisk parting of the fur with the fingers of both hands and a picking off by hand or directly by mouth of extraneous items on the skin. These items are often swallowed.

Analysis of 220 grooming events in the Cape baboons indicated that grooming of female by female was easily the most common (Fig. 4–16), and females also spend much longer (averaging 6½ minutes per grooming event) in grooming than do males (averaging 1⅓ minute)—patterns of grooming behavior which are also true of baboons in Kenya. Adult males spend longer in grooming the estrous females with whom they are in temporary consort relationship than they do in other grooming situations. Females with infants, as already noted, are particularly attractive to other females, who frequently approach and groom them. In S group, Cape, it was also noted that, in female grooming pairs, females nearing or at full estrus were the groomed animals nearly four times as frequently as the converse. This is probably a direct reflection of the increase in all kinds of social activities by an estrous female, and of some alteration in her status, but the significance of the observation is not yet clear.

Infants are frequently groomed by their mothers and by other females, and young animals frequently groom each other, although it is not known whether the young males tend to be groomers markedly less often than the young females. Although a female low in the dominance order may be groomed by other females superior to her, she may also be driven away fairly often by another female.

The nature of the dominance relations within a group are reflected in the amount of grooming attention an individual receives; a dominant, central male, for example, is more frequently groomed than a peripheral or less dominant male. Nevertheless, it seems that all members of the group to

Fig. 4–16. One adult female grooms another while an infant-one clings to her mother's belly. S group, Cape.

some extent receive and give grooming attention, and the function of grooming is assumed to be not simply a cleaning of the body surface but a continuing reinforcement of the social bonds.

LIPSMACKING AND "GREETINGS" BEHAVIOR. Lipsmacking occurs, as we have stated, when a would-be groomer approaches another animal and also during grooming. It also occurs as a prelude to or accompaniment of a series of gestures which are provisionally described as "greetings" behavior. This behavior was most clearly and frequently seen in S group, Cape, as a sequel to the release near the group of a tame young male baboon (about four years old). Once the aggressiveness of the a male had subsided, baboons of about the same age and of both sexes approached the stranger on many separate occasions and exhibited several reactions to him, which were sometimes reciprocated by him. These included standing on hind legs; embracing the hindquarters of the other when it was standing on all fours, one arm placed between the hind legs, the other holding the buttocks from the rear; putting the muzzle down to the genital-pelvic area; placing the mouth up to the mouth.

On the few occasions when this behavior was seen at other times in S group, it seemed that the initiative usually came from the more dominant ·animal of a pair. Thus, the a male was observed to rise briefly on his hind legs, while lipsmacking, as some adult females passed close to him. Also, female B, dominant over female E, approached lipsmacking up to E, held the latter briefly with her hands between E's thighs and put her head right down to E's pelvic region. In the beginning the initiative always came from the animals in the group toward the tame baboon although later he initiated some exchanges. Such behavior was seen rarely in the Kenya groups also, and it is doubtful if all the components in the sequences described for the Cape occur there. Adult females have been observed standing on hind legs in front of adult males, and when two adult males are approaching each other lipsmacking, one may put its nose to the genital area of the other and even grasp the other's penis.

This type of performance is clearly very similar to that which females, or sometimes males, would carry out after approaching a black infant. The usual sequence was to walk, lipsmacking, toward mother and infant, then stand on hind legs, pick the infant up by its back legs, and briefly touch or "kiss" the infant's backside. Such performances frequently occurred in S group in the winter of 1961 when there were five black infants in the group. What is common to the situation of both the tame baboon and the infants is that they were all newcomers to the group. These performances were not accompanied by any vocalization. Although the adult male (Lone) who joined the SV group, Kenya, did not elicit greetings behavior, such sequences occurred in that group occasionally when a female approached a young infant.

Another kind of "friendly" behavior that is difficult to describe fully, because it is usually fleeting and has no very pronounced gestures in it, occurs between baboons, often young animals, when they are several yards away from each other. The communicating animal briefly flattens its ears against its head, but without any other facial or gestural element common to threat behavior, and looks in the direction of the other animal. Among chacma baboons the animal also utters an intermittent "chattering" call (Table 4–5, item 11), a call that is also heard between young animals playing with one another. Also, when one or more baboons have become separated from the main body during the day range and when they begin to approach the group again, a loud, rather high-pitched bark (Table 4–5, item 12) is heard, usually from the smaller party, in both Kenya and the Cape. Apart from the apparent pitch difference and apart from the nature of the situation, it is difficult for the observer to distinguish this from an alarm bark.

When most of the members of a group are feeding quietly and at ease fairly close to one another, as among the thick bush areas in part of the Cape Reserve, a chorus of grunting (Table 4–5, item 13) may occasionally be heard without any other signs of communicative behavior. In Kenya this chorus inevitably occurred once or more as the group settled into its sleeping place for the night. What initiates these vocalizations is not clear, although sometimes they appear to start with the deep and repeated grunting of the *a* male, to which the others may have been responding.

PRESENTING AND MOUNTING. Among baboons, presenting is usually done by a subordinate animal, and mounting is usually done by a dominant one. The presenting animal approaches and stands in front of the other, sometimes backing toward it, tail raised or turned to the side. Distances at which the presenting animal approaches its rump to the other have varied considerably, from two feet or less up to several yards. The farther away the presenter stands, the more nervous it usually appears, sometimes looking back at the other, and sometimes jumping away as if in expectation of attack. In all the groups studied the act of presenting is one of the clearest indications of relative status. In S group, Cape, where at the time of study the *a* male was clearly and consistently dominant over the other males in the group, he presented to no other animal. On one occasion when the *a* male was sitting near the top of a gully up which the group was moving, seven different animals (five adult females, and males *2* and *3*) presented briefly to him in six minutes. The *a* male responded (by mounting) to only one of the seven.

At least four different situations may occur where presenting is seen, and there are some postural variations in each:

(1) A female in estrus stands in front of a male and turns the tail to the side, lifting it slightly. The male varies his response according to her state

of estrus and probably according to her dominance rank among the females. If she is in full estrus, he may pick over her perineal fur, or mount and copulate. Otherwise, he may briefly put his nose to her genital region, or he may ignore her altogether.

(2) Females not in estrus tend to present much more briefly, as do juveniles who may present to a dominant male, peering at him, then scamper away.

(3) Adult females, when approaching another adult female with an infant-one, would sometimes lower their hindquarters toward her, looking over their shoulder at her and lipsmacking (Fig. 4–17). The mother sometimes responds by briefly grasping the presenting female's hindquarters, but never mounts her.

(4) An adult male may walk quickly over to another adult male and lower his hindquarters, which action usually elicits grasping of the posterior. This sequence is rarely followed by mounting, in fact, it is usually the *dominant* male who initiates this sequence, and whose rump is grasped. It is probably significant that the hindquarters are lowered substantially more in situations (3) and (4) than in normal presenting, and that mounting is unlikely to follow this lowered-hindquarters type of presenting. In any case, presenting and mounting among adult males was most frequently ob-

Fig. 4–17.   An adult female approaches a mother slowly, with hindquarters lowered. SR group.

served in a group in which there was tension or excitement among the adult males, as in the Kenya SR group in the presence of estrous females or feeding experiments. When in consort with an estrous female the male Dano of this group trotted over to Mdomo and mounted him, dismounting after two pelvic thrusts. Both males lipsmacked vigorously during the mounting. Mounting at other times is a very brief gesture, and quite infrequent in groups where the male dominance relations are stable.

CONCLUSION

Within the limited sampling of baboon habitats represented by the Kenya and southern Africa studies, the range of ecological variation has been considerable. Animals that can maintain themselves successfully in mountainous riverine and savannah habitats are by definition adaptable, and the evidence for the diversity of feeding within the general vegetarian scheme indicates a readiness to make use of local natural resources, just as they make use of man-created food opportunities.

It is also of special interest that baboons seem capable of organizing themselves into groups, or, as among hamadryas, forming loose associations between groups in a variety of ways that indicates the flexibility of the basic social scheme. For example, a viable social structure appears in groups as diverse in composition as SR, Kenya, S Cape, C Cape, and so on, and with group members totaling from a low of 9 to a high of 185. Going beyond our own data to that of Kummer and Kurt (1963), we find an even more significant variation, but one which is still based upon the same sort of dominance relationships.

Adaptability in coping with a diversity of physical conditions, therefore, is inseparable from the adaptability revealed in the pattern of social organization. The remarkable success of the pattern of interaction between aggressiveness and protectiveness must, it seems, be attributed to the propensity these animals have evolved for learning continually during development to modify, extend or inhibit the form and frequency of their expressive and communicative behavior. This is essentially a perceptual, social learning, and it appears to be founded on a fairly narrow range of vocal and other expressive gestures and postures.

When we look back over the data synthesized in baboon studies in DeVore and Hall 1965 and in this chapter, the basic similarity of the findings from two widely separated regions of Africa should not lead us to neglect the possibility that some major differences may be found in the behavior patterns characteristic of what are sometimes designated as two separate species. For example, although the dominance relations look fundamentally alike in the two regions, the details of mating behavior show some differences. The rhesuslike copulation series characteristic of the Cape chacma does not seem to have its counterpart in the Kenya groups, and in mating behavior detail, the female vocalization of the chacma and the male "grin" have not been reported in Kenya. The attack-threat behavior patterns seem essentially alike, but it is not yet evident that the full sequence of what is provisionally termed greetings behavior is ever shown in the Kenya animals, although some components of it undoubtedly occur. In evaluating such comparisons, however, it is necessary to emphasize that the full extent of behavior variations within the regions of study has not yet been sufficiently sampled. Particularly where quantitative differences in behavior are in question, the difficulty in making significant comparison is very great.

chapter five
THE NORTH INDIAN LANGUR[1]

Phyllis Jay Dolhinow

INTRODUCTION

Man and monkey have shared the forests, villages, and cities of India for thousands of years. One kind of monkey, the langur, has also had an important part in the traditions and epics of India in its role as the monkey deity Hanuman. The common langur monkey of North India, *Presbytis entellus*[2], and very closely related forms distributed widely throughout India, are related to the langurs of South and Southeast Asia.

Before the present study very little information on the Indian langur monkey was available. Pocock's interest in the behavior of langur monkeys was incidental to his main concern with langur taxonomy and distribution (1931, 1934, 1939). In 1928 McCann described his occasional contacts with langurs and added brief communications from other observers of langur behavior. This article was followed in 1933 by another short account of encounters with langurs in several areas of India. The literature includes articles on other types of southeast Asian langurs (Pocock 1934; Hill 1937; Phillips 1935; Washburn 1944; and Stott and Silsar 1961).

Ayer's book *The Anatomy of Semnopithecus entellus* (1948) deals exclusively with anatomical problems and is the most complete published physical description of langurs. The Indian langur is a quadrupedal monkey with a specialized stomach adapted to digesting large quantities of relatively unnutritious, mature leaves. Because of their ability to digest mature leaves

---

[1] This chapter is a revision of "The Common Langur of North India" written in 1963 and published in 1965 (DeVore 1965a).

[2] References in the literature to Indian langurs of the species *entellus* include the use of three generic names: *Pithecus* (Wroughton 1918, 1921; McCann 1928, 1933; Pocock 1931; Zuckerman 1932), *Semnopithecus* (Hutton 1867; Cunningham 1904; Hill 1936; Heape 1894; Prater 1948; Champion 1934; Pocock 1939), and *Presbytis* (Simpson 1945). Following Simpson (1945) the generic name of *Presbytis* will be used in this chapter. Although there is great diversity in the literature regarding the generic and subgeneric nomenclature, *entellus* is commonly used in reference to the species observed in this field study.

and to live for months without drinking water, langurs can live in extremely dry areas that are inhospitable to monkeys unable to subsist on mature leaves during summer months. Many langur troops survive the summer with no water other than that from leaves and bark. Whenever possible, langurs supplement their leafy diet with fruit, vegetables, buds, and sprouts, but they were not observed eating meat or insects or digging in the ground for roots.

Other studies of langurs have been completed in the years since the fieldwork on which this chapter was based ended in 1960. A team of research workers from the Japan Monkey Centre including K. Yoshiba, D. Miyadi, S. Kawamura, and Y. Sugiyama, and M. D. Parthasarathy from South India, observed social change in troops of South Indian langurs living in a dry deciduous forest near Dharwar in Mysore State (Sugiyama 1964, 1965a, 1965b). The results of this two-year study, from June 1961 to March 1963, provide striking contrasts with the patterns of social behavior among the North Indian langurs. S. Ripley completed a one-year study of the closely related Ceylon gray langur in 1963 (Ripley 1965, 1967), and F. Poirier observed the Nilgiri langur (*Presbytis johnii*) in the Ootacamund area of the Nilgiri hills in South India (Poirier 1968a, 1968b, 1969).

If our understanding of langur behavior were based only on the North India study reported in the earlier version of this chapter, we would grossly underestimate the variability that characterizes *entellus* langur social relationships. Variability is obvious in many aspects of behavior, from the amount and kinds of aggression among animals, to variations in group size and composition and in local population densities. Our study of behavioral variability is far from complete, and much further work is needed before the causes and correlations of variability in social behavior will be understood. It is clear that the amount of usable and suitable space for living, the availability of food, water, and appropriate sleeping places, as well as predator pressure, all influence the behavior and structure of a troop of monkeys. Severe crowding tends to increase aggression and may change the usually relaxed relationships among troops to tense antagonistic meetings with fights over boundaries or sources of food and water. The behavior of the members of the social group with one another is also subject to change under varying conditions. When there is excitement and fighting, or tension in anticipation of aggression, certain activities that normally occupy hours of the day decrease in frequency or are not seen at all. Young animals spend less time playing and adults and young alike spend much less time in mutual relaxed grooming. The implications of these changes are very important. One of the most sensitive relationships in the troop, and one of the most crucial, is that of mother and infant. There is ample evidence from both the laboratory and the field that this relationship is exceedingly sensitive to what goes on around the pair. A mother's tension and fear is readily communicated to her infant and her behavior may change markedly with the degree of her concern for what is going on around her in the troop. All these factors influence what and how the infant learns, and some of the differences in adult behavior must be related to experiences in the first few years of development.

The present confusion of taxonomic relations of very closely related forms of langurs and the degree and nature of their differences in behavioral adaptations to various environments are problems that require intensive investigation if they are to be resolved. At the present time even boundaries of langur distribution are poorly defined. There are two major forms of the Indian gray langur, one in the north and the other in the south. It is possible to tell them apart on sight by the way the tail is carried and by the pattern of fur length and distribution on the crown of the head. The southern langur has a conelike peak of fur on the top of its head, and its tail is carried up with the end pointing down and behind the monkey. The North Indian langur has a round-appearing head with no peak of fur. Its long tail is carried up over the back toward its head in a large single loop (see photographs). Because there have been no surveys in the area of possible overlapping distribution we do not know if these forms are found together in any area. The Godavari and Tapti rivers traditionally have been considered the boundary between rhesus and bonnet macaques (Southwick, Beg, and Siddiqi 1965) and they may also mark the division between North and South forms of langurs. A thorough survey of central India will be necessary to determine the extent, if any, of overlapping between North and South forms of both langurs and macaques. The Ceylon langur is clearly another variation of *entellus* and is probably not distinct on more than a subspecific or racial level. The Ceylon langur carries its tail differently from the way either Indian variety does and has a tuft of fur that comes to a pointed peak on the top of its head.

The Field Study

The common langur of North India was observed for more than 850 hours during an 18-month field study in India, from October 1958 through April 1960. Approximately 75 percent of the total hours of observation were undertaken in excellent conditions where the majority of group members could be seen at one time. Observations were concentrated on 4 troops, although more than 1000 langurs were counted in 39 troops. From November 1958 through November 1959 3 troops were observed in the forest near Orcha village, Bastar District, Madhya Pradesh. Survey trips into the surrounding hills made it possible to determine the size and composition of other troops. From December 1, 1959, to March 30, 1960, intensive observation was undertaken on one troop near Kaukori village, 14 miles from Lucknow in Uttar Pradesh. A brief resurvey of the Kaukori troop was made in February 1963. Another langur troop of approximately 20 langurs and one adult male rhesus macaque was located near the village of Halwapura,

Fig. 5-1. In a typical North Indian habitat the Kaukori troop moves leisurely in the late afternoon to tall trees it will use for the night.

3 miles from the Lucknow-Hardoi road and approximately 5 miles from Kaukori. In 1959 only the one troop of langurs at Kaukori was recorded for an area of approximately 8 square miles. Kaukori is on the densely populated Gangetic Plain where peasant agriculturalists utilize every available plot of land that can be irrigated and will support a crop. In contrast, Orcha is a tribal area of minimum cultivated acreage and very low population density. Kaukori and Orcha are located on the map in Fig. 5-2. Table 5-1 presents an ecological comparison of the two locations. Langurs were also observed in other areas representing a wide range of environments. In 1959-1960 brief observations were undertaken at Mussoorie and Agra in North India, and at Nagpur, Kondagaon, and Narayanpur in central India. The latter three places were checked periodically during 16 months to determine whether a birth season existed and if any major change in social behavior occurred in the course of the year.

Detailed observation of many aspects of social behavior was difficult and often impossible in the Orcha forests because of restricted visibility. However, it was possible to observe many basic patterns of langur behavior and details of the ecological adaptations of three troops over a period of 16 months. Orcha was surrounded by reserved forest with a full complement of wild animals, including predators. Both animal population and forest have remained essentially undisturbed by man. After an initial 12 months at Orcha in 1958–1959, observations were shifted to Kaukori in North India. Conditions at Kaukori were excellent for the investigation of special problems that required the prolonged following of well-known individuals. The Kaukori troop could be located and followed at any time of day and the

Fig. 5–2. Study areas of langur monkeys in North and Central India.

TABLE 5-1   Ecological comparison of Orcha and Kaukori troops of Indian langur monkeys

Characteristics	Orcha (Abujhmar Hills)	Kaukori
Land under cultivation or dwelling	3 percent (maximum)	98 percent
Human population density	8 per sq. mi.	650 per sq. mi.
Source of langur's food:		
Man's crops	1 percent	90 percent
Forest	99 percent	10 percent
Surrounding land in virgin condition or in forest regrowth	90 percent	1 percent
Summer hot season	Moderate (90–100°)	Severe (100–118°), with wind and dust storms
Annual rainfall	80 in.	30–50 in.
Monsoon rainfall	Approx. 60 in.	20–40 in.
All-year water supply	Rivers and streams good	One large reservoir; wells barely adequate and not easily accessible
Altitude	2500 ft.	400 ft.
Coldest months	December–February	December–February
Winter showers	1–4 in.	Less than 1–4 in.
Associated fauna	Man, domestic animals, jackal, hare, mongoose, hyena, tiger, leopard, wild dog, deer, sambar, wild pig, peacock, jungle chicken, etc.	Man, domestic animals, jackal, mongoose, hyena
Possible predators	Man, wild dog, tiger, leopard	Man (rarely)
Reaction to man	Flight	None, except flight when chased
Range overlap	With langur and rhesus troops	With one rhesus troop
Intergroup relations	With many langur and rhesus troops	With langur and rhesus troops
Use of "whoop" vocalization	Intertroop positional cue, and intragroup use	Intratroop use only
Time on ground in day	Approx. 30–50 percent	Approx. 70–80 percent
Sleeping trees	Many	Many
Birth season	Insufficient data but births in most months	Birth concentration in April–May
Observation conditions	Poor	Excellent

group members could be kept in sight. This made it possible to investigate many details of maternal behavior, socialization, and dominance – details that could not be observed at Orcha because it was difficult there to maintain contact with the animals.

Although Orcha and Kaukori provided very different contexts for group life, there appeared to be no major differences in basic patterns of social behavior between the two environments. Excellent observation conditions at Kaukori made it possible to record long, detailed sequences of all patterns of social behavior, and by contrast, Orcha forests made it more appropriate to study group ecology and intergroup relations. These two areas, Kaukori and Orcha, presented different research opportunities and it was essential to undertake suitable problems of investigation and observations in each location.

In February 1963 brief observations were made on langur groups living near Dharwar in Mysore State at the location of the Japan Monkey Centre's research site. Several groups living along roadsides between Bangalore and Dharwar, and between Bangalore and Ootycamund were recorded, although detailed observations of behavior were not possible. Since that time additional troops of North Indian langurs have been located and surveyed. During 1964 and 1965 the National Center for Primate Biology at the University of California, Davis, undertook a survey of the rhesus macaque populations in North India. In the course of this survey 34 troops of langurs were observed, 30 in forest environments and 4 either in cultivated countryside or near a town.

Study Methods

Interactions were recorded as they were observed and photographs were taken of all major forms of behavior. Fieldnote entries included time, actors, and interaction. Troop movement and weather conditions were recorded daily.

Before a troop was chosen for observation each area was surveyed to estimate population concentration and average troop size and composition. Time was limited and the purpose of this study was to investigate behavior characteristic of normal free-ranging troops; therefore, norms of troop size and structure were first determined in order to avoid the selection of either extremely large or small troops, or troops of unusual age or sex composition. An average observation day lasted from 6 to 8 consecutive hours. Most observations were made during daylight, but evening and night activity was also recorded every 4 or 5 days. Night observations were limited to nights of full moon when it was possible to recognize individual monkeys.

Observation techniques varied with the reaction of monkeys to my presence and with visibility conditions. In Bastar forests systematic observations were difficult. Langurs in that area seldom saw man and their immediate reaction to my presence was flight or concealment in treetops which made it impossible to follow an individual for more than a few hours. Observations were made from semiconcealment but it was extremely difficult to remain hidden from them for more than an hour. The monkeys were very alert and within that time one of the group usually noticed me. Forest troops gradually became used to my presence and this made it possible to follow them at a distance of about 50 feet. However, if any sudden movement in the brush startled them, they immediately fled from sight.

Troops living in towns presented special observational problems. Frequently part or all of the group climbed over walls and into yards or houses and out of sight. When this happened people usually chased the troop. Because city langurs are chased frequently and are often crowded into areas much smaller than their natural range, their behavior is tenser and more aggressive than the behavior of monkeys living in less crowded and less stressful surroundings of forest and open field. A similar change in social behavior among rhesus monkeys is recorded by Southwick, Beg, and Siddiqi (1965). Rhesus macaques living in a city also fight more frequently and severely and are tenser than forest rhesus.

The Kaukori troop was accustomed to man, although it did not tolerate people too close unless food was offered or the people were not paying attention to the animals. For the first 10 days of observation I stayed approximately from 50 to 100 feet from the monkeys and did not try to conceal my presence. I wore the same colored clothing each day and avoided making any sudden movements. When the troop moved I followed them at a distance, stopping each time they did. After approximately 10 days I moved to the edge of the troop where I sat within 10 feet of the least shy animals. I was careful not to look directly into a langur's face since this is a form of threat. Within another week I moved to the center of the troop and from that time until the end of the study I recorded behavior from within the troop. Whenever an animal threatened I turned or moved away – as is characteristic of subordinates – and never returned the threat. My rapport with the troop depended in part on my subordinate position and refusal to interact with troop members. No animal, regardless of how subordinate he or she was to the rest of the adults in the group, found it necessary to avoid me. No attempt was made to feed or provision the animals or to keep them nearby. It is always difficult to determine just how much the observer influences the behavior of the animals being watched, especially since many of the monkeys' reactions are communicated very subtly. In so far as I could tell, there came a time when members of the troop were used to my presence and did not try to avoid me or spend time watching my movements, and their behavior appeared normal. That is, when their behavior was the same as that observed from a greater distance or when I was watching them and they did not appear to know it. Occasionally an infant or juvenile would come dashing up and grab my notebook, the hem of my skirt, or my arm or sometimes my legs in an invitation to join the play. Difficult as it was, these invitations were ignored and after a moment or two the young langur returned to more willing companions.

The greatest initial problem in this study, as in most studies of primate groups, was the identification of individual animals. Variations in size, coat color, scars, temperament, and reactions to other animals made it possible to recognize many of the troop after only a few days of observation. After 2 months it was not difficult to identify most animals at from 50 to 75 yards, by the way they moved and by their general appearance. Being able to

Fig. 5–3. North Indian langurs may spend part of the day in trees, as this group is doing. In the foreground a mother sits with her newborn infant shortly after its birth.

recognize each individual and knowing their personalities was extremely important in this study and it was possible mainly because the observation conditions at Kaukori allowed close and prolonged contact with each langur.

ECOLOGY

Because the distribution of arboreal monkeys is closely related to forest distribution, many localized populations of monkeys have become distinct on at least a racial level. Minor variations in coat color, fur length, and other characteristics have been the basis for a confusingly large number of classifications of langurs (Napier and Napier 1967).

The common langur of North India is probably the most ground-living of the Indian langurs. It is distributed over most of North India with the exception of the western deserts. Pocock (1939), Blanford (1888–1891), and McCann (1928) discuss langur distribution in detail. Groups of langurs live in areas ranging from dry scrub, with only occasional low trees, to thick forests. They are extremely adaptable and their ability to live on mature leaves without drinking water enables them to occupy many vegetation zones. The langur is found from sea level to altitudes of more than 8000 feet at Nanital. Here the troops occupy deciduous oak forests and are reported to move to lower elevations during the coldest months of winter (Southwick, personal communication). I have observed langurs at Mussoorie. Hingston (1920) reports langurs to the timberline at 11,000 feet in the foothills of the Himalayas, and notes that they have been observed in fields of snow. He also reports that these langurs appear to be less quarrelsome than langurs living at lower altitudes, although when the langur is driven to lower elevations during the winter, "pitched battles are said to occur between the two species."

Population Structure and Ecology

The Troop and Ecology

The effect of ecological conditions on troop size and population density among monkeys is poorly understood. It appears that among langurs drinking water may be much less important than type of vegetation as a limiting factor on the number of groups an area can support. In regions where part of the year is completely without any rain, langurs do not drink for several months. At least a minimal quantity of new growth is available all year through in the Orcha forest area but this is not true in many other areas. In some of the dry central and northern regions of India little if any rain falls in winter, and these areas are covered only with low scrub that provides a minimum of new vegetation during extremely dry summer months. The presence of suitable sleeping trees may also affect langur troop distribution.

Troops in central India tend to be larger and are spaced to take advantage of artificial water sources used to irrigate crops. Orcha troops are smaller and the population is spread evenly throughout the forest (Table 5–2).

In areas of water scarcity troops that otherwise might not have any contact with one another come together at wells and reservoirs. When several troops use the same areas around water reservoirs their interactions are peaceful. Smaller troops tend to stay at a distance while larger ones drink, but on 7 occasions more than 2 troops were observed to drink from the same pool at once. Very small streams in the Orcha forests do not flow during the summer months but pools of water remain. Several troops may use the same pool at different times but if they arrive together the smaller usually waits or moves to the opposite side to drink.

Troop Size

During the 1958–1960 study more than 1000 langurs were counted in 39 troops. The smallest was of 5 and the largest consisted of more than 120 monkeys. Troop sizes (recorded in Table 5–2) averaged from 18 to 25 in the Bastar forests and from 25 to 30 in the dry central regions. The largest was observed one-half mile from a water reservoir in the late winter and in the dry season. Large aggregates of langurs at water reservoirs are usually temporary, and although several troops may mingle for an hour along the edges of a river or a pool, they separate and return to their own home ranges.

The langur troops of this study appeared to be relatively closed social systems; that is, monkeys rarely changed troops and there was little evidence that recent shifts from one troop to another had taken place. The membership of a troop remains constant except for deaths, births, and the departure of a few adult males that leave to live as nontroop males. The reasons a male leaves or is forced to leave are not known. A total of 53 nontroop males were observed living alone or in groups of from 2 to as many as 10 males. The sizes of these all-male groups were:

Number of Groups	Number of Males
9	1
2	2
4	3
1	4
1	6
1	8
1	10

A group of males not associated with the bisexual troop occupies a range that overlaps the ranges of adjacent bisexual troops, but the males usually avoid using these overlapping areas whenever the troop is nearby. Three such males lived next to the Kaukori troop, and since there were no other langurs for several miles in any direction, it is most likely that these males originally lived in the troop. It was not possible to determine the troop origin of any of the nontroop males either in this area or in others in the amount of time available. An observer would have to be present at the time of separation from the troop to know the exact origin of each male. One of the three males living near the Kaukori troop tried on two separate occasions to follow the troop. Each time he did he was held off by adult males and one old female in the troop. The fighting that took place during these skirmishes was the most severe of any fights seen during all my observations on North Indian langurs. The nontroop male may not have been a stranger in that he lived near the troop for at least the period of study, but he was forcibly denied access to the troop.

Thirty of the 34 troops located in the 1964–1965 survey in North India lived in forest habitats. The troops were composed of from 3 members to more than 35. Four additional troops were located in cultivated countryside and of these only one had more than 30 members. One adult male was recognized as living apart from a bisexual troop but there were probably many more nontroop males than this incomplete survey indicated. The observer usually assumes that a single male may be a member of a nearby troop unless it is possible to spend several days in one area. This assumption is made because restricted visibility in forest areas makes it quite possible to see single animals and not the rest of the troop to which they belong.

Two adult rhesus macaques, a male and a female, were living with the langur troop as "full time" members during the period of the study. Villagers said that the rhesus had been with the group for several years before observations started. This may or may not be accurate, but they were there at the start and finish of the study period. Three other langur troops each had one rhesus member. The relationship between rhesus and langurs will be discussed later in this chapter.

Troop Structure

Troop compositions for three Orcha and the Kaukori troops are presented in Table 5–2. Although the number of males and females of all ages in the total population is probably approximately equal, there appear to be more adult females than adult males. In troop counts, on an average, there are usually between 1.5 to 2 adult females for each adult male.

Females are counted as adult from the time they give birth to their first infant, at about $3\frac{1}{2}$ to 4 years of age, whereas a male is not socially an adult until he has reached full physical development and his canines are

TABLE 5-2   Composition of Kaukori and Orcha troops

Group	Inf.-1		Inf.-2		Juvenile		Subadult		Adult		Total		Total Both Sexes	Nontroop Males	Rhesus Macaque	
	M	F	M	F	M	F	M	F	M	F	M	F		M	M	F
Kaukori	1	–	8	5	5	5	2	3	6	19	22	32	54	3	1	1
West Orcha	–	–	4	1	2	2	–	1	3	5	9	9	18			
North Orcha	1	1	3	–	–	5	1	2	6	9	11	17	28 ⎫			
East Orcha	–	1	1	1	–	–	–	1	2	4	3	7	10 ⎭ 1			

M = male
F = female

fully erupted, at about 6 to 7 years of age. Thus females enter the adult category as long as 4 years before a male is considered adult.

Daily Round

Activity starts just before dawn when a few monkeys shift position in the sleeping trees. As soon as the first light can be seen movement increases, infants squeal, and adults move from branch to branch. When the sun is visible groups come to the ground or sun in the treetops. Early morning is spent eating and moving; estrous females are active and the young play. Midday usually is spent quietly, with adults resting and grooming while infants and juveniles play or rest with their mothers. Feeding and movement start again in the late afternoon and continue until the groop moves to trees it will use for the night. The troop prefers large sleeping trees and usually sleeps in different trees each night. Settling down takes from 45 minutes to one hour and during this time there is constant moving and grunting from adults and squealing from some of the young. Females and young form small sleeping groups with adult males scattered throughout the group. Night is spent quietly with occasional waking and slight shifting of positions. During the hours of darkness, whenever a monkey shifts its position or disturbs a neighbor, there is a wave of urination and defecation from adjacent monkeys. As soon as it is light another daily cycle of activity begins.

The daily activity cycle lasts from approximately 8 hours during monsoon months to more than 12 hours during long summer days. Hours of sleep at night vary from more than 15 in the monsoon to 10 in the summer. The 2 to 4 quiet midday hours are spent resting and grooming or eating. A troop usually moves from 1 to 2 miles a day, less during the monsoon months and more during summer.

Group Range

Langur troop ranges vary in size from approximately ½ square mile to as large as 5 square miles, with average sizes of from 1 to 3 square miles. Fig. 5-4 presents the daily routes of the Kaukori troop for 80 consecutive

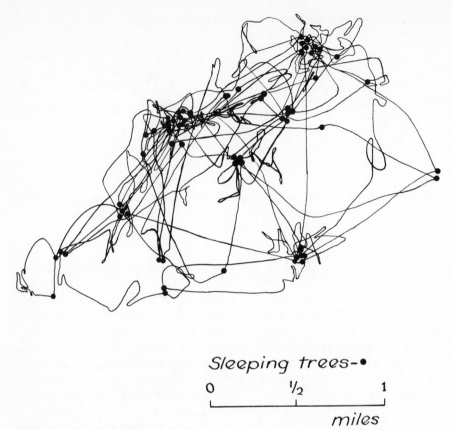

Sleeping trees-•

0   ½   1

miles

Fig. 5–4. Daily travel of the Kaukori langur troop for 80 consecutive days.

days. From this figure the range map in Fig. 5–5 was drawn, illustrating boundaries of the troop range and within this, the core areas, those used most frequently. As was also true of Orcha troops, the core areas coincided with preferred sleeping trees (more than 80 percent of which were in core areas) and sometimes with water sources, those of the Kaukori troop being centered around water reservoirs, since during 5 months of the year all other wells and irrigation systems were dry. In Fig. 5–4 the plotted daily routes of the Kaukori troop show that parts of the home range are seldom used except as the troop travels from one core area to another. Fig. 5–6 summarizes range size for 4 Orcha troops, and Fig. 5–7 illustrates the annual home range and core areas for 3 Orcha troops. In this forest area there is a definite seasonal use of different parts of the range. Dark areas in Fig. 5–7 represent concentrated use during dry months when cultivated areas are overgrown and forest is relatively open, with visibility restricted in grassy open areas and in fields, and greatly improved in the forest and on the forest floor. Lighter areas represent core areas in wet monsoon months when the forest becomes almost impenetrable and the fields are cleared of plants that obstruct crops. Treetops are covered with a thick canopy of vines that provides some shelter from the rain, and during wet months troops move less during the day in order to remain under this vine canopy.

Rhesus macaque troops use part of the area included in langur troop ranges at Kaukori and Orcha. On 9 occasions members of Orcha rhesus and langur troops were observed eating together in the same trees and on 15 occasions they ate together in harvested fields. The 2 groups mixed peacefully, with neither threats nor aggressive behavior. The Kaukori rhesus seldom came in contact with the langurs, since the rhesus remained in the village.

Langur troop home ranges may overlap extensively, but core areas do not. Areas of range common to several troops are usually occupied by only one group at a time, although when two troops are in the same area they do not threaten each other. Fighting between two troops was never observed; if they both happen to be nearby, the larger group usually takes precedence and the smaller remains at a distance until the larger moves

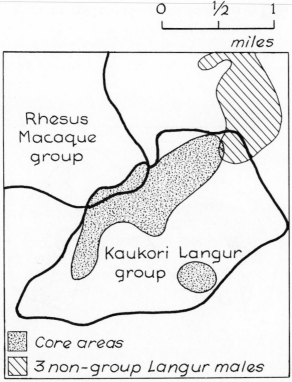

Fig. 5–5.   Home range and core areas of the Kaukori langur troop, and home range of three nontroop langur males.

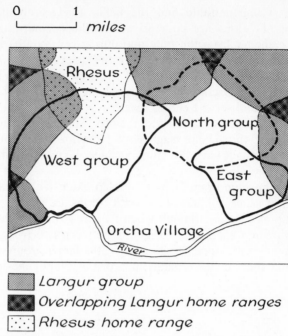

Fig. 5–6. Overlapping home ranges of langur and rhesus troops.

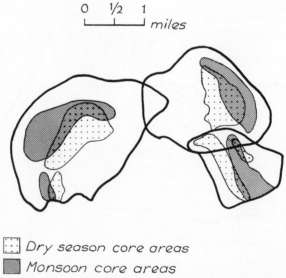

Fig. 5–7. Core area shifts in langur home ranges during summer and monsoon season.

away. Langur troops seldom come together in the forest or in open fields since groups are separated very effectively by their daily routes and patterns of range use. An effective spacing mechanism that allows troops to determine each other's position is a deep, resonant "whoop" vocalization produced by adult males when a troop is about to move suddenly or for a long distance.

Interspecific Relations

Kaukori is extensively cultivated with almost no ground cover of brush or scrub; as a result there are few animals other than man and his domestic animals. No instance of predation on langurs of the Kaukori troop was observed. Because the Orcha area includes large reserved forests where hunting is not permitted, it is very rich in wildlife. In addition to man and domestic animals, there are tiger (*Felis tigris*), leopard (*Felis pardus*), wild dog (*Cuon dukhunensis*), wild pig, hyena (*Hyaena striata*), bear (*Melursus labiatus*), many large ungulates, sambar (*Cervus aristotelius*), bison (*Bos gaurus*), spotted deer (*Cervus axis*), barking deer (*Cervulus muntjac*), hare, mongoose, squirrel, small forest mammals, peacock (*Pavo cristatus*), red jungle fowl (*Gallus ferrugineus*), and other birds.

Fig. 5–8.   An adult female langur is being groomed by an adult female rhesus macaque, a member of the langur troop.

No instance of predation was ever observed in Orcha, and the impression received from this study was that it would be unlikely that a healthy langur would fall prey to a ground-living carnivore even though several predators, such as leopards and jungle cats (*Felis chaus*) climb trees. It seemed that the langur's skill in climbing and jumping would make it very difficult to catch and kill even an injured langur if the monkey reached the safety of trees. However, when Schaller studied the tiger and leopards of North India he found that of 335 tiger feces collected in central Kanha Park 20 or 6.2 percent contained langur hair. Of 22 leopard feces collected in the same area 6 or 27 percent contained langur hair (Schaller 1967: 281, 311). Clearly, the impressions of the langur study were incorrect, which points up how cautious the observer must be about assumptions of events that do not happen during a study. The chances of seeing a kill are so small that even with many hours of observations no kill was even suspected. It is most unlikely that the many tiger and leopards of the Orcha area would have different hunting preferences or patterns from those in Khana Park. It is more reasonable to assume that predation occurred and that langurs and rhesus fell victims to the large cats. It is probable, of course, that injured or sick monkeys were more frequent prey than were healthy primates.

On three occasions langurs gave alarm barks and moved up into the treetops when a leopard appeared; twice the group gave an alarm but continued eating in the trees; and once a langur group slept within 100 feet of a tiger and his kill.

Although forest langurs did not associate with any other species for more than a few hours at a time, langurs were extremely alert to sudden movement or warning calls of other species in the forest. Several times a herd of spotted deer grazing nearby barked and alerted the langurs, and peacock often alerted the monkeys since these birds, with their exceptionally keen eyesight, saw man and predators long before the langurs noticed them.

Langurs and Rhesus Macaques

Two rhesus macaques, an adult male and female, lived with the Kaukori troop of langurs. They ate, moved, slept, and interacted with the langurs, and did not leave the troop for more than a few hours at a time during 4 months of observation. The rhesus were probably originally members of a troop of macaques that lived in Kaukori village and whose home range in part overlapped the range of the langur troop, but at no time during observations did the 2 rhesus try to reenter the village rhesus troop. Villagers asserted that the rhesus had been associated with the langurs for several years.

The male rhesus was approximately 10 years old (an estimate by Kenji Yoshiba of the Japan Monkey Centre) and bore many scars from past battles. His right ear was badly mangled and he had numerous healed scars on his arms, hands, face, and body. At the time observation began in December, he was recovering from a deep 5-inch wound on the right side of his face. He was a rather thin adult but very well muscled. He appeared lean only when compared to other adult males in the rhesus troop, although he was as large in body length and height as any of the rhesus males. The female rhesus was a young multiparous adult, probably about 5 years old (estimate by Yoshiba). She had no scars or other signs of old wounds, and at the end of the 4 months of observation was in advanced pregnancy. In contrast to the male, she was very plump. The 2 rhesus were usually not far apart and if the troop was resting they sat next to each other and groomed or slept.

Each rhesus was dominant over all the langurs in the troop. Whenever the male approached a group of langurs, they dispersed and would not stay close to him for more than several seconds at a time. It was seldom that he could sit near them without their walking away, although when he was relaxed, they sometimes only turned away and were careful not to look directly at him. Usually, however, within a few minutes they walked off one by one. The female rhesus was also dominant over all the langurs including the most dominant males, but she was more careful than the fearless male rhesus and she did not provoke fighting unless he was nearby and could support her in the event that a langur returned a threat. She did not, however, deliberately provoke the beta male langur. When she passed alpha, she looked directly at him but walked quietly past while he calmly looked directly at her in return. If she was alone, alpha did not move away; if the male rhesus was with her, she often deliberately threatened alpha. Although the male rhesus did not threaten in such encounters, his presence near her made the most dominant langur move slowly but definitely away from the rhesus pair. Together the rhesus could rout the entire langur troop and on occasion they broke up disputes involving a number of the adult langurs.

The male rhesus always commanded a distance surrounding him into which langurs never dared enter. When he was irritable or tense, this distance was as large as 25 feet in radius; even when he was relatively relaxed, all young langurs and most adults avoided sitting or walking very close to him. Regardless of his mood, langurs avoided looking directly at him, and when the male rhesus passed or sat by them, they either got up and walked away or were careful to face away from him and not to exchange glances. A direct look was one of the most effective and frequently used means by which the male rhesus cleared away nearby langurs when he was relaxed. Staring was not always sufficient to move a few of the more dominant langur males, but staring combined with a threat gesture always succeeded in making them move away. Juveniles were frightened by both rhesus, but whenever the male approached they scattered quickly and if they were still associated with their mother ran directly to her and grabbed her fur.

The male rhesus was easily annoyed by noisy threats among the langurs, and he chased and scattered langurs when they were active near him. Often a slap on the ground or on the branch called attention to his irritation and they either moved away or stopped threatening each other.

Neither the male nor female rhesus hesitated to dominate a langur whenever position or food was desired. As soon as the rhesus saw peanuts, which villagers sometimes tossed to the langurs, the rhesus took complete control of the supply. When the nontroop male tried to enter the troop, it was primarily owing to the activity of the male rhesus that the strange male was repelled as decisively as he was, although he acted in combined attack with the langur males and one old female langur.

The female rhesus was extremely short-tempered and irritable compared to female langurs. If female langurs threatened each other near her, she chased them and quickly stopped the interaction. Although the rhesus was unquestionably dominant over all the adult female langurs, she was less clearly so over the adult males. No female, with the exception of one very old female langur, ever fought with the rhesus female or returned her threats. Her behavior toward dominant male langurs was not so confident as toward adult females, unless, as mentioned, the male rhesus was nearby.

Rhesus diet differs in part from that of the langur, and the rhesus often ate from different though adjacent fields. Langurs eat leafy young and mature plants, leaves, and shoots, while rhesus prefer to eat whole grains and young leaves or buds. The rhesus digestive system is not able to utilize mature leaves, a specialization characteristic of the subfamily to which the langur belongs. Often both rhesus and langurs were in the same field but generally ate from different parts of the same plants. Although many items of their diets were the same, many were not, and this substantially reduced food competition between the two species.

There were some relaxed social relationships between the two kinds of monkeys in the troop. The female rhesus groomed certain of the adult langur females and one large female infant langur. In almost every instance, it was the rhesus that initiated grooming by walking up to the langur; if the latter did not move away, as most did when she approached, she would groom it. Three of the adult female langurs regularly reciprocated as soon as the rhesus stopped grooming and turned her back. Several others allowed the rhesus to groom but did not return the attention. The large female infant was exceptionally relaxed near the female rhesus and would occasionally run to her without being solicited and begin to groom. These sessions lasted until the infant moved away to play. Interestingly, the mother of the infant was not one of the females that allowed the rhesus even to sit near her, and she was never seen to groom or be groomed by the rhesus.

The female rhesus went into estrus once in the first month of observation and was repeatedly mounted by the male rhesus. On one of the days when she was sexually receptive, the langur troop was on the edge of the village near the rhesus troop and two adult males from the village rhesus troop approached the female rhesus. Her consort drove the two intruders away with fierce threats, tree shaking, and constant loud vocalizations. The langurs appeared to be somewhat interested in the entire series of events and some sat around the contesting males in a large circle watching attentively. None of the langurs was directly involved in the action and none made any attempt to interfere. Still other langurs sat within 30 feet of the commotion and ate calmly. The female rhesus rushed back and forth between the two males, screaming and extremely agitated, but she returned to her consort each time. Later in the day when the langur troop and the two rhesus moved back into the fields, one of the adult males from the rhesus troop followed at a distance. The male rhesus with the langur troop chased him back to the edge of the village but the male continued to follow the pair for almost two days, staying about 100 yards behind the troop. During these two days the male rhesus consort was exceptionally tense and paced back and forth between the female and the edge of the troop.

As members of a langur troop, the rhesus responded appropriately to a wide range of gestures and vocalizations that are not included in their own species' repertoire. In addition, they responded immediately to langur warning barks and the langurs to theirs. The two rhesus threatened langurs by slapping the ground or by shaking branches, and the langurs would return the threats in the form of slapping the ground or making biting movements into the air. A sharp bark or a series of grunts from a female langur that had been chased by the rhesus would serve to renew the attack by the rhesus until the langur stopped her nonsubmissive conduct. No submissive gesture was offered to the langurs by the rhesus, but the langurs sometimes presented in submission when they were threatened from a short distance and this appeared to stop an attack by either of the rhesus.

A dominant rhesus male often carries his tail upright, but although the male rhesus was able to dominate all the langurs, he did not walk through the troop holding his tail in a vertical position. Instead, he usually carried it out behind his body parallel to the ground. The female rhesus carried her tail either pointed toward the ground or at a slightly higher angle but never higher than the male held his tail if he were in sight.

The sharp, quick, stacatto motions so characteristic of rhesus were a striking contrast to the smooth, calm, slow, and often graceful movements of the langurs. Personality differences between the two species are similarly striking. Rhesus seemed much quicker to respond to a social stimulus and did so with stronger, more forceful gestures. They were far more vocal than were the langurs, even when relaxed.

Female Reproductive Cycle

A female langur is sexually receptive for the first time when she is approximately 3½ years old. The first sexual cycles of three females in Orcha groups were irregular and spaced at longer than normal intervals.

Fig. 5-9. An adult male langur, sitting beside an estrous female, mildly threatens another langur.

A female is sexually receptive only during estrus, a period of from 5 to 7 days midway between menstrual periods, which occur approximately once every thirty days. The period of estrus corresponds to the portion of her monthly cycle when ovulation is most likely to occur. Perineal swelling or "sexual skin" does not occur in langurs, and the female initiates sexual behavior by displaying three gestures not associated as a sequence in any other social context—only the sexually receptive female simultaneously shakes her head, drops her tail to the ground, and presents to adult males.

Sexual behavior actually plays a very small part in the life of an adult langur female. She is pregnant for approximately 6 months and is not sexually receptive again until her infant is weaned at from 10 to 12 months of age; she will not give birth again for another 6 to 8 months. Births in many areas are spaced at approximately 2-year intervals. When a female is not in estrus, adult males show no sexual interest in her. She is the sole initiator of sexual activity and she is not mounted unless she solicits the male.

In a troop containing many females there may be one female in estrus during most days of the year. Table 5-3 indicates the number of days during which more than one female was in estrus. Although estrous females do not solicit males in any particular preference order, there is a tendency for the

TABLE 5-3   Number of estrous females in Kaukori troop

Number of Estrous Females	Number of Days Per Month				Total Number of Days
	December	January	February	March	
1	14	8	9	8	39
2	1	6	4	3	14
3	—	8	3	1	12
4	—	—	2	1	3
5	—	—	—	1	1
Total days per month	15	22	18	14	

most dominant males to consort with estrous females at the height of their receptivity at the middle of their estrous period, the time at which ovulation and conception probably occur.

Birth Season

In central India births are concentrated in the months of April and May. In February and March of 1959 only 4 newborn infants were observed in groups along the roadside, but in April, 27 newborn infants were counted within 6 of the same troops. In contrast, troops in Bastar forests contained newborn infants in most of the months of the year, although it is possible that there may be a concentration of births during some months, a concentration that was not clearly observed because of the difficulties of repeated surveys.

Infants born in April and May, the first months of summer, begin to eat solid food early in the beginning of the monsoon, June and July, when tender leaves and shoots are plentiful. During April and May wells dry and irrigation systems are empty, crops are harvested, and very little new growth is available as food. The infant is dependent on its mother's milk during these first months of life until new leaves are available. Since there is rainfall and new growth throughout the year at Orcha, the necessity for a specific birth season is probably not so acute as in dry regions of central India.

Among the langurs of South India most births occur in January and February, at the beginning of the dry months (Kawamura, personal communication). Langurs in some areas of the South do not drink water for several months, but there, as in northern India, the young begin to eat leaves at the time when such new growth is available after the start of the monsoon season. However, a definite correlation between an infant's dietary need for available new growth of leaves and the spacing of births has not been established.

BEHAVIOR

Social Organization: The Troop

The context for langur social life is a stable well-organized troop composed of monkeys of all ages and both sexes. Such a group is the basic unit of the species. Although it is possible to survive living apart from a social group, only a small number of adult males live outside bisexual troops, and probably only for part of their adult life.

The Kaukori troop in North India was a very well-structured social organization whose behavior was regular, quite predictable, and stable. The overall impression from five months of observation was one of conservative patterns with very infrequent change. The activities of individuals did vary over this period of time but most such variations were relative to changes in reproductive cycles and the changes in activities associated with different stages in maturation.

In general, all adult females were preoccupied with care of their own young and showed great interest in the young of other adult females. Adult males did not display an interest in infants. Instead if they were not relaxed or grooming, they were active in dominance interactions, in controlling the patterns of troop movement, and, when there were sexually receptive females, in copulation. The young langurs, when not with their mothers, spend hours each day in play. This is an oversimplified cross section of life, but each age-sex category of animals within the troop had activities that characterized the way they spent most of their time. There were striking variations in interests and activities depending on the personality and inclinations of individuals.

Many important activities crosscut the entire troop, and were not the prerogatives of any special group of monkeys. The most obvious of these was grooming, a pastime that occupied as much as five hours a day for adults. Grooming was pleasurable, as witnessed by the complete relaxation of the participants and the willingness of animals to groom and especially to be groomed. Grooming is also very practical; particles of dirt, burrs, and parasites are removed from the fur, and when the animal is wounded, the cuts or abrasions are picked clean of foreign matter.

The troop has strong social traditions that are part of the daily lives of all members. An important part of social behavior is the dominance structure that unites all the group males and in a looser way females as well. This structure helps perpetuate a system of predictable relationships among familiar animals. The network of dominance ranking is not rigid and many changes take place over time, some minor and some very important. However, on a short-term basis, from day to day and week to week, the dominance relationships of the Kaukori langurs were stable.

Dominance rank among the Kaukori animals was not easily determined by the observer. There is no single criterion that will be accurate in unraveling this structure, since dominance is not a unitary feature of life. It is exceedingly complex and is often expressed very subtly with many nuances dependent on the feelings of the monkeys at the moment. It took several months to ascertain the power structure among the adults in this troop because dominance was expressed in many ways: by controlling access to desired foods, copulating with estrous females, being groomed more than having to groom, walking first or alone on a narrow path, and by other events in daily behavior. Individuals that are able to take precedence over another or others may not do so at all times. Whether this is because they do not want to or are not motivated to assert themselves at a particular moment is something the observer cannot tell. However, it became very clear that sometimes an individual may not contest or take something that at another time he or she does easily; it is not that they cannot, but that they do not. Other langurs apparently have no difficulty in knowing when an animal is or is not interested in an object or an event since most of the time it was also clear to the observer.

For example, if two adult males are resting quietly and a ripe mango happened to fall midway between them, the less dominant of the two males will glance at the more dominant. If there are no signs of the latter's going to pick up the fruit, the subordinate will pick it up. But if the more dominant alerts and shows any movement or intention to take the fruit, then the subordinate of the two males makes no attempt to get it and often gives no indication that he even saw the fruit fall. It would be grossly misleading in the event the lower-ranking male picked up and ate the fruit to say that he was dominant on the basis of this interaction. It is possible to make general ranking of animals within a troop only after observation of many interactions in a wide variety of situations. Dominance is as much a potentiality as an openly asserted act or event, and at most times animals act and interact freely, apparently regardless of dominance status. This free interaction is especially true among relaxed adults when dominance ranking is not obvious. Attributes of adult male and female ranking will be discussed later in this chapter.

Although an arboreal monkey may not be under as great pressure to remain in a social group as is a ground-living monkey, there are advantages in a group way of life. Many langur activities are most effectively carried out by the group, and among these the most important is the socialization or rearing of the young. The presence of a newborn langur infant is an extremely strong cohesive factor for all females in the group. Intense female interest assures the young of the protective environment and care they need until they are independent members of the group. The importance of contact with age-mates in a social setting for the development of a normal adult monkey has been demonstrated many times in laboratory analysis (Mason 1965).

Without a stable social troop these contacts would be most unlikely, whereas in an average-size troop of from 20 to 25 langurs the young monkey will always have a peer group.

The troop provides protection because each member is alert and warns all others as soon as danger is sighted. The chances that a monkey will be warned are many times greater if he lives in a troop than if he lives alone. A common social tradition is shared among the members of a group and this includes knowing where to find food and water, which trees are best for sleeping, and where it is safest to run from predators.

Patterns of interaction within a troop can be conceptualized as a more-or-less clearly defined network of social relationships among its members. Not all possible interactions that might occur are actually observed. Some age-sex categories seldom interact with each other, as for example, there are no interactions between small infants and adult males. It is not sufficient to count the number of times any given animal or class of animals does something because of the great variability among interactions with respect to the length of time they last and the intensity of involvement of the participants.

Just how much an animal is involved in an interaction is sometimes impossible to determine when the only clues are observable reactions. A langur can sit within 30 feet of a fight and appear not to be aware of its occurrence. If it were possible to monitor the watching animal's internal reactions such as his heartbeat rate, blood pressure, and body temperature, it would be possible to measure his involvement.

Socialization: The Ontogeny of Social Behavior

Social maturation is an orderly and well-integrated process characterized at each step or stage by new and changing relationships and behavior patterns. As a monkey matures it learns social skills within a group of age-mates and gradually assumes patterns of behavior characteristic of each level of social development. The social maturation of the individual is the concern of the entire troop. A monkey's life in a social group consists of constantly reacting to and interacting with other monkeys; it is a complex adaptation of learning to get along with others. Patterns of social behavior described in the following sections are based primarily on observations of the Kaukori troop in North India.

Fig. 5–10 illustrates the major vocalizations and gestures that appear at different stages of maturation. Some elements of communication are replaced, whereas others are supplemented to produce the communication system of the adult. Although elements are designated as if discrete, in reality they interact to form a rich and highly varied series of socially meaningful behavior patterns.

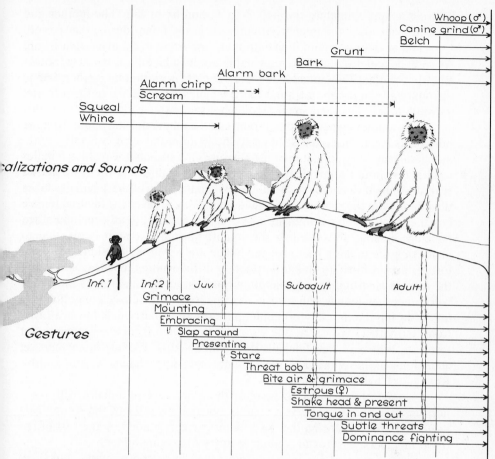

Fig. 5–10. Major vocalizations and gestures of langurs as they appear during maturation.

Maternal Behavior and Infant Social Development

The mother–newborn-infant relationship is the strongest and most intense bond in the life of a langur monkey (Jay 1962). At birth the brown-colored infant is dependent on its mother for nourishment and transport, but it is by no means completely helpless. It is a clinging, vocal animal, capable of controlling to a great extent its relations with the mother. It can cling to her unaided within a very few hours after birth and can secure itself so tightly under her body that she can run on the ground or make long jumps through the air without dislodging it. The infant is able to move independently of its mother long before the change in its natal coat color, which remains dark brown for approximately the first 3 to 5 months of life. The mother and adult females are most intensely interested in the infant during that period.

The birth of an infant is an extremely important event in a langur troop. As soon as females in the troop, whether adult, subadult, or juvenile, notice a newborn infant they immediately cluster closely around its mother. She is surrounded by a group of from 4 to 10 females, all reaching out gently and trying to touch, lick, and smell the newborn infant.

The mother inspects, licks, grooms, and manipulates the infant from the hour of its birth. When it rests quietly she grooms and strokes it softly without disturbing or waking it. She usually turns her back to the waiting females until a few hours after birth, when the infant is dry, and then she allows several of them to handle it. Within a few minutes one of these females takes the infant from the mother's arms and holds it; as soon as the favored female has the infant in her arms, she inspects it minutely, gently manipulating, nudging, licking, and smelling the infant. Special attention is directed to inspecting the infant's head, hands, and genitals. At the first sign of discomfort in the newborn it is taken by another of the waiting females, although if the mother is sitting nearby she often reaches out and intercepts the infant before another female can take it. As many as eight females may hold the infant during the first day of its life, and it may be carried as far as 50 feet from its mother. However, she can take her infant from any female in the troop regardless of that female's dominance status. Passing infants among adult females is reported by Kawamura for South Indian langurs and by Ripley for Ceylon langurs.

Among North Indian langurs, if there are several females with newborn infants, they usually stay together near the center of the troop. This protected area, where mother and infant are surrounded by the rest of the troop members, assures maximum security for young infants.

Female langurs are intensely interested in newborn infants, but there is considerable variation in skill among mothers in caring for their infants and among females in holding infants of other females. Not all females can keep an infant quiet and content, and a few are awkward and clumsy, but the majority experience little or no trouble caring for infants. Competent females are casual but firm, apparently unaware of the movement of the often very active infant; whereas less capable females are uncertain of themselves and constantly readjust the infant and shift it from side to side. Very few females are inept with infants. Those who are inept hold the infant too tightly, upside down, or away from their body while they inspect it constantly. These same females also appear to have difficulty making a newborn cling and nurse in a comfortable position.

Females vary in their behavior toward the newborn of another female. Some females pull the infant away from their breasts each time it succeeds in finding a nipple, whereas many females let the infant press against their chests and at least take a nipple in its mouth. Of all females observed holding another female's infant, less than one fourth deliberately helped it locate and grasp a nipple.

The length of time a female can hold a newborn infant before it is uncomfortable, that is until it squirms and squeals, is one measure of her ability to handle it. Older females that have given birth to many infants appeared, in general, more efficient mothers than very young females. These older and more experienced mothers are usually more confident and expend little effort keeping a newborn calm and quiet. This suggests that experience is important in developing skill in caring for an infant. Because langur mothers allow other females to hold their infants, no langur female is completely without experience in infant care.

The female's temperament or personality also affects her interest and aptitude for holding infants. A tense, nervous, and easily irritated female frequently startles the infant with quick or unpredictable motions; whereas a calmer and more relaxed female makes few sudden movements.

Dominance does not appear to be important in the daily life of a langur female and her status is seldom apparent in her relations with other group members. It is unlikely that her status has great significance for the development of her offspring. Infants of all females in the group have free access to other infants in play and can venture into any part of the group without

TABLE 5–4 Stages of langur development

Stage Designation	Age	Basis for Stage Designation
Brown infant	Birth to from 3 to 5 months	Brown coat color; color change from approximately 3 to 5 months of age
White infant	From 3 to 5 months to 12 to 15 months of age	From end of color change to just after weaning
Juvenile	From 15 months to 4 years (M) or to 3 years (F)	Intense peer orientation; play is major activity
Subadult	From 4 years to complete physical maturation at approximately 6 to 7 years (M); from 3 to 4 years (F)	No play, orientation to adults, dominance important. Male marginal, little participation in troop activities; female displays intense interest in infants and spends much time grooming with adult females
Adult	At approximately 6 to 7 years for male, canines fully erupted, full muscular development; female from birth of first infant at approximately 3½ to 4 years	Entrance into adult dominance hierarchies; assume all roles of the adult

being threatened by the adults. Some females tend to stay together more than other females, but these preference groupings are temporary and not exclusive. Infants of females that avoid one another may spend time each day playing. It is probable that the dominance status of a langur mother is not as influential on the development of her young as is her temperament.

Brown Infant

Several obvious characteristics of a newborn or very young infant set it apart from all older monkeys. Brown natal coat color is present during the first few months of life when the infant most needs nourishment from its mother and protection by all adult monkeys. The young infant's movements are uncoordinated, hesitating, and awkward. When it needs help in steadying itself or in clinging, it grasps its mother's fur or touches her and she immediately supports it or helps it to cling. These movements of a brown infant as it touches the mother, shifts positions, clings, or falls appear to stimulate her so that she holds, adjusts, and fondles the infant.

The second month of life is one of rapidly increasing muscular coordination and greater independence from the mother. The infant spends several hours a day exploring its rapidly expanding environment of new and strange objects to lick, smell, and touch. The infant is carried but it is no longer necessary for the mother to pull it under her body when she is about to walk. At about 2½ months the infant wanders as far as 10 feet from her but makes frequent trips back to nurse briefly. If they are sitting in a tree, the mother often restrains the infant by pulling it back by a leg or its tail. As the infant goes farther from the mother it encounters age-mates and adult females, and instead of hopping from one place to another it starts to focus its attention on other monkeys. By 3 months the infant supplements its milk diet with solid food. It samples plants the mother picks and takes bits of food she drops.

The 2 months of color change from dark brown to light gray coincide with important alterations in the social relations of an infant. It is often left by its mother with another female and her infant so that one female may sit with several infants at once. Active interest of adult females in the infant declines rapidly as the infant turns white, although they continue to be tolerant of the infant when it climbs and tumbles on them. The mother is less solicitous but she is constantly aware of the infant's whereabouts and continues to care for and protect it. If the group is alerted suddenly, she signals her tensions by a grunt, patting the ground, or moving quickly toward the infant, and it rushes to her.

The social unit for the protection and care of a newborn langur in the Kaukori troop does not include adult males. They are indifferent to a newborn and seldom are within 15 feet of one. The loudest squeals of a newborn seldom draw the attention of an adult male, and when an infant needs help it is the mother or another adult female which gives assistance. If an adult male accidentally frightens an infant, the mother instantly threatens, chases, and often slaps the male.

White Infant

This is a period of increasing independence from the mother and gradual orientation to age-mates. The young monkey learns to produce vocalizations and gestures and to use them in appropriate situations. As a brown infant the monkey displayed motions that communicated only the most general states of discomfort or comfort.

By 5 months of age the infant runs 20 or 30 feet from its mother, but she is constantly aware of its whereabouts and moves to it whenever the troop is alerted. Because she vocalizes to draw its attention, her protective role is more obvious than when she had only to take a few steps to reach it. When the troop is relaxed, she no longer signals short moves away from her infant while it is playing, and it must follow her if it wishes to nurse or cling.

By 6 months of age the infant plays with age-mates for several hours every day. Play groups may include as many as 16 young monkeys but the average size is from 2 to 4, and half of all play groups have only 2 members. Running, jumping, chasing, wrestling, and tail-pulling are supplemented by more complicated forms of play, sometimes oriented toward objects such as twigs, potsherds, cloth, or any other object – not necessarily a movable one. Play is usually interpersonal and exploratory. When the young are approximately 8 months old the males are much larger and stronger than the females of the same age and large play groups divide into smaller ones composed of young of approximate size. Mothers seldom intervene even when play gets very rough. Squeals and grunts are heard during active play but sharp or loud outcries and continuous squealing are rare since these cries may draw the attention of adult females that break up the play group by slapping the ground or mildly threatening the playing animals. Adult interference in play is more frequent in small langur troops with few young members, where play groups must include infants and juveniles of unequal size and strength. When the infant is 10 months old it spends from 4 to 5 hours a day in group play and frequently travels with age-mates when the troop moves.

Many basic patterns of social behavior, including dominance and sex, first appear in play groups. As a context for learning to get along with other monkeys the play group provides social environment in which experimentation and mistakes go without punishment or the threat of danger from other monkeys. Dominance among very young langurs is poorly defined and is not established until they are older.

Whereas adult females orient themselves, as mothers and protectors, to all the young in the troop, it is the infant itself which must draw the attention of adult males. The male infant has no contact with adult males until he

Fig. 5–11.   A young infant-two drops down from his mother as she stops.

Fig. 5–12.
A small infant-two reaches hesitatingly away from his mother and keeps one hand tightly grasping her fur. (*Courtesy John Wiley & Sons, Inc.*)

Fig. 5–13.
An infant-two reaches for a piece
of food in her mother's mouth.

is approximately 10 months old. At this time the infant first approaches the
adult in a very special manner. The infant runs, squealing tensely, to the
moving adult and veers away just before it touches him. Gradually the infant
appears to gain confidence and touches the male's hindquarters. Within a
week after this the infant approaches and mounts by pulling himself up over
the adult's hindquarters. Infant mounting of an adult male is similar in form
to a male mounting a female or to dominance mounting between two adults,
but when displayed by the infant it does not appear to be either sexual or
dominance behavior. In a few weeks another element is added and the infant
runs around to face and embrace the adult. Touching, mounting, and embracing occur thereafter either as a series or as separate events. One adult
male may be mounted and embraced by as many as 4 infants and juveniles
in rapid succession. As shown below, this approach is displayed by the
young male until he is approximately 4 years old.

The female white infant, in contrast to the male, has almost no contact
with adult males. Instead she spends more time than does the male infant
grooming and being groomed by adult females. The year-old infant, male or
female, takes an active role in the group including grooming and alerting the
entire troop with an alarm bark if danger approaches.

Weaning is an extremely stressful period for both the infant and the
mother. In addition to physical rejection, an important part of weaning is
the emotional rejection of the infant. The mother, who has been the major
source of protection and security, becomes hostile and denying. The infant
can no longer run to her when it is frightened by an adult. The infant must
resolve its conflicts with other members of the troop unaided by a protecting
mother.

Early rejection is mild. At first the mother avoids her infant by moving
away and taking long jumps from one tree to another, making the infant

Fig. 5–14.   Large play groups are composed of both infants and juveniles. (*Courtesy John Wiley & Sons, Inc.*)

maneuver long distances down to the ground and up into her tree. When the infant catches her she merely turns away, and only if the infant persists in trying to nurse does she hold it off with outstretched arms. The infant squeals and its movements are quick, tense, and jerky. It crouches, peers into her face, and runs around to stay in front as she turns from it.

For weeks the mother alternates between temporarily rejecting her infant and allowing it to cling. After several months of rejection even 30 minutes of being followed and harassed by the infant may not wear down her resistance, as did the shorter periods of infant persistence during the early weeks of weaning. Even the most boisterous weaning tantrum seldom draws attention from other adults. If a female is disturbed by the noise and movement, she moves or threatens the mother. Rather than return the threat, the mother usually moves and is followed by the infant.

Severity of rejection varies among females. Some females make little effort to reject their infants until after the resumption of estrous cycles. A rare female may be very positive in her rejection and after only a few weeks may strike the infant whenever it approaches. The mother's temperament is also important in determining the amount of hostility or physical force used in rejection. Irritable females are usually also more irritable with their infants. Older multiparous females appear to reject their infants with less effort than do younger adult females, but there are exceptions. A very few females bite infants during rejection, but when they do, the infant is not visibly hurt.

Fig. 5–15. A male white infant is about to mount an adult male.

In the last month of weaning even the most permissive mother persists in her rejection of the infant in an effort to break the remaining ties of infant dependency. Her tension, increasing irritability, and flight from the infant suggest that this final period is unpleasant and strenuous for her. The daily routine of both mother and infant is affected by the intensity of their encounters and antagonism between the two reaches a peak toward the end of weaning. The infant strikes her and screams as it jumps about her and shakes the branches.

When the infant is approximately 14 months old it no longer receives preferential treatment by adults and is threatened if it disturbs an adult. Active running and wrestling of male infants bring them in contact with adults more than the quieter play of female infants. As a result adults threaten and chase male infants more than twice as often as they do female infants (see Table 5–6).

Juvenile

The juvenile is weaned and independent of its mother. When the juvenile is approximately 2 years old its mother gives birth to another infant and the remaining social ties with her last infant are severed. Sibling relationships did not exist in the Kaukori troop. Juveniles moved freely among adults and it was not possible to determine the mother of any particular juvenile. The young juvenile assumes an independent life in the troop, where its protection is provided by its own alertness and that of the troop, and not by the mother's signaling changing group tensions.

The juvenile stage is a period of expanding social relationships with

TABLE 5-5   Touching, mounting, and embracing of adult males by male infants and juveniles

	Frequencies by Age Groups		
Form of Approach	Small White Infant	Large White Infant	Juvenile
Touching the hindquarters	4	19	33
Mounting	0	10	38
Embracing	0	12	61

adults and subadults. A gradual familiarization with the range of the troop's activities is important in the development of behavior patterns that will characterize adult social interaction. The juvenile learns by experience which patterns of behavior, vocal and gestural, are appropriate in a given situation and learns also the consequences of inappropriate responses. Since the juvenile is no longer protected by its mother, it is essential that the juvenile develop those forms of social behavior that allow a maximum of integration into troop life with a minimum of interindividual aggression.

Threat gestures directed by a large juvenile to an adult are responded to with threat or aggressive behavior. Similarly, a submissive gesture may avert impending adult threat or aggression. The juvenile's skill in displaying gestures and its longer consistent sequences of gestures and vocalization contribute to effective communication of the juvenile's emotional state, whereas the infant intersperses recognizable social gestures with random, often play, behavior.

Play is the most time-consuming and important activity of the juvenile. Since play groups are composed of other juveniles and infants, the young

TABLE 5-6   Dominations by adults of white infants and juveniles

Domination by Adult Males of White Infants and Juveniles

Number of Dominations of Each Age Group

Sex	Small White Infant	Large White Infant	Juvenile
Female	0	5	5
Male	0	11	50

Domination by Adult Females of White Infants and Juveniles

Number of Dominations of Each Age Group

Sex	Small White Infant	Large White Infant	Juvenile
Female	3	9	21
Male	2	18	41

juvenile is assured of a context in which inappropriate motions will not result in serious injury, although physical contact in rough play among older male juveniles may turn into fighting if one juvenile is hurt or repeatedly chased by other large juveniles. Early patterns of dominance appear in this context and although dominance among juveniles is mostly a function of relative size, the juvenile gains experience and becomes familiar with both dominant and subordinate situations.

Dominance interactions do occur among female juveniles but less frequently than among males. The female spends only a few hours a day in play and when she plays it is more likely to be with other female juveniles and infants and in a less active manner than males play. Play groups among large juveniles are almost entirely unisexual, since large female juveniles seldom play with large male juveniles for more than a few minutes at a time.

The social bonds of the juvenile with adult females are much less intense than are those of the infant with adult females. Grooming remains the main form of relaxed social contact between the juvenile and the adult female. The male juvenile seldom initiates grooming with an adult unless it is to placate the adult or to get closer to food.

The male juvenile mounts and embraces adult males more than four times as often as does the male infant (see Table 5–5). The following are typical of adult male and male juvenile interactions:

KAUKORI

December 17, 11:15 A.M. A juvenile male ran screaming to alpha male, embraced, grasped alpha's fur, and then groomed him. The adult returned the embrace and put his face into the fur of the juvenile's shoulder but the juvenile tried to move to one side in order not to remain directly in front of the adult. In a minute alpha ran off and the juvenile ran squealing after him but did not attempt to touch the adult again.

KAUKORI

March 2, 4:05 P.M. A large juvenile male ran squealing to beta male and groomed beta while he lay on a branch. The juvenile continued to give a steady weak squeal and when he stopped squealing beta raised his head and looked at the juvenile, which started squealing again. As long as the juvenile squealed, beta did not look directly at him.

The juvenile may occasionally take part in the harassing of older males when the latter are in consort with estrous females. The following two examples illustrate this behavior:

KAUKORI

January 17, 9:00 A.M. An estrous female was sitting beside a juvenile male when an adult male came to her in response to her presenting. The adult male slapped the branch at the male juvenile, which ran down the tree and sat 15 feet away watching the pair. The juvenile continued to move about after a few minutes of sitting quietly and watching. He squealed constantly. When another adult male came to the base of the tree and threatened the copulating male, the latter chased the juvenile out of the tree, hitting him and threatening the second male, which withdrew 20 feet and sat.

KAUKORI

February 4, 12:15 P.M. An estrous female was mounted by alpha male. After the copulation alpha chased away another adult male that had been circling them vocalizing and gesturing. A male juvenile ran up to the female while alpha was chasing the other adult male and the female hit the juvenile. Alpha returned and mounted. The juvenile ran up to the female's head and hit at the male but the female pushed the juvenile away. Alpha did not react to the juvenile and did not dismount. When alpha completed the second copulation, he dismounted and the female ran after the juvenile, hit it, and grimaced. (The grimace is a pulling back of the corners of the mouth as in a broad grin; the teeth may or may not be exposed.)

Subadult

The subadult stage is a period of transition from the juvenile stage (in which peer-group orientation, play, nonaggressive behavior, and reliance on the group for direction are important) to the adult stage (in which dominance hierarchies, sexual behavior, maternal behavior for the female and leadership of the troop for the male replace the dependency of the young monkey). During the years when not yet physically mature the subadult cannot compete with adults for a position in the adult dominance hierarchy. However, it orients toward adults and experiences and participates in many forms of adult activity without entering dominance competitions. The subadult stage is characterized by display of predominantly submissive gestures to adults. Although the subadult female is subordinate to all adult females, the subadult male is not. Contacts between subadult males and adult females are tenser than female-subadult/adult-female interactions. From approximately 4 to 5 years of age the male repeatedly tries to assert his dominance over as many adult females as possible.

During the 2 years from approximately 4 to 6 years of age, the subadult male has fewer close social bonds within the troop than at any other time in his life. He is, in many respects, an extremely marginal participant in group life. Although he has free access to any part of the troop, he spends most of the day on its edge and near adults only when they are resting quietly.

Subadult males compete with each other over many things, such as food, right of way on paths, positions in sleeping trees and on the ground, and estrous females. Dominance interactions among them are characterized by more physical contact, aggressive threats, and active chasing than when adult males compete. In one instance, a large subadult male in the Kaukori troop was mounting an estrous female when a second subadult ran up to the pair, jumped on the first, and knocked him off. The larger subadult turned, threatened, slapped, and then chased his attacker.

Subadults are subordinate to all adult males and this lower ranking is not disputed. When adult males fight, the subadult stays a safe distance away, sometimes more than 50 feet. When he is near an adult male the younger male is obviously tense and gives the impression of being quite nervous, especially when there is activity and the adult is behaving aggressively. In an extremely stressful situation — as for example after an adult threatens and does not make it clear that he has relaxed and will not continue to threaten — the subadult may approach nervously and embrace the adult in a gesture of subordination and placation. Probably because mounting is a gesture of dominance among adults, the subadult no longer displays this approach as he did when he was a juvenile. His larger size, increased strength, and in general his potential to interact on a par with adults is associated with the loss of freedom to touch an adult male in the form of mounting. This is an interesting instance of a change, associated with maturational changes, in the meaning or social significance of a behavior.

Subadult males very seldom come into contact with subadult females except in relaxed mutual grooming. When the subadult female first becomes sexually receptive the subadult mounts her whenever he is solicited, but this too is not frequent. Subadult male contacts with adult females are more frequent than with subadult females, and these consist mainly in relaxed mutual grooming. Most adult females are able to dominate small subadult males. Larger subadults, however, gradually can and do dominate most adult females in the troop. Only if a female attacks a large subadult or adult male suddenly, without warning, or if the male is relaxed when she threatens him, is she able to succeed in dominating him. When two or more females combine their attack they can always dominate even the largest subadult male. Because the subadult male works his way up through the female dominance hierarchy before he really takes his place in the adult male ranks of power, there is often tension when subadult males and adult females are near each other. If an adult female is threatened by a more dominant adult, she may turn and threaten a smaller subadult male instead of the more dominant animal. This redirected aggression often initiates fights involving subadult males and adult females, and when the fighting draws the attention of a nearby adult male, the interaction may then grow to involve the adult male as well.

Infants have almost no contact with subadult males, whereas juveniles may approach, touch, mount, and/or embrace subadult males in the same manner as they approach adult males. During the contact there is less tension than with adult males and occasionally the subadult initiates wrestling or plays with the juvenile.

Although the subadult female has a much wider radius of activity than has the subadult male, she does not have persistent or strong ties with any segment of the troop. Her approaches to adult males are hesitant and she moves away from adult females when they are irritable or aggressive. In any case, experience has shown her that she will always lose out in a contest with any adult.

As a subadult the female is very interested in young infants and when she has the opportunity she plays and wrestles gently with them. Her relationship with adult females is usually relaxed and they often spend 4 or 5 hours a day together sitting quietly or alternately grooming one another. Subadult females readily join grooming groups of adult females but may hesitate to approach certain of the more irritable adult females if they are sitting alone.

It appears that the subadult female is subordinate to all adult females, but dominance interactions are seldom observed. At the first hint of fighting the subadult usually moves away. She is seldom the focus of aggression until her initial estrous period, when she actively solicits mounting by adult and subadult males. Her sexual behavior is like that of the adult female, which is discussed in the following section.

Adult Female

The most important and time-consuming role of the female langur, and her primary focus as an adult, is maternal behavior, motherhood. She raises one infant after another from the time she assumes adult roles at the age of about four years until the time she dies. More than two thirds of her life may be spent nurturing and protecting her infants. She is pregnant approximately one quarter of her adult life. She lactates and cares for a dependent infant one third of her adult life and weaning occupies a major part of the rest of the time. Weaning may overlap the resumption of her estrous cycles, but often it does not. Her dominance status, associations with adult males, female companions, and daily activities are to a large extent a function of her status as a mother and the phase of her reproductive cycle.

The next most important and time-consuming activity, one which she engages in throughout her life, is mutual relaxed grooming. A female has close grooming ties to many of the members of the group and by far the majority of her contacts with adult males occur in mutual relaxed grooming.

Adult Female Dominance Relations

It is not possible to assign a female langur to a position in a so-called female dominance hierarchy except within a general level of dominance that includes females of approximate rank. The female dominance hierarchy is relatively unstable and poorly defined. A female's dominance position fluctuates with the different stages of her reproductive cycle. However, by far the majority of interindividual contacts do not involve expressions of aggression or contest. At least one third of the activities involving what may be labeled dominance are of a very subtle nature, and it is only after the troop members are well known that any hierarchical arrangement can be described. Dominance is often a matter of degree with no clear wins or losses so that an adult female may be dominant in one situation and not in another.

The most accurate way of conceptualizing the female dominance structure is as a series of levels. An individual female occupies a rank within a general level and she is dominant over those females in a lower level and submissive in most situations to the females in the next higher level (see

TABLE 5-7 Adult female dominance hierarchy of Kaukori troop

Female	Dominates Frequency	Most Frequent	Dominated Frequency	Most Frequent	Present to Female	Associated Young
A	30	G, D	10	C	2	F Infant*
B	21	K	6	C, L	3(A, H)	Pregnant
C	19	A, B	6	A, B	1(A)	Pregnant (?)
D	13	K	10	A, C, I		F Infant*
E	10		4	B		F Infant* Pregnant (?)
F	8	G, C	11	D, G	5	M Infant*
G	8		8		2(F)	Pregnant
H	8		6	A		M Infant*
I	6	A, B, D	3	A, B, J	1	M Infant*
J	4	D	3	A, C		Pregnant
K	4	G	12	B, D		M Infant*
L	3	B	2	B	1(D)	Pregnant
M	2		8	I	1	F Infant** Pregnant (?)
N	2		8			Small M Juvenile
O	3		9			M Infant*
P	2		10			M Infant*
Q	2	10	10			F Infant*
R	1		8	I, D, E		——
S			8			F Infant*
Subadults						
a	3 (in alliances)		12			——
b			13			——
c			10			——

*   White
**   Black

Table 5-7). In Fig. 5-16 adult female dominance hierarchies for three Orcha troops and the Kaukori troop are presented. The hierarchies of small groups usually represent a cross section of age grades and tend to be linear with most or all females of unequal status. Relationships among females within one level are often not predictable and may vary within short periods of time. Actual fighting during dominance interactions is rare. Encounters are quick and usually last only from 5 to 30 seconds. Female dominance interactions, unlike those among males, are seldom preceded by a gradual building up of tension. In female fights physical contact is usually limited to slapping, and no adult female was visibly wounded in such fighting. In 150 encounters recorded for one troop, biting was observed in only three.

Table 5-8 summarizes aggressive gestures and vocalizations in order of increasing aggression. These are displayed by both males and females, although the female often skips some elements during her quick interactions. Table 5-9 summarizes the submissive gestures and vocalizations used by both adult females and males. Table 5-10 summarizes gestures and vocalizations given by dominant and subordinate adults and juveniles; it illustrates that most gestures and vocalizations are used by many individuals in a troop and that few of these elements of communication are specific to a very limited number of situations. The gestures and vocalizations in these tables are rather crudely defined and clearly many are composites of smaller units involving sometimes difficult to perceive but important changes in posture and in features of the face, and the like.

The very dominant female is not necessarily involved in dominance interactions more frequently than other females; she is able to take food whenever she wants and when she is irritated or tense she is often avoided by less dominant females. Female L is relaxed and direct in her movements,

Group	North Orcha	West Orcha	East Orcha (linear hierarchy)	Kaukori
Most	A C B	A B	A B	A D B C ⓔ E
Medium	D E F	C D E	C D	K G F I M O ⓛ H ⓙ N P ⓠ
Least	GH I a b	a	a	R S b a c

General Levels of dominance

Upper case = Adult female
Lower case = Subadult female
◯ = Dominated one or more of the five most dominant females

Fig. 5-16. Adult and subadult female dominance hierarchy in four langur troops.

TABLE 5–8 Aggressive gestures and vocalizations

	Male		Female	
	Gesture	*Vocalization*	*Gesture*	*Vocalization*
I N C R E A S I N G	Stare	Silent or low grunting	Tense	Grunting or silent grunting
	Tense, turn to face other monkey	Silent or grunting and belching	Stare	
	Intense stare	Silent or grunting and belching		
	Slap ground	Silent or grunting and coughing	Slap ground Grimace	Grunting or barking
	Grimace	Silent or grunting		
	Crouch, then stand suddenly	Barking, grunting or sharp coughing	Bite air, toss head	Barking, grunting or coughing
	Slap ground and bob head	Grunting, coughing or barking		
A G G R E S S I O N	Bite air, toss head	Silent or grunting		
	Lunge in place	Silent or grunting and barking	Lunge in place	Barking, grunting or coughing
	Chase	Silent or grunting or barking	Chase	Silent or barking
	Hit or slap	Silent, grunting or barking	Hit or slap	Silent or barking and coughing
	Bite	Silent, grunting or barking	Bite	Silent or barking and coughing
	Wrestle	Silent	Wrestle	Silent or barking and coughing
S U B S I D E N C E	Monkeys separate and dominant sits	Grunting or belching		
	Hand on other monkey	Silent	Female sits	Barking, grunting
	Subside	Belching or canine grinding		

TABLE 5-9   Submissive gestures and vocalizations (in order of increasing subordination)

Gestures Male and Female	Accompanying Vocalizations	
	Male	Female
Avoid visual contact	Silent	Silent
Turn head; look away	Silent, or belching or soft grunting	Silent or soft grunting
Turn back	Silent or soft grunting	Silent or soft grunting
Move tongue in and out of mouth	(Rare) silent or soft grunting	Grunting or soft squealing
Grimace	Silent or grunting	Grunting or soft squealing
Embrace	(Rare) coughing	Grunting, coughing or soft squealing
Walk away	Silent, or grunting, belching	Silent or grunting
Present	Silent or grunting, belching	Grunting or squealing softly
Combine turning back, presenting, and grimace	Grunting or coughing	Grunting, coughing or squealing
Run away	Silent or grunting	Silent or grunting and coughing

and does not take part in many interactions, but she can walk up to females A and B and take a piece of food lying at their feet.

Reduction of all dominance interactions to numbers of successes and failures is only a partial reflection of the dominance structure and needs qualification. The patterns of social interaction most difficult to evaluate, such as avoidance, are extremely important in understanding dominance. Table 5-7 summarizes dominance interactions during four months' observation of the Kaukori troop. Approximately one third more dominations occured than are shown, but interactions are tabulated only if it was possible to identify all participants. In this troop, as in most, the observer gets the distinct impression that a major aim in the life of the observed is to do half of what happens behind a bush or out of sight. Approximations of the realities of a troop's structure must be made with the difficulties of continuous observation in mind.

All females listed in Table 5-7 were well known as individuals. Differences in personality, age, and experience were certainly importantly

TABLE 5-10 Vocalizations and gestures given by dominant and subordinate adults and juveniles

Vocalization	Given by Adult Male	Adult Female	Juvenile	Subordinate	Dominant	Equals
Whoop	X	X		X	X	X
Belch	X	r*		X	X	X
Canine grind	X			X	X	X
Cough	X	X		X	X	X
Bark	X	X	r	X	X	X
Grunt	X	X	X	X	X	X
Squeal	r	X	X	X	X	X
Scream		X	X	X		X

Gesture						
Grimace	X	X	X	X	X	X
Stare threat	X	X			X	X
Bite air	X	X		X	X	X
Slap ground	X	X	X	r	X	X
Bob head	X	X		r	X	X
Lunge in place	X	X			X	X
Chase	X	X	X		X	X
Slap	X	X	X	X	X	X
Bite	X	X	X		X	X
Wrestle	X	X	X		X	X
Dominance pause	X	X			X	X
Embrace	X	X	X	X	X	X
Present	X	X	X	X		X
Tongue in & out	X	X	r	X	r	X
Hand on monkey	X	r			X	
Turn back	X	X	X	X	r	X

* r = rare, seldom observed

related to differences in status. Some females are relaxed and slow to threaten; others are irritable and appear to have little tolerance for disturbance.

On 13 occasions 2 or more females formed temporary alliances, which lasted from 30 seconds to approximately 2½ minutes, during which these females simultaneously chased and threatened another monkey or other monkeys. The target of the alliance can be either a dominant or subordinate male or female. The alliances are apparently spontaneous and voluntary. The members of the alliance act with little reference to each other and often the alliance fractionates and individual females turn their attention in separate directions. It is often impossible to determine just what triggers a fight or chase; so that rather than to say these alliances are spontaneous, a more accurate way of describing them is to admit that the observer was not aware of the events that initiated them. The following are two illustrations of female alliances:

1. Five adult females were eating in a thick lentil-bush patch when one of the five, a dominant female designated as 1 in this sequence, brushed against female 2, which turned and slapped the ground toward 1. Female 3 sitting nearby ran toward 1, which backed away five feet. When she saw 3, 2 also ran toward 1; and female 4, sitting near, stood and ran toward 1. Then 1 ran out of the field followed by 2, 3, and 4. The three females chased 1 30 feet and disbanded; 1 sat, barked toward 3 and 4, which grimaced and slapped the ground, and 2 returned to the lentil patch, followed shortly by 4 and 1.

2. Four other females were involved in this sequence. 1 was sitting in a tree eating a mango and dropped half of it. When female D ran and picked it up, 1 threatened her by slapping the branch on which 1 was sitting, and 3, also in the tree, jumped down toward female D. Female 1 then jumped down, and female 3, which had been sitting nearby, ran toward female D. D, with the mango in her hand, ran into the mango tree grove followed by females 1, 2, and 3. Female 1 ran directly up to female D and slapped her; female D in turn slapped 3. Then 1, 2, and 3 again chased female D, which dropped the mango and ran up into a tree; 2 grabbed the mango and ran off into another tree.

More than half of all temporary alliances recorded led to a long sequence of action. The sequence diagrammed and described in Fig. 5-17 is one of the more complicated examples of a long, connected series of dominance interactions. The female designated as 1 is one of the most dominant females, and 2 and 6 are also relatively dominant. The entire sequence lasted 1 minute and 10 seconds.

Female 1 was being chased in a mango tree by 2; one then chased 3, which presented to 1; 4 chased 1, which ran down out of the tree and over to 3 and back to the tree, where 4 chased 1 again; 1 ran up into the tree and was chased down by 3; and 3 ran after 1, which repeatedly presented while running; 6, which had been sitting in the tree, jumped down and presented to 1, and then ran up to 7, which was sitting nursing her infant. Seven presented to 6; 1 displaced both 5 and 3 and ran back into the tree; 3 displaced 1 and 4 and at this point the female rhesus monkey came running and scattered all the female langurs (diagrammed in Fig. 5-17).

Since it appeared that most adult females did not display strong patterns of individual preference for each other, the impression was that chance dictated the nearness of a female that might join in a combined threat or alliance. Because very few alliances were observed, this impression of random coalitions may be an artifact of sampling, and it is possible that females that groomed together more than they did with others may have joined each other in alliances. This is a possibility that should not be ruled out.

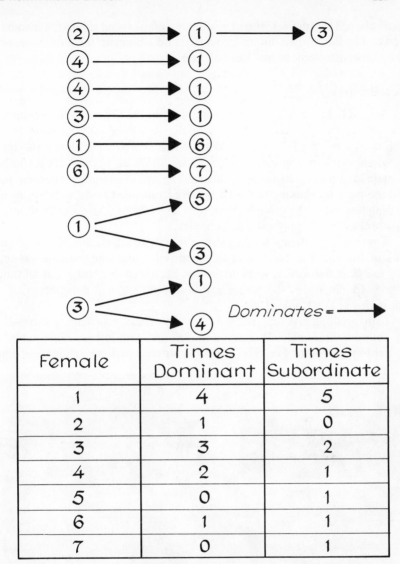

Fig. 5-17. A complex series of dominance interactions among adult females.

Male dominance is never contested by a female unless a male accidentally frightens an infant or a female with an infant. When this occurs he is liable to immediate attack from the mother of the infant and nearby females. On one occasion male D walked under a mango tree in which a small infant was playing. The mother was sitting under the tree, and as the male approached she reached up to grasp the infant who became alarmed and jumped up to the next branch just out of her reach. The female jumped up and chased the male 150 feet while the infant raced around squealing in the tree. The adult male ran up into a tree and whooped; the mother stopped and he turned to look at her and belched.

Sexual Behavior

Sexual behavior plays a very small part in the life of an adult female in terms of the time it takes. She is sexually receptive only when in estrus, which lasts from 5 to 7 days a month, when she is not pregnant or lactating. When she is not in estrus adult males show no sexual interest in her. A female langur exhibits no perineal swelling and must indicate her receptivity to the male by shaking her head rapidly from side to side, presenting, and dropping her tail to the ground. No female associated with a small infant was ever observed to display estrous behavior.

There is a tendency for an estrous female to solicit the most dominant males in the troop at the height of receptivity, the time ovulation is most likely to occur. However, some males are preferred over others at all times. Table 5–11 illustrates the frequency of copulation and solicitation of the adult males in the Kaukori troop.

Consortships last only a few hours. Once a consortship is formed the male follows the female. When she moves away he follows slowly, and he very seldom threatens her. The consort pair may be harassed by less domi-

Fig. 5–18. Juvenile and subadult males harass adult male copulating with estrous female.

nant males which run about the pair and bark, threaten, and slap at the consort male. The aggressive behavior of a harassing male is directed almost exclusively to the consort male and not to the estrous female. In general, the dominance status of a female increases when she is in consort with any male, and the dominance of the female with a high-status male is potentially greater than that of the female with a low-status male.

Adult Male

Adult Male Roles

Every adult male in a troop can be described as a unique personality with variations in behavior that set him apart from all other males in the troop. These differences are not obvious, however, until the male is well known and has been observed in a wide variety of interactions. Differences in maternal behavior as well as in tolerance for other monkeys has been described for adult females, and variations in behavior of similar magnitude are characteristic of individual males. These variations are expressed in intensity of reactions, in gestures, vocalizations and associates. Most adult males in the Kaukori troop could be identified without their being seen by the way they produced certain vocalizations. For example, the "belch" always was given by one male in a rapid series, while other males emphasized various parts of the sound or the volume. Variations of this magnitude were distinguishable only after long familiarity with the individuals. It is possible that changes in dominance status may affect the degree of confidence or of hesitation with which many forms of vocalization are given, but only one major change in a male dominance hierarchy occurred during the

TABLE 5-11 Frequency of sexual behavior

Adult Male	Solicited by Estrous Females	Copulates
A	47+	18
B	35	10
C	19	6
D	55+	10
E	6	—
F	4	4
Subadult Male		
G	17	14
H	3	2
Total	186+	64

period of observation and neither of the two males involved noticeably changed their vocal patterns.

Adult males are leaders and coordinators of troop activity. Their roles are extremely important in the maintenance of troop unity and stability as well as in determining the troop's use of its range and its relationships to other langur troops.

Males initiate and determine the direction of troop movement. Male leadership determines the part of the range that will be used, where the troop will feed, and which trees it will use for sleeping. In addition, the male coordinates intergroup relations by producing the whoop vocalization that is an effective means of informing all nearby troops of the location of his troop. Adult males also coordinate intragroup activity and troop movement by whooping, and thereby minimize the possibility of leaving any member behind. The "whoop" call is necessary for quick and effective gathering of a group in forests where visibility is poor and it is difficult for monkeys to stay in sight of one another. Adult males maintain internal troop stability by establishing and asserting a stable male dominance hierarchy that structures the relationships of adult males within a troop. Since adult males are dominant over adult females, the stable, linear male hierarchy is far more effective than is the poorly defined female hierarchy in determining the outcome of disputes which arise among troop members.

In the Kaukori and Orcha troops the adult males invariably initiated a display characteristic of Indian and Ceylon langurs. This display consists of whooping and dashing about in trees and on the ground, jumping against branches and trunks with all four feet, and taking great leaps from place to place. The males whoop during the entire series of leaps. Females occasionally begin to run about, banking off objects with all four feet, but in the Kaukori troop the females did so only after the display was initiated by the adult males. The cause of this display was not always obvious, but frequently it was due to some disturbance outside the troop such as an airplane overhead or a shot in the distance. Langurs in Ceylon displayed this sequence more frequently than did the North Indian langurs, and in Ceylon females as well as males began the display. It was elicited as often by dominance interactions within the group as by events outside the troop. Here too the animals would attempt to jump down deliberately as hard as possible on limbs, which might break and fall crashing to the ground. Often very large limbs were broken in this manner. When the commotion settled, males sat quietly belching and grinding their canines. Gradually the tension subsided and the troop was quiet.

Dominance

A relaxed dominant adult male langur does not stand out in a group when there is no tension. Most of the time, adult males mingle freely with most of the group members. There are usually very few aggressive interactions to indicate which males are more or less in control. It is interesting that those adult males in the Kaukori troop that were most noticeable in the early months of this study because of their activity were actually among the less dominant males.

In general, dominant males are less active in minor disputes than are subordinate males. The high-status male is able to take positions, food, and estrous females from other males, and when he is tense and irritated he is surrounded by a wide area of potential threat or personal space. Beta males are usually more active in dominance interactions than are alpha males whose high dominance status is uncontested by the rest of the troop.

Adult male dominance is established and maintained with a minimum of aggressive behavior. Subtle pauses and hesitations predominate in dominance interactions. If a male pauses in passing another male as the two are alongside one another, this is a dominant gesture, and the sitting or stationary male either looks away or moves. The dominant animal may pause without breaking the rhythm of his stride. If the passing animal pauses when he is slightly in front of the stationary male, this indicates that the passing male is subordinate.

Physical nearness is often used as a subtle gesture. Repeated approaches to a less dominant male is a subtle means of forcing him to move. A characteristic gesture of dominant males seldom displayed by a female is placing the hand on a less dominant monkey. Often it signifies the end of an interaction, a dismissal by the more dominant animal.

Another subtle indicator of status is the size of a male's "personal space," an intangible area surrounding him into which another monkey cannot enter without the danger of threat from the animal within the space. The more dominant the male, the larger the area of space he can maintain for his exclusive use. When a male is relaxed, however, he does not maintain an area of personal space. Females may also maintain a surrounding area into which another female or a less dominant monkey may not enter without making his or her intentions clear to the more dominant animal. To conclude, in all instances the potential threat is only to less dominant members of the troop.

Slight postural shifts and the direction of visual focus are two extremely subtle movements that communicate a potentially changing emotional state and an awareness of surrounding activity or tension. Subtle movements by an adult of low-dominance status are usually ignored by a more dominant animal unless the high-status animal is irritated and tense.

Often a seemingly inappropriate mixture of dominant and subordinate gestures and/or vocalizations is used in a sequence. To illustrate: a female may grimace, slap the ground, tense her body, and then present to a more dominant female; a female lunges in place, slaps at another adult female, then presents. Such ambiguous sequences are commonly observed in females of intermediate dominance status. In part this reflects a rapid alteration of mood, but it is also observed in interactions between two animals of very different status. The total impression of a series, and the reaction of the animal to which the sequence is directed, depends upon which gestures are emphasized or repeated and which are displayed last. The effect of an aggressive threat followed by presenting is usually submission. In many other combinations the effect is not so easily predicted since it depends on the relative dominance status of the animals involved and the sequence of gestures displayed.

The Dominance Hierarchy

Table 5–12 summarizes the dominance hierarchy for the six adult and two subadult males in the Kaukori troop. The degree of marginal participation is indicated by the horizontal dimensions of this diagram. The status of each individual is well defined and constant for relatively long periods of time. Adult males act independently during a dominance interaction and do not support each other or form even temporary alliances. Ranks in Table 5–12 take into consideration more than the numbers of dominations credited to each male. Table 5–13 illustrates several aspects of adult-male social behavior important in the assessment of adult-male dominance. Table 5–14 details the dominance interactions which were recorded.

There are many measures of dominance, and when individual interactions are analyzed, it is clear that the critical measure is not necessarily the total number of times a male dominates or is dominated. The number of completed copulations does not always correspond to the dominance rank of the male, as illustrated in Table 5–12. Male G, a subadult, was not successful in competition for estrous females; estrous females solicited him only when no other male responded. Male D was a preferred sexual partner by almost every female in the group.

Presenting and grooming among males reflect ranking. A subordinate male presents to a dominant male, which usually responds by grooming him for two or three seconds. In a typical sequence the subordinate male ap-

TABLE 5–12 Adult male dominance hierarchy of the Kaukori troop

---

A (Slate)
B (Rip)
                    C (Slit)
D (Patch)
                              E (Rat)
                                        F (Mangle-ear)

- -

G (Subadult)
H (Subadult)

TABLE 5-13   Summary chart for adult male dominance (Kaukori)

Male	A	B	C	D	E	F	G	H
Dominates other males	48	57	20	4	6	1	1	1
Dominated by other males	4	16	10	31	16	12	24	7
Solicited by estrous females	47+	35	19	55	6	4	17	0
Copulations	18	10	6	10	0	4	14	2
Groomed by adult females	33	66	65	26	32	16	18	10
Presented to by adult males and grooms them	1	5	1	0	2	0	0	0
Relative size (1 = largest)	3	2	1	4	2	5	6	7
Relative age (1 = oldest)	4?	3	2	5?	1	6?	7	8

proaches a male of higher status, presents, and is then groomed for a few seconds. After this the dominant male places his hand on the lower-status male and tension decreases as both males relax. The subordinate usually moves a short distance away probably to avoid any further physical contact.

Only one major shift in a male hierarchy was observed in 16 months of fieldwork. Male A, Rip, in the Kaukori troop lost his top ranking to male B, Slate, A moving down into the position that B had occupied at the beginning of the study. Exceptionally little aggressive fighting occurred in the 2½ weeks during which these 2 males changed positions in the dominance hierarchy. The first indication of increasing tension between them was constant belching whenever they were within 50 feet of one another. The following examples are typical of the interactions between A and B males, and of these males with other troop members during the first week of the change.

TABLE 5-14   Dominance interactions for adult males (Kaukori)

Domi-nator	Dominated								Uniden-tified Male	Total
	A	B	C	D	E	F	G	H		
A	—	16	2	6	4	5	6	1	8	48
B	4	—	8	19	3	3	9	4	7	57
C	0	0	—	4	9	0	5	1	1	20
D	0	0	0	—	0	1	1	0	2	4
E	0	0	0	2	—	1	2	1	0	6
F	0	0	0	0	0	—	1	0	0	1
G	0	0	0	0	0	1	—	0	0	1
H	0	0	0	0	0	1	0	—	0	1
Total	4	16	10	31	16	12	24	7	18	138

January 20. Rip (originally male A) moved past male D and belched 2 times. Male D ran off; then Rip moved to male C. At this, male C swung his hindquarters to Rip in a gesture of subordination, and Rip groomed male C for 3 seconds. C walked away, sat, and Rip once more moved close to C. Again, C walked away, this time grimacing, and sat 10 feet away from Rip. Rip belched once, and C stood and walked farther. Rip again belched toward C, but the latter remained seated 25 feet from Rip.

Rip approached C, but C turned his back and shook his head toward Rip. Then Rip groomed C. All the nearby adults had tensed but not moved when Rip approached C, but when he put his hand on C, they all relaxed and continued eating or resting.

Rip then moved to male F, then being groomed by female D. Rip gave 2 very slight shakes of his head; F jerked aside, and the female groomed Rip.

January 21. Slate (originally male B) moved to Rip. Rip stood and walked off 10 feet and sat by a female. The female immediately started to groom Rip. Slate belched two times directly to the female grooming Rip and she stopped grooming, presented, and moved to groom Slate.

Rip came down a tree, belched two times and walked past male E. E was being groomed and stood when Rip hesitated but Rip moved on, deliberately passing within 10 inches of him. E did not volunteer to move from Rip and Rip did not assert his dominance.

Slate walked to within 4 feet of Rip and sat. Slate tossed his head once but Rip did not move. One minute later Rip walked slowly away. Neither male vocalized.

Rip was tearing about whooping. Monkeys were scattering in all directions. Several juvenile males and male H ran after Rip squealing and tried to mount but could not catch up with him. When Rip initiated this running and vocalizing, Slate, at the other end of the mango grove, did the same. When the monkeys calmed down after 20 minutes of very active interaction, Rip sat quietly. Slate dashed down from a tree with one final whoop but Rip remained sitting. Both males sat belching.

Slate walked out onto a large branch where Rip was sitting. Rip got up slowly and moved 5 feet away and sat quietly. Slate sat where Rip had been sitting.

A striking characteristic of langur dominance and one well documented by this major change in status is the rarity of violent aggressive fighting. No single battle resulted in the replacement of the former top ranked male. Instead, a gradual process with no single fight but a constant pressuring of A by B eventually resulted in A's dropping in status. The percentage of dominant responses by the former A male decreased while the percentage of dominant responses by the former B male increased. As soon as the new ranks were established these males seldom threatened each other or other adults.

Comparisons with South Indian Langurs

The ecological variability of the areas of langur distribution in India includes habitats of from dry scrub to thick wet forests, from sea level to 11,000 feet altitude. Many of these areas are shared with rhesus monkeys, but langurs can also live in much drier areas than are hospitable to rhesus macaques. The South Indian langurs at Dharwar were able to go 4 to 5 months at a time without drinking water (Yoshiba 1968), a remarkable ability made possible by anatomical and physiological specializations quite different from those of rhesus monkeys (Bauchop and Martucci 1968).

Scientists from the Japan Monkey Centre (K. Yoshiba, S. Kawamura, and Y. Sugiyama) and an Indian scientist (M. D. Parthasarathy) worked in the forest near Dharwar in Mysore State from June 1961 until March 1963 observing langurs (Sugiyama 1964, 1965a, 1965b; Sugiyama and Yoshiba 1965; Yoshiba 1967). This team documented some exceptionally interesting differences in social behavior between the northern and southern forms of the common Indian langur. Table 5–15, taken from a summary article by Yoshiba (1968), describes the three major study areas of Orcha, Kaukori, and Dharwar. Some major differences draw attention immediately. The population density in Dharwar is very high, and not only is this a very crowded area it is also one in which the basic structure of individual troops varies from the typical multi-male bisexual troop characteristic of most of the north. In the Dharwar forests one-male troops are more common than multi-male troops.

Social changes are frequent and dramatic at Dharwar, where "inter-troop encounters are daily affairs. More than once a day (on the average) a troop meets another troop at the overlapping area of the ranges of the two troops" (Yoshiba 1968). This frequency of encounters permitted the ranking of adjacent troops as subordinate or dominant to each other depending whether they withdrew from meeting or avoided situations wherein they might meet. In general, all troops, even those categorized as submissive, will defend their core areas even when it is a more dominant troop that approaches these important locations. These daily meetings seldom led to severe fighting between multi-male or one-male bisexual troops.

The stimuli for social change within the troop came from the en-counters of bisexual troops with all-male groups. The relationships between bisexual troops and all-male groups were much more aggressive than the interactions between two bisexual troops. At times the males from a bisexual troop are forced out and replaced by males that have not been living with a troop. This is accomplished only with great fighting that often results in serious wounds, and when the shift of male membership is made, in per-manent social change.

More than 10 major social changes were documented in 2 years of observation at Dharwar, and in every case they were caused by contact between male groups and bisexual troops. Changes were of several kinds including shifting of troop leadership, and division of troops in which one segment was led by a male who had formerly been in an all-male group. Regardless of the stimulus for a change, one invariable result was an in-crease in sexual behavior after the change occurred.

TABLE 5–15 Comparisons of three study areas

Characteristics	Orcha	Kaukori	Dharwar
Summer conditions	Moderate	Severe	Severe
Winter conditions	Moderate	Severe	Moderate
Annual rainfall	80 in.; 75% in monsoon	30–50 in.; 70–80% in monsoon	30–50 in.; 90% in monsoon
Natural vegetation	Moist deciduous forest	Dry scrub forest	Dry deciduous forest
Human influence	Very weak	Very strong	Rather strong
Other wild animals	Tiger, leopard, and so forth, abundant	Almost none left	Tiger and so forth, survive in decreased number
Langur population density	7–16 per sq mi	7 per sq mi	220–349 per sq mi
Troop size	22 (average)	54	16 (average)
One-male troop	Less common	—	Common
Nontroop male	Very few	A few	Many
Sex ratio of adult troop members	6 females to a male	3 females to a male	6 females to a male
Home range of a troop	1.5 sq mi Seasonal change of core areas	3 sq mi	0.072 sq mi No seasonal change of core areas
Percent of time on ground per day (approx.)	30–50%	70–80%	20–40%
Weaning age	11–15 months		20 months
Infant male/adult male relations	More tense with less contact and characteristic approach		More relaxed
Juvenile male/adult male relations	More tense with characteristic approach		More relaxed
Subadult male's position in troop social life	Extremely marginal with less contact with the adult male		Near that of the adult male with more contact
Harassment of sexual behavior	By adult and subadult males of the troop		By females of the troop or nontroop males

TABLE 5-15—*Continued*

Characteristics	Orcha	Kaukori	Dharwar
Male dominance hierarchy	Clearly defined and constant among the adults and subadults of the troop		Not clear between the adult and the subadult of the troop; defined but unstable among nontroop adults
Female dominance hierarchy	Observed but poorly defined		Seldom observed
Intertroop relations	Peaceful and tolerant with less frequent encounters		More aggressive with frequent encounters
Relation between the troop and nontroop males	Very aggressive; nontroop males more easily expelled		Very aggressive, with occasional success of nontroop males in entering the troop
Frequency of major social changes		Low	Very high

When a leader male was forced out of his troop all the males that had been members of the troop including 1-year-olds left with him. On 3 different occasions the "new leader of the troop seriously injured all the infants and some of the 1-year-old juvenile females" (Yoshiba 1968:236).

Males that became new leaders of bisexual troops did not attack infants once they had been established as top male. Yoshiba (1968:236) estimates "that at Dharwar forest such social change may occur in a troop with a frequency of once every three to five years."

Summary

This chapter has described briefly the social life characteristic of four troops of North Indian langurs. Their daily routine, social organization, and some major events in the lives of males and females were illustrated.

A langur's life is spent either in or very near the safety of trees. The social organization of a North Indian langur troop is not oriented primarily to protection of the individual by group action except that each langur's alarm call may alert each other animal to an awareness of danger. Unlike macaques or baboons, a langur protects himself as an individual most effectively by dashing up into the nearest tree instead of depending for protection on large adult males with well-developed fighting prowess. Sexual dimorphism is not pronounced among the North Indian langurs, but adult males can be distinguished from adult females by the male's slightly larger size and more robust body build. Relations among adult male langurs are relaxed. Dominance is relatively unobtrusive in langur daily life and most of the activities that occupy an individual's time are minimally related to the status of the actor in the power structure of the troop. Aggressive threats and fighting are exceedingly uncommon.

This relaxed nature of the North Indian langur's life is one of the first characteristics an observer notices. The daily pattern of activity of a langur group is in rather sharp contrast to the more boisterous, noisy interactions of a rhesus troop, a contrast easily discerned since throughout North India langurs are frequently found side by side with macaques. Often the two different kinds of monkeys live together in the same area and use many of the same food and sleeping trees as well as the same sources of water. When they are observed mingled together in trees or fields differences are even more striking. Rhesus are more intense, quicker moving, more easily provoked to threat, more aggressive, and more vocal than are langurs. The stocky, short, muscular rhesus moves slowly and cautiously in the trees, whereas the slim, greyhound-like langur takes long elegant jumps with apparent ease. Dominance interactions among northern langurs are infrequent and subtle, while among rhesus interactions are frequent and fighting is often severe. As a consequence rhesus bear many scars of past fights, whereas langurs seldom show signs of any serious wounds.

As the results of additional field studies are made available it is increasingly clear that there is great variability in behavior among Indian langurs. The causes and correlations of this variability are not clear. Detailed ecological studies will provide information important to patterns of langur dispersion, numerical density, and group dynamics. Yoshiba (1967) suggests that South Indian langur group ranges are related to food availability and Sugiyama (1964) also emphasizes that the behavior of Dharwar langurs is greatly influenced by the high population density.

Few Old World species have more variation within their behaviors than *entellus*, and generalizations written in 1963, revised now in 1969, may be just as much in need of revision in another six years time — providing there is continuing research.

part 2

ANALYSIS OF BEHAVIOR: SPECIAL TOPICS OF PRIMATE BEHAVIOR

Many of the more recent studies of primate behavior focus on those species that display striking behavioral, morphological, or habitat specializations, and on species within which there is evidence of substantial behavioral variability in different habitats. Questions raised in the formative years of primate studies can be asked again, now directed to a much wider range of data on many primates never before studied. Recent research has suggested additional questions, which in turn guide the planning of new field studies and laboratory investigation. Answers, more often than not, are only partial and do not offer simple solutions, but rather, lead into complexities of still other questions demanding still more information for their resolution.

As the questions become more and more complex, the network of investigation expands to include related sciences and new techniques, and the circle of researchers grows to include specialists with many different skills. Primate studies are no longer the concern of the zoologist, the psychologist, or the anthropologist working alone, because to perform all the necessary tasks of understanding behavior is beyond the specialization of any one person. A major result of the increasing complexity of primate studies is the development of research teams composed of a number of technically trained people in addition to the observers who concentrate on what they see the animals doing. Whatever specialists or specialized techniques are needed, the work must always be based on careful detailed observations, descriptions, and film recording of the actual behavior of the animals. When surgery is used to change brain structure or to implant telemetric devices to deliberately alter the animals' behavior, it should be done only after the species is very well known and each individual in the particular social group is familiar. Control animals must constantly be observed as a baseline against which to measure the changes in behavior of the monkey or ape that has been operated on. The tedious yet intriguing work of recording exactly what the animal does remains an indispensable part of any study or research project, and a great deal of training and skill goes into the ability to perceive and record, to select from the continuum of actions and reactions, those elements that are significant and should be preserved in the written record. The need for practice extends also to photography. It takes much skill to know just when to release the shutter so that the essential moment of some complex action is recorded. The photographer of motion films must be sufficiently alert to patterns of behavior to realize when important interaction is starting; if he is not, all he records is the ends of important sequences.

Remarkably few of the early generalizations concerning primate behavior, even on a very broad level, have survived intact after the scrutiny of the last few years. Some of these generalizations emerge and are discussed in the following chapters of the book, but all present generalizations should be regarded as suggestive and tenative. Each awaits further data for support or revision. As is often true, the more we know the more difficult it is to make simple statements that do not need to be qualified with each new study. Unlike the fossil record of primate evolution, in which many of the pieces are lost forever, the record of the living primates may be filled in by the necessary fieldwork. This, of course, is time-consuming and often difficult, and man is destroying some of the evidence (the animals and their habitats) as fast if not faster than it can be studied. We proceed with the awareness that if only one species in a large genus is studied it may turn out to be the most unusual one, or that the troop observed may be the only one living in a particular kind of habitat. If more complete investigations of the species were made, a very different picture might well emerge.

Every field study goes through an initial period of planning and selection. The kind of primate most appropriate to the interests of the investigator must be carefully selected, and then a search for the most appropriate study location must be made. Such a location should provide observation conditions that allow the observer to see enough specific kinds of behavior to gather the information needed. Sometimes clear visibility of the animals must be sacrificed in the interest of gaining more special ecological information. Perhaps the investigator's concern is with the relationships among several species that live in one area. It is not always possible to find a location that permits easy visibility of individual animals and at the same time offers examples of the additional types of interactions the research demands. Every field study, then, must be undertaken in the place optimal for the research problems of the observer. If, for example, a study is planned of the effects of predation on social structure, then it is necessary to search for a study site that includes a full complement of predators in sufficient numbers to resemble the situation before man entered the area (and decreased predator populations by hunting.) It also means that the number of tourists should be as low as possible since their presence greatly affects hunting success and the general behavior of large cats. A frequently visited game park with reduced numbers of cats will not suffice. Here again the selection of study location must be coordinated with the research problems of the investigator.

It has not always been possible to locate studies where optimal data collection is possible, and in assessing reports of species behavior, limitations of observation conditions and habitat characteristics must be taken into account. This introduces many difficulties in comparative studies when scholars who did not participate in the original research projects must rely on the literature for information. Authors of field reports have seldom discussed the disadvantages of their study locations and the effects these disadvantages may have had on the behavior observed.

Added to these problems are further difficulties in the way various researchers describe and label the behavior they see. There is remarkably little agreement among investigators on the vocabulary of behavior units; not only do different observers call what may be the same items of behavior by different names, but in addition each worker makes his own interpretation of the meaning and emphasis of each recorded unit. What one observer calls drastic another might refer to as little out of the ordinary; a crisis to one might be a routine event to another. With these problems in mind the reader must weigh each article he reads.

Although Chapter 6, "Field Studies of Old World Monkeys and Apes," was written five years ago, many of the problems outlined still pertain. Some of the questions raised at that time are much closer to being answered, but for the most part the last five years of research has enable us to rephrase rather than to resolve most of the issues that were of concern. Emphasis on the relationship of field and laboratory research continues to bear repeating since feedback between the two makes each more efficient in behavior analysis.

Hall raised issues in Chapter 7, "Social Learning in Monkeys," that are central to most investigations of primates. His emphasis is on social learning—on what and how the primate learns and on how critical it is to design research tasks that are appropriate for each species. Further, he emphasizes the necessity of undertaking investigations under the conditions required for normal learning, since the deprivations inherent in learning in anything but a natural context manifest themselves in what is learned and in general behavior. Hall states that "learning is a product of evolution and is therefore an integral feature of a species' adaptations, and not something that can be studied independent of them." The social group is the context of primate life from birth to death, and studies of learning must begin there. Hall illustrates many of his points with material taken from his field study of the patas monkey. His uncompromising point of view has raised controversy, but it is readily documented by the results of his own research and that of many others.

In Chapter 8, "Aggressive Behavior in Old World Monkeys and Apes," Washburn and Hamburg survey primate aggression as it is learned, rewarded, and manifested in a social system. The setting for aggression, as for learning, is the small social groupings of animals that know each other well. Agonistic behavior is a daily part of life, and the patterns and level of aggression in a group are directly related to the composition of the group and the stresses acting upon it. Competition and aggression among social groups are viewed as necessary for dispersal and control of local population density and as related to availability of food, water, and other commodities of the environment. The biological basis of aggression is discussed as it relates to the patterns of behavior characteristic for different species and for males and females.

Every field study has concerned itself in part at least with how animals communicate, both visually and gesturally. A communication act normally includes a sequence of motions, postures, gestures, and sounds, and it is rare that any single item carries the entire message. A great deal of information is conveyed among animals by the way an animal sits or how it moves. Monkeys and apes communicate with one another constantly, even when they are not making stereotyped, easily recognizable gestures or vocalizations, much as humans pass information among themselves. In both man and monkey the sender may not be aware that he is making a statement about his emotional state and in fact, the receiver may process the information with a similar lack of awareness. Although he may respond to them, man tends not to be conscious of many of the cues that signal emotional states, such as a slight shift of glance, the way the head is held, visual focus, or a refusal to look at something or someone, or even pupilary reactions. Often, a subtle nuance in tone of voice or a pause or hesitation will indicate to the listener that the person speaking means something other than the words imply; the spoken message may easily be modified or qualified to contradict the dictionary meaning of words.

There are many modes of communication among nonhuman primates. Facial gestures are more important in some species than in others, and some features of the face are used for emphasis in signaling; for example, the white eyelid patches that adorn many species can be made to flash when the eyebrows are raised rapidly. Posture or stance can convey information about the animal, from its abject submission to its aggressive certainty, and tail position indicates status and intention among some species. In general, sounds convey much less information than do postures and gestures, but sounds do serve to call attention to the sender, and then the observing animals can see its focus of attention or note further signals. The vervet monkey has developed alarm calls specific to the type of predator, signaling whether the predator is avian or terrestrial, but this is unusual. Sometimes anatomical features function to emphasize the sounds a primate produces, as for example, the hyoid bone development in howling monkeys of South America, which increases the resonance of their very loud calls enabling them to carry over long distances.

A nonhuman primate communicates its emotional state as it is at the moment of signaling; there is no reference to the past or to the future, and the number of items that can be communicated are vastly fewer than those within the most limited human ability. Sight seems to be far more important than hearing in determining what is communicated, although more structural analysis of sounds from different species may show that the animals are qualifying messages by varying qualities of the sound far more than we are aware of with our unaided ears. The physical structure of sounds has been analyzed and sounds, apparently identical to the human ear, have been found to be structured in different ways. Some fieldworkers have noted that they could identify individual animals by their voices, and it is certain that the animals can do at least that. Moods can also be conveyed to humans by some individual animals, when the listener is very familiar with them.

The development of human language has greatly enlarged the communication ability of the human primate (Lancaster 1967), and one of the most important features of human language is the ability to name objects. This kind of information has been added to the emotional content of communication in man and is made possible by the development of areas and connections within the brain that are for the most part characteristic of man alone and not part of the nonhuman primate brain.

Shirek-Ellefson's Chapter 9, "Social Communication in Some Old World Monkeys and Gibbons," surveys major aspects of primate communication as they occur in social groupings. She draws upon her field study of *Macaca irus* to illustrate many subtleties of communication. This chapter, like preceding ones, emphasizes the social life of primates and concludes that the meaning of acts of sharing information can only be understood in this context. Most communication is among animals that live together closely every day of their lives. Even intergroup communication is usually among known troops, although the resounding long-distance calls given by some species, such as the gibbon, may be heard by distant troops.

In "The Development of Primate Social Relationships and Motor Skills Through Play," Chapter 10, Dolhinow and Bishop discuss some major features of one of the most important activities of all young primates. Remarkably little attention has been paid to this crucial activity, which takes up so much of a young primate's time. Not all behaviors are learned or practiced in play, but the list of important activities that appear early, are practiced again and again, and gradually achieve a high level of performance, includes many skills that are vitally important in the individual's later life. The young primate learns gestures and interaction patterns that are species-specific, as well as those general to most kinds of primates. Laboratory studies have demonstrated that depriving a monkey of play with peers can bring about severe and permanent damage in its future social life, and field studies suggest the range of skills that are affected.

Chapter 11, "Evolution of Primate Behavior" by Washburn and Harding, concerns the relationship between biology and behavior in the course of primate evolution. It outlines the major changes that have occured in the evolution of our species and emphasizes the essential feedback relationships between structures and behaviors that have made the human way of life possible.

"Primate Patterns," Chapter 12, summarizes briefly some of the major themes of primate social behavior. At this point in time, variations are more striking than are common features that apply to all species. The more we know, the more we qualify, although we are probably more overwhelmed with the details of differences now than we will be in the future.

chapter six
FIELD STUDIES OF OLD WORLD MONKEYS AND APES[1]

S. L. Washburn, Phyllis C. Jay, and Jane B. Lancaster

For many years there has been interest in the evolutionary roots of human behavior, and discussions of human evolution frequently include theories on the origin of human customs. In view of the old and widespread interest in the behavior of our nearest relatives, it is surprising how little systematic information was collected until very recently. At the time (1929) Yerkes and Yerkes collected data for their book on the great apes (Yerkes and Yerkes, 1929) no one had devoted even one continuous month to the systematic study of the behavior of an undisturbed, free-ranging nonhuman primate. Apparently scientists believed that the behavior of monkeys and apes was so stereotyped and simple that travelers' tales or the casual observations of hunters formed a reliable basis for scientific conclusions and social theorizing. As a part of the program of the Yale Laboratories of Comparative Psychology, Yerkes encouraged a series of field studies of the chimpanzee (Nissen 1931), the mountain gorilla (Bingham 1932), and the howling monkey (Carpenter 1934). These first studies proved so difficult that Yerkes could write, in the introduction to Carpenter's study, "His is the first reasonably reliable working analysis of the constitution of social groups in the infra-human primates, and of the relations between the sexes and between mature and immature individuals for monkey or ape" (Carpenter 1934:4). Zuckerman (1932, 1933), quite independently, had realized the importance of field observations and had combined some field work with physiology and the older literature to produce two very influential volumes. From this beginning, only Carpenter continued to make field studies of behavior, and his study of the gibbon (Carpenter 1940) is the first successful study of the naturalistic behavior of a member of the family Pongidae. Hooton (1942) summarized what was then known about the primates,

[1] Supported by USPHS grant MH 08623. We thank Anne Brower, John Ellefson and Lewis Klein for reading the preliminary version of the manuscript and for helpful criticisms.

particularly stressing the importance of behavior and the work of Carpenter and Zuckerman.

The war stopped field work, and no major studies were undertaken for some 15 years. Then, in the 1950s, investigators in Japan, England, France, Switzerland, and the United States independently started studies on the behavior of a wide variety of free-ranging primates. For the history of science it would be interesting to examine the reasons for this burst of parallel activity. Field studies were undertaken at more or less the same time, and publications start in the late 1950s and accelerate rapidly in the 1960s. This trend is still continuing and is well shown by the pattern of frequency of citations in a recent review by Hall (1965b). The review cites the papers of Bingham, Carpenter, Köhler (1925), Nissen, Yerkes, and Zuckerman, but there are no references to additional field studies in the period 1941–1951, and most of the references are to papers appearing in 1960 or later.

The increased interest in primates, and particularly in the behavior of free-ranging primates, has given rise to several symposiums, and results of the new studies have been published almost as soon as they have been completed. Data from the recent field studies are included in volumes edited by Buettner-Janusch (1962, 1963–1964), Washburn (1963a), Napier and Barnicot (1963a) and, especially, DeVore (1965a). The volume edited by DeVore is devoted entirely to recent field studies and their evaluation. It includes accounts of the behavior of five kinds of monkeys, of chimpanzees, and of gorillas. Each chapter is by the person who did the field work, and in addition there are eight general chapters. Two new journals also are devoted to primates: *Primates*, published by the Japan Monkey Centre, and *Folia Primatologica*, which has completed volume 3. Carpenter's field studies and general papers have been reprinted so that they are now easily available (Carpenter 1964). Southwick (1963) has published a collection of readings in primate social behavior, and Eimerl and DeVore (1965) contributed a volume on the primates to the Life Nature Library. Field studies have recently been reviewed by Jay (1965b), and proceedings of a symposium have been organized and edited by Altmann (1967a). This abundance of published material makes it hard to believe that only a few years ago a course on primate social behavior was difficult to teach because of the lack of easily available. suitable reading material.

THE NEW FIELD STUDIES

Obviously, with so much new data a complete review is impossible, and readers wishing more information and bibliography are referred to Jay (1965b) and to the symposiums previously noted. Here we wish to direct attention to the nature of the recent field studies and to a few of their major contributions. Perhaps their greatest contribution is a demonstration that close, accurate observation for hundreds of hours is possible. Prior to Schaller's field work, reported in 1963 (Schaller 1963), it was by no means clear that this kind of observation of gorillas would be possible; previous investigators had conducted very fragmentary observations, and Emlen and Schaller deserve great credit for the planning and execution of their study. A field study of the chimpanzee that seemed adequate in the 1930s now seems totally inadequate, when compared to Goodall's results (Goodall 1965). Today a field study is planned to yield something of the order of 1000 hours of observations, and the observer is expected to be close to the animals and to recognize individuals. A few years ago observations of this length and quality were thought unnecessary, if not impossible.

The importance of studies in which groups are visited repeatedly and animals are recognized individually may be illustrated by the problems they make it possible to study. For example, during one season of the year chimpanzees "fish" for termites by breaking off sticks or stiff grasses and sticking the prepared implement into a termite hole (Goodall 1964), and this whole complex of nest examination, tool preparation, and fishing is learned by the young chimpanzee. It can be seen at only one time of the year and can be appreciated only by an observer whose presence no longer disturbs the animals. Habituation to the observer is a slow and difficult process. Goodall (1965) reports that after 8 months of observations she could approach to no closer than 50 meters of the chimpanzees and then only when they were in thick cover or up a tree; by 14 months she was able to get within 10 to 15 meters of them. The problem of tool use in nonhuman primates has been reviewed by Hall (1963c), but the essential point here is that the amount of throwing and object manipulation in the monkeys (Cercopithecidae) was greatly exaggerated in travelers' tales, which were uncritically accepted, and it took years of observation in a favorable locality to reveal the complexity of this kind of behavior in the chimpanzee (Lancaster 1965).

PREDATION

Another example of the value of continued observations is in the study of deliberate hunting by baboons. In three seasons of field work and more than 1500 hours of observation DeVore had seen baboons catch and eat small mammals, but apparently almost by chance, when the baboon virtually stepped on something like a newborn antelope and then killed it (DeVore and Hall 1965; DeVore and Washburn 1961). But in 1965 DeVore saw repeated incidents of baboons surrounding, hunting, and killing small mammals (DeVore, personal communication).

The whole matter of predation on primates has been difficult to study. Rare events, such as an attack by an eagle (Haddow 1952–1953) may be very important in the survival of primates, but such attacks are seldom observed, because the presence of the human observer disturbs either the predator or the prey. We think that the present de-emphasis of the importance of predation on primates arises from these difficulties of observation and from the fact that even today most studies of free-ranging primates are made in areas where predators have been reduced or eliminated by man. Most predators are active at night, and there is still no adequate study of the nocturnal behavior of any monkey or ape. Predation probably can best be measured by studying the predators rather than the prey.

Recognition of individual animals is necessary for the study of many problems, from the first stages of the analysis of a social system to observations of social continuity or constancy of group membership; such observations are exceedingly difficult under most field conditions. For example, understanding of the dominance system implies repeated recognition of a number of animals under sufficiently various conditions so that the patterns of interaction become clear. Again, to be sure that a group has lost or gained a member, the observer must know the whole composition of the group.

Long-continued observations have proved to be important in many unexpected ways. For example, rhesus monkeys have been observed in several of their many very different habitats, and it has been found that young rhesus play more in cities than in some kinds of forest and play in the forest more at some seasons than at others. These differences are due in part to the amount of time which must be spent in getting food; the same forest troop may play more when fruits are available and hunger may be rapidly satisfied than at times of the year when the diet is composed of tiny seeds which take a long time to pick. Extracting the small seeds of sheesham pods during the months when rhesus troops spend most of their time in the sheesham trees takes many hours of the day (Jay and Lindburg 1965). What might easily have been described in a short-term study as a species-specific difference of considerable magnitude turns out to be the result of seasonal and local variations in food source. It is essential to sample behavior in several habitats to gain an understanding of the flexibility of the built-in behavior patterns of a species, flexibility which precludes the need for development of new forms of genetically determined behavior to cope successfully with different habitats.

The long-term study in which many groups of a species are observed in different, contrasting localities, and in which at least some groups are known so well that most of the individuals can be recognized, will correct many false notions and will make valid generalizations possible. Although so far there have been only a few major investigations of this sort, some important generalizations seem possible.

ENVIRONMENT AND SOCIAL BEHAVIOR

Nowhere is the extent to which the behavior of a species is adaptable and responsive to local conditions more apparent than among groups of rhesus living in India. Rhesus occur naturally in such diverse environments as cities, villages, roadsides, cultivated fields, and many types of forest ranging to altitudes of over 2400 meters. Contact with man varies in these habitats from constant and close to rare and incidental.

Where rhesus groups are subjected to pressures of trapping, harassment, and high incidence of infectious disease, groups are tense and aggression is high. These pressures are found in areas where there is most contact and interaction with man, such as in cities and at places of pilgrimage. The animals are in generally poor physical condition, and numerous old and new wounds are evidence of a high rate of intragroup fighting. Tension among groops occupying adjacent areas of land is similarly high where there is insufficient space for normal movement and behavior, and where there may be intense competition for a limited supply of food and water. This is in sharp contrast to those groups living away from man where normal spacing among groups can be effected by the means evolved by the species. In the latter environments, such as forests, the rhesus are in excellent physical condition and what aggressive behavior occurs functions to maintain stable social groups and relationships among the members of the group; wounds are substantially fewer, and disease appears to be rare.

There has been considerable controversy in discussions of the relationships among social groups of the same species as to whether or not the geographical area occupied by a group should be called a territory or a home range. The point we wish to emphasize is that, within one species, populations living in different habitats may act quite differently toward neighboring groups. Populations may be capable of a wide variety of behavior patterns ranging from exclusive occupation of an area which may be defended against neighboring groups to a peaceful coexistence with conspecifics in which wide overlap in home ranges is tolerated. Because local populations of a species may maintain their ranges in different ways it is necessary to investigate all variations in group spacing in diverse habitats before attempting to describe characteristic behavior patterns for any species.

Not unexpectedly, population and group composition reflect these differences in habitat and stress. Groups living on the Gangetic plains, where trapping, harassment, and disease are important factors, are smaller, and the proportion of young members is also significantly smaller (Jay and Lindburg 1965; Southwick, Beg, and Siddiqi 1961a, 1961b). The long-term effects of pressures on different rhesus populations in northern and central India are now being investigated by a team of anthropologists of the National Center for Primate Biology.

A city presents a very different set of challenges to a rhesus group than does a forest. Often there are no trees to sleep in; living space must be shared with man and his domestic animals. Food is not available in the form common to other habitats, and monkeys may have to depend on their skill in stealing food from man. Often the food has been prepared by man for his own consumption, or it consists of fruits and vegetables pilfered from houses, shops, and streets. Garbage is picked through and edible portions are consumed. It is essential that the monkeys learn to differentiate between those humans who represent a real threat to their safety and those who are safe to approach. They must react quickly and learn to manipulate doors, gates, and other elements of the physical environment unique to their urban habitat. This is a tremendously different setting from that in which most rhesus live. City rhesus are more manipulative, more active, and often more aggressive than are forest rhesus. Clearly, the same species develops quite different learned habits in different environments.

ANNUAL REPRODUCTIVE CYCLE

The belief, which has been widely maintained, that there is no breeding season in monkeys and apes gave rise to the theory that the persistence throughout the year of groups, or highly organized troops, was due to continuous sexual attraction. The evidence for a breeding season has been reviewed by Lancaster and Lee (1965) who found that in many species of monkeys there is a well-marked breeding season. For example, Mizuhara has presented data (personal communication) on 545 births of Japanese macaques of Takasakiyama. There were on the average approximately 90 births per year over six consecutive years. The average length of the birth season was 125 days, but it varied from 95 to 176 days. The majority of the births occurred in June and July. Copulations were most frequent in November to March and were not observed during the birth season, and in spite of this the highly organized group continues as a social unit throughout the year.

The birth season has been studied in other groups of Japanese macaques, and in general the situation is similar. There is no doubt that both mating and birth seasons are highly restricted in the Japanese macaque. The birth season is spring and summer, but its onset and duration vary considerably. If observations were limited and combined for the whole species, as they were in early studies, the birth season would appear to be much longer than in fact it is for an individual group, and it is the events within the local group, not averages of events for the species, that bear upon the role of sexual attraction in holding primate society together.

Under very different climatic conditions, in India, rhesus macaques also have a birth season, but copulations were observed in all months of the year, although probably not with equal frequency (Southwick, Beg, and Siddiqi 1961a, 1961b). Among rhesus on a small island off Puerto Rico births occur from January to June, and copulations are restricted to July–January (Koford 1965). These data confirm the point that a birth season will be more sharply defined in a local group than in a species as a whole. There is a mating season among rhesus introduced on the island, but only a peak of mating in the same species in their native India (Southwick, Beg, and Siddiqi 1961a, 1961b). It is clear that survey data drawn from many groups over a wide area must be used with caution when the aim is to interpret the behavior of a single group. Since the birth season is an adaptation to local conditions, there is no reason to expect it to be the same over the entire geographical distribution of a species, and under laboratory conditions rhesus macaques breed throughout the year.

No data comparable to those for the macaques exist for other primates, and, since accurate determination of mating and birth seasons requires that reasonable numbers of animals be observed in all months of the year and that groups be observed in different localities, really adequate data exist for only the Japanese macaque. However, Lancaster and Lee were able to assemble data on 14 species of monkeys and apes. They found that probably the most common situation is a birth peak, a time of year at which births tend to be concentrated, rather than sharply limited mating and birth seasons. This is highly adaptive for widely distributed species, for it allows the majority of births to occur at the optimum time for each locality while maintaining a widely variable basic pattern. The birth season may be a more effective adaptation to extreme climatic conditions. There may be a birth peak in the chimpanzee (Goodall 1965), and there may be none in the mountain gorilla (Schaller 1963), but, since we have no more data than are necessary to clarify the reproductive pattern in a single species of macaque, we can conclude only that, while birth seasons are not present in either gorillas or chimpanzees, a peak is possible in chimpanzees, at least for those living near Lake Tanganyika.

Prior to the recent investigations there was a great deal of information on primate reproduction, and yet as late as 1960 it was still possible to maintain that there were no breeding seasons in primates and that this was the basis of primate society. Until recently the question of seasonality was raised without reference to a birth season as distinguished from a birth peak, or to a limited mating season as distinguished from matings throughout the year with a high frequency in a particular period.

FREQUENCY OF MATING

Obviously many more studies are needed, and one of the intriguing problems is the role of potency. Not only does the frequency of mating vary through the year, but also there appear to be enormous differences in potency between species that are reproducing at a normal rate. In nearly 500 hours of observation of gorillas, Schaller (1963) saw only two matings, fewer than might be seen in a troop of baboons in almost any single morning. The redtail monkey (*Cercopithecus ascanius*) mates rarely (Haddow 1952, 1953), but the closely related vervet (*Cercopithecus aethiops*) does so frequently. To a considerable extent the observed differences are correlated with structure (Schultz 1938), such as size of testes, and all these species seem to be reproducing at an adequate and normal rate. There is no evidence that langurs (*Presbytis entellus*) are less successful breeders than rhesus, but the langurs copulate less frequently (Jay 1965a).

Now that more adequate data are becoming available, the social functions of sexual behavior should be reinvestigated. The dismissal of the theory that sexual attraction is *the* basis of primate society should open the way for a more careful study of the multiple functions of sexual behavior. The great differences among the primate species should provide data to prove or disprove new theories. In passing it might be noted that the human mating system without estrous cycles in the female and without marked seasonal variations is unique.

SYSTEMS OF MATING

Mating systems, like the presence or absence of seasonality in breeding and the frequency of copulation, are extremely variable in monkeys and apes. Eventually the relation of these variations to species adaptations will be understandable; at present it is most important to note that monkeys do not necessarily live either in harems or in promiscuous hordes as was once assumed. Restrictive mating patterns such as the stable and exclusive pair-bond formed between adult gibbons (Carpenter 1940) and the harem system of the Hamadryas baboon (Kummer and Kurt 1963) are comparatively rare. The most common mating pattern of monkeys and apes is promiscuity more or less influenced by dominance relationships. In species in which dominance relations are not constantly at issue, such as langurs (Jay 1965a), chimpanzees (Goodall 1965), or bonnet macaques (Simonds 1965), matings appear to be relatively promiscuous and are often based on the personal inclination of the estrous female. When dominance relationships are constantly at issue, as in baboons (Hall and DeVore 1965), Japanese macaques (Tokuda 1961–1962), and rhesus macaques (Conaway and Koford 1964; Southwick, Beg, and Siddiqi 1965), sex often becomes one of the prerogatives of dominant rank. In such species dominant males tend to do a larger share of the mating than do more subordinate animals, but it is only in unusual situations that subordinate animals are barred from the mating system altogether. Mating systems probably support the general adaptation of the species to its environment. In most baboons and macaques the tendency for a few males to do much of the mating may be partly a by-product of natural selection for a hierarchy of adult males which dominates the troop so that in a dangerous terrestrial habitat external dangers will be met in an orderly way. Selection is not only for a male which can impregnate many females but it may also have favored a dominance-oriented social organization in which sexual activity has become one of the expressions of that dominance.

DOMINANCE RELATIONSHIPS

Long-term field studies of monkeys and apes in their natural habitats have emphasized that social relationships within a group are patterned and organized in very complex ways. There is no single "monkey pattern" or "ape pattern"; rather, there is great variability, both among different species and among different populations of the same species, in the organization and expression of social relationships. A difference in the relative dominance of individuals is one of the most common modes of social organization in monkey and ape societies. Dominance is not synonymous with aggression, and the way dominance is expressed varies greatly between species. In the gorilla, for example, dominance is most often expressed by extremely attenuated gestures and signals (Schaller 1963); a gentle nudge from the dominant male is more than enough to elicit a submissive response from a subordinate, whereas, in baboons, chases, fights, and biting can be daily occurrences (Hall and DeVore 1965). In many primates there is a tendency for the major age-sex classes to be ranked in a dominance order; for example, in baboons, macaques, and gorillas, adult males as a class are usually dominant over adult females, and females are dominant over young. This may not always be true, for in several species of macaques some females may outrank some adult males (Simonds 1965), although groups dominated by a female (such as the Minoo-B troop of Japanese macaques) are extremely rare (Yamada 1963). Dominance relationships may be quite unstructured, as in the chimpanzee (Goodall 1965), where dominance is expressed in interactions between individuals but where these relationships are not organized into any sort of hierarchy. A much more common situation is one in which dominance relations, among males at least, are organized into linear hierarchies that are quite stable over time, as in baboons (Hall and DeVore 1965), langurs (Jay 1965a; Ripley 1967), and macaques (Altmann 1962; Itani, Tokuda, Furuya, Kano, and Shin 1963). Sometimes these dominance hierarchies are complicated by alliances among several males who back each other up very effectively (Hall and DeVore 1965) or even by an alliance between a male and a female (Simonds 1965). Although dominance varies widely among monkeys and apes both in its form and function, it is certainly one of the most important axes of social organization to be found in primate societies.

GENEALOGICAL RELATIONSHIPS

Recognition of individual animals and repeated studies of the same groups have opened the way to the appreciation of other long-continuing social relationships in monkeys and apes which cannot be interpreted in terms of dominance alone. Long-term studies of free-ranging animals have been made on only two species of nonhuman primates, Japanese macaques, which have been studied since 1950 by members of the Japan Monkey Centre, and Indian rhesus macaques living free on Cayo Santiago, Puerto Rico, the island colony established by Carpenter in 1938. In these studies, when the genealogy of the animals has been known, it has been obvious that genetic relationships play a major role in determining the course and nature of social interactions (Yamada 1963; Imanishi 1960; Koford 1963b; Sade 1965). It becomes clear that bonds between mother and infant may persist into adult life to form a nucleus from which many other social bonds ramify. When the genealogy of individual animals is known, members of commonly observed subgroupings, such as a cluster of four or five animals grooming or resting together, are likely to be uterine kin. For example, members of a subgroup composed of several adult animals, both male and female, as well as juveniles and infants, may all be offspring of the same female (Sade 1965). These relations continue to be very important in adult life not only in relaxed affectional relationships but also in dominance interactions. Sade saw a female rhesus monkey divert the attack of a dominant male from her adult son and saw another adult female protect her juvenile half-sisters (paternity is not determinable in most monkey societies). There is a very high frequency of grooming between related animals, and many animals never seek grooming partners outside of their own genealogies.

It should be stressed that there is no information leading us to believe that these animals are either recognizing genetic relationships or responding to any sort of abstract concept of family. Rather these social relationships are determined by the necessarily close association of mother with newborn infant, which is extended through time and generations and which ramifies into close associations among siblings. We believe that this pattern of enduring social relations between a mother and her offspring will be found in other species of primates. Because of their dramatic character, the importance of dominance and aggression has been greatly exaggerated compared to that of continuing, positive, affectional relations between related animals as expressed by their sitting or feeding together, touching, and grooming. Much of this behavior can be observed easily in the field, but the extent to which it is in fact an expression of social genealogies has been demonstrated only in the studies cited above.

Positive, affectional relations are not limited to relatives. Male Japanese macaques may take care of young by forming special protective relationships with particular infants (Itani 1959), but whether these males have any special relationship to the infants as either father or brother is uncertain, and the mating system is such that paternity cannot be known either to the observer or to the monkeys. MacRoberts (1965) has recorded a very high frequency of care of infants by males in the Gibraltar macaque. In addition, he has demonstrated that these positive protective relations are very beneficial to the juvenile. Two juveniles which had no such close relationship were forced to be peripheral, were at a great disadvantage in feeding, and were groomed much less than other juveniles in the group.

The status of the adult can be conferred on closely associated young (frequently an offspring when the adult is female), and for this reason the young of dominant animals are more likely to be dominant. This inheritance of rank has been discussed by Imanishi (1960) for the Japanese macaque and by Koford (1963b) for the rhesus. Sons of very dominant females seem to have a great advantage over other males both because their mothers are able to back them up successfully in social interactions and because they stay with their mothers near the other dominant animals at the center of the group. They may never go through the stage of being socially and physically peripheral to the group which is typical for young males of these species. A male cannot simply "inherit" high rank; he must also win this position through his own abilities, but his chances of so doing are greatly increased if he has had these early experiences of associating with and being supported by very dominant animals.

There could hardly be a greater contrast than that between the emerging picture of an orderly society, based heavily on affectionate or cooperative social actions and structured by stable dominance relationships, and the old notion of an unruly horde of monkeys dominated by a tyrant. The 19th-century social evolutionists attributed less order to the societies of primitive man than is now known to exist in the societies of monkeys and apes living today.

COMMUNICATION

Research on the communication systems of monkeys and apes through 1962 has been most ably summarized and interpreted by Marler (1965). Most of the data represent work by field observers who were primarily interested in social structure, and the signals, and their meanings, used to implement and facilitate social interactions were more or less taken for granted. Only in the last year or so have communication systems themselves been the object of careful study and analysis (see, for example, Altmann 1967a). Marler has emphasized both the extraordinary complexity of the communication systems of primates and the heavy dependence of these systems on composite signals (Marler 1965). Most frequently it is not a single signal that passes between two animals but a signal complex composed of auditory, visual, tactile, and, more rarely, olfactory signals.

Communication in some monkey species is based on a system of intergrading signals, whereas in others much more use is made of highly discrete signals. For example, most vervet sounds (described by Struhsaker 1967a) are of the discrete type, there being some 36 different sounds that are comparatively distinct both to the human ear and when analyzed by a sound spectrograph. In contrast, Rowell and Hinde have analyzed the sounds of the rhesus monkey (Rowell and Hinde 1962; Rowell 1962) and found that of 13 harsh noises, 9 belonged to a single intergrading subsystem expressing agonistic emotions.

As more and more study is done on primates it will probably be shown that their communication systems tend to be of mixed form in that both graded and discrete signals are used depending on the relative efficiency of one or the other form in serving a specific function. In concert this use of both discrete and intergrading signals and of composites from several sensory modes produces a rich potential for the expression of very slight but significant changes in the intensity and nature of mood in the signaling animal. Marler has emphasized (Marler 1965) that, except for calls warning of danger, the communication system is little applied to events outside the group. Communication systems in monkeys and apes are highly evolved in their capacity to express motivation of individuals and to facilitate social relationships. Without this ability to express mood, monkeys and apes would not be able to engage in the subtle and complicated social interactions that are a major feature of their adaptation.

SOCIAL LEARNING

Harlow and Harlow's experiments (Harlow and Harlow 1965) show the importance of learning in the development of social life; however, monkeys and apes are so constituted that, except in the laboratory, social learning is inevitable. They adapt by their social life, and the group provides the context of affection, protection, and stability in which learning occurs. No one factor can explain the importance of social behavior, because society is a major adaptive mechanism with many functions, but one of the most important of these functions is the provision of a rich and protected social context in which young mature. Field observations, although mainly observations of the results of learning rather than of the process itself, provide necessary clues as to the nature of the integration of relevant developmental and social factors. These factors can then be estimated and defined for subsequent intensive controlled research in a laboratory or colony.

It has become clear that, although learning has great importance in the normal development of nearly all phases of primate behavior, it is not a generalized ability; animals are able to learn some things with great ease and other things only with the greatest difficulty. Learning is part of the adaptive pattern of a species and can be understood only when it is seen as the process of acquiring skills and attitudes that are of evolutionary significance to a species when living in the environment to which it is adapted.

There are important biological limitations which vary from species to species and which do not reflect differences in intelligence so much as differences in specializations. For example, Goodall (1964) has observed young chimpanzees learning to fish for termites both by their observation of older chimpanzees and by practice. It takes time for the chimpanzee to become proficient with these tools, and many mistakes are made. Chimpanzees are not the only primates that like termites, and Goodall has observed baboons sitting near chimpanzees watching and waiting while the latter are getting termites. The baboons are just as eager as the chimpanzees to eat termites but are unable to learn how to fish for termites for themselves.

It is likely that there are important variables among groups of a single species that make it possible for the acquisition of new patterns of behavior or the expression of basic learned species patterns to vary from group to group and from one habitat to another. For example, the nature of the integration and operation of a social unit vary in the extent to which it depends on the personalities of individuals in the group – this is another dimension of our understanding of how social behavior may affect species survival. Particularly aggressive adult males can make the behavior of their groups relative to that of adjacent groups with less assertive males substantially different. For example, a group with very aggressive males can control a larger geographic area than is occupied by a group with much less aggressive males. The tenor of life within a group may be tenser or more relaxed depending on personalities of adults in the group.

Imprinting has traditionally been distinguished from other learning processes by the fact that in imprinting the young animal will learn to follow, to be social (Collias 1962), without an external or immediate reward (Sluckin 1965). However, among monkeys and apes, simply being with other animals is a reward, and learning is reinforced by the affectional, attentive, supportive social context of the group (Hall 1963a). Butler was the first to use the sight of another monkey as a reward in psychological experiments (Butler 1954). The field worker sees sick and practically disabled animals making great efforts to stay with their group. Among ground-living forms, animals that have lost or broken limbs or are so sick that they collapse as soon as the group stops moving, all walk along as the troop moves. Instances of wounded rhesus macaques' moving into langur groups after the rhesus have left or been forced out of their own group have been recorded. Clearly, it is essential for the young monkey or ape to mature in a social setting in which it learns appropriate skills and relationships during early years and in which it continues to learn during adulthood. "Where the individual primate is, in temporary isolation, learning a task without reference to any other member of its species, the learning is not normal" (Hall 1968).

FUTURE PRIMATE STUDIES

At present many long-term studies are in process and major films are being edited (Goodall on chimpanzee and DeVore on baboon). There will be about twice as many major accounts available in 2 years as there are now.

Since it is now clear that detailed descriptive studies of undisturbed free-ranging primates can be made, and since available data show that there are substantial differences in the behavior of the different species, more species should be investigated. So far studies have concentrated for the most part on the larger ground-living forms which are easier to study. There is no study of *Cercocebus,* little on *Colobus* (Ullrich 1961), and nothing on the numerous langurs *(Presbytis)* of southeast Asia. New World monkeys have been investigated very little, and there are numerous genera that have not been the subjects of a major field study. Also, since local variation is important, forms such as the chimpanzee and gorilla should be studied in more and contrasting localities.

Once the general characteristics of the behaviors of several species are known, then interest can shift to topics such as detailed ecology, birth, infant behavior, peer groups, affectionate behaviors, sex, or dominance, to mention only a few. The behavior of a whole species is a large problem, and description has to be at a very general level when the goal is a first general statement. A problem-oriented study permits choice of species and elaboration of techniques. A further advantage of the problem-oriented approach is that it allows the close coordination of the field work with experimental work in the laboratory. Fortunately, no division has developed between those doing the field work and those involved in the experimental analysis of behavior. Many scientists have done both controlled experiments and field studies. The interplay between naturalistic observation and controlled experiment is the essential key to the understanding of behavior (Mason 1965). The character of the natural adaptation of the species and the dimensions of the society can be determined only in the field. Many topics, such as geographic range, food, predation, group size, aggression, and the like, can be seen only under field conditions. But the mechanisms of the observed behavior can be determined only in the laboratory, and this is the more complicated task. The relation of a field study to scientific understanding is like the relation of the observation that a man walks or runs to the whole analysis of locomotion. The field worker lists what the animals eat, but this gives no understanding of nutrition. The kinds of interactions may be charted in the field, but their interpretation requires the laboratory. Field workers saw hours devoted to play, but it was Harlow's experiments that showed how essential this activity was to the development of behavior. As the field studies develop it is to be hoped that they will maintain a close relation to controlled experiment. It is most fortunate that the present studies are being carried on by anthropologists, psychologists, and zoologists. An understanding of behavior is most likely to come from the bringing together of the methods and interests of many sciences, and we hope that the field studies remain a part of general behavioral science and do not become independent as workers and problems become more and more numerous.

Even now, in their preliminary state, the field studies can offer some conclusions that might be pondered by students in the multiplicity of departments now dividing up the study of human behavior. Behavior is profoundly influenced by the biology of the species, and problems of perception, emotion, aggression, and many others cannot be divorced from the biology of the actors in the social system. Early learning is important, and an understanding of the preschool years is essential to an understanding of behavior. Play is tremendously important, and a species that wastes the emotions and energies of its young by divorcing play from education has forfeited its evolutionary heritage — the biological motivation of learning. Social behavior is relatively simple compared to the biological mechanisms that make the behavior possible. Ultimately a science of human behavior must include both biological and social factors, and there is no more reason to separate the study of human behavior into many compartments than there would be to separate the field studies from the intellectual enrichment coming from the laboratory.

chapter seven

SOCIAL LEARNING IN MONKEYS

K. R. L. Hall

If the psychologist is to make a general contribution to the study of primate social behavior distinguishable from that of zoologically or anthropologically trained investigators, it should be in a specialty of his field, namely, the study of learning.

The principles of learning, as we know them from the standard texts, still have a somewhat narrow species and situational reference; it is not very clear, therefore, how naturalistic social studies and experimental studies could be integrated. Many psychologists nowadays are sensitive to this problem, especially as it relates to human social behavior and the learning processes involved in that behavior. They now know that by aiming at experimental precision in the laboratory, before they have discerned some of the major problems that should be the objects of this precision, they are in danger of setting up a science of learning that is valid only in the experimental situation.

This awareness no doubt has been brought about by the influence of several trends in the other sciences concerned with behavior, as well as from within psychology itself. Psychologists studying the behavior of children, human social groups, and behavioral deviants, have not found the learning principles derived from the laboratory entirely adequate. Their problem is similar to that of the fieldworker on the nonhuman primates. Beset by the complexity of his data, he is conscious of the need to progress beyond the essential descriptive and classificatory phases by the systematic use of experiment and by integrating his findings with those of general behavior theory. The artificiality of laboratory learning situations has been frequently emphasized. Leuba (1955:28) for example, said: "Most human learning and development . . . does not occur when people are deprived of food or water for twelve or more hours, are given electric shocks, are under the influence of powerful neurotic drives, or are in any situations even remotely resembling these." A similar emphasis is given by Bandura and

Walters (1963:109), from their studies of children's social learning: "generally speaking, it has been assumed, rather than demonstrated, that reinforcement principles apply within complex social settings and that they govern the social behavior of human beings in precisely the same manner as they regulate the responses of human and animal subjects in highly structured nonsocial laboratory experiments."

The influence of ethology in changing and expanding the view of learning has been, and continues to be, profound. It stems in part from a general biological orientation (the implications of which have not always been appreciated by psychologists) to the effect that learning is a product of evolution and is therefore an integral feature of a species' adaptations, and not something that can be studied independently of them. "This characteristic of animals, whereby their abilities for the learning of particular types of responses are related to their ecological requirements, is one of the fundamental points of emphasis of ethological studies" (Etkin 1964:186). As a principle of great importance, derived from comparative studies of bird behavior, Hinde and Tinbergen (1958:255) point out that ". . . many of the differences between species do not lie in the first instance in stereotyped behavior sequences but consist in the possession of a propensity to learn." Learning processes serve and are an extension of the processes of evolution. Comparative studies of primate learning only at their peril can ignore the biological frame of reference. Assessments of relative learning ability or "intelligence" in the nonhuman primates are usually irrelevant and inaccurate because they have ignored it, as well as for other reasons of method (Hall 1963b). For the lower vertebrates, as for the primates: "Evolution, through selection, has built the biological base so that many behaviors are easily, almost inevitably, learned" (Washburn and Hamburg 1965a:613).

The group is the environment in which the young primate does his learning, and in which traditions of feeding, ranging, and the like are perpetuated; it is not unreasonable then to suppose that the starting point for learning studies on the primates—both observational and experimental, both in nature and in captivity—must be the social group rather than the isolation cage. The wild infant monkey or ape does not survive except in the group. Survival requires it to live socially and, preeminently, to learn socially and to do so not only in the long period of its dependence and growth, but also throughout its life. The emphasis here is not new, being a part of biological and anthropological thinking, but its implications have not been manifested in many primate behavior studies.

ADAPTABILITY OF WILD MONKEYS

If we are to increase the relevance of the experimental studies of primate learning we must consider first some of the major characteristics of the social life of wild primates to which learning is likely to have contributed most importantly. Interspecific differences are now known to be enormous, and it is as ridiculous to suppose that the talapoin goes through more or less the same processes of learning as the baboon as it would be to equate the ground-nesting plover and the great tit in this respect. However, there will be an obvious basic common factor in most primate species, namely the setting provided by a group around the mother-infant relationship, although in some species the social surroundings will be numerous and diversified (as presumably in a large baboon group), whereas it will be very limited in the family-party type of organization.

In field studies, as Washburn and Hamburg (1965a) point out, the problem is that observations are primarily of the results of learning rather than of the process itself. Variations in the life of the group can be imposed by the investigator and the results observed, as in several of the Japan Monkey Centre studies. Unplanned variations, such as habitat destruction or drought, can significantly test the limits of adaptability of a population, but a sufficient experimental control is difficult to impose in natural conditions. We must therefore continue to rely mainly upon working out our ideas about the social learning processes by analysis of field data, with the intention of formulating these in such a way that they can be elaborated in the captivity situation. It is perhaps still necessary to point out that, when we try to analyze species-characteristic behavior and behavior of populations or groups of a species, it is always to be understood that genetic factors constitute a major source of variation, but that learning must operate in some manner and in some degree whatever the situation and however stereotyped a behavior may appear to be.

If we list some of the main features of the environment and the activities in terms of which we usually work our field data, it is fairly obvious that we are concerned with certain conformities or traditions. These are in part defined by the physical adaptations of the species, in part imposed by local conditions, and in part the product of intergenerational spread and continuation of habits that have proved serviceable in the local situation:

Feature (physical-social)	*Conformities*
Home range.	Routes of feeding areas, sleeping places, areas of contact with neighboring groups, day-ranging patterns, seasonal changes, danger areas.
Food and water.	Discriminations of edible from inedible, ripe and unripe, manipulatory techniques, water locations.
Day-activity pattern.	Times of resting, times of activity, coordination of movement, rest, feeding, and so forth.
Other animals (different species).	Discriminations of predators or noxious creatures, tolerances of large mammals, warning cues from other animals.
Other, extragroup animals (same species).	Tolerance distances between groups, habituation or avoidance or territorial display.

Feature (physical-social)	*Conformities*
Size and composition of groups. Sexual reproductive characteristics. Individual social development. Group social interaction pattern, and social interaction code.	Processes of socialization continually involve familiarization with individuals in the group, appreciation of social distance, and social discriminations derived from conflicting impulses (*approach:* friendly, sexual; *avoidance:* fear, threat; and so forth).

What are the social learning processes chiefly involved in these conformities? These processes evidently differ between species; but they may, at this level of analysis, suggest a different way of looking at the problem. An accepted way of classifying learning processes is to break them down into forms that differ in kind and in complexity. Habituation, for example, is a very "primitive" means of adjustment, whereas insight or perceptual learning sometimes involves a restructuring of relationships (as in the chimpanzee instrumentation problem-solution) that appears to be of greater complexity and suggests an ability in the nervous system to mediate several alternative relationships between stimulus and response. Of these several processes each has its reference experiments, and they are experiments that usually have been derived from a particular type of situation of a nonsocial kind. When we are studying the complex interactional system of a monkey group, which comprises a kind of natural unity, we may suppose that any or all of the defined learning processes may be operating in the adjustments of individuals to their environment. However, we may also question whether we can obtain an entirely satisfactory account of conformities and adjustments in this social matrix by superimposing on it those learning categories. The doubt is precisely due to the nonsocial character of most of the experimental references. In Gestalt terms, for example, the social situation of a group of baboons cannot adequately be broken down into the sum of the performances of the individuals in it (conditioned response of A, plus conditioned response of B, and so on). It is clear that the social unity is the group, that the behavior of the individual is only comprehensible in terms of its membership in that group to which it has belonged since birth, and that changes in the composition of the group will in some degree affect all members of it. Where there is some kind of role differentiation and rank relationship in a group, as in baboons and patas monkeys, the Gestalt quality of the group is exceedingly clear. In a simple form, we have been able to demonstrate this repeatedly in our laboratory group of patas by observing the effects on the group of removing individuals, one at a time, and introducing or reintroducing individuals (Hall and Mayer 1967). The dynamic equilibrium of the whole pattern of relationships is temporarily disturbed and is then restored, but in a modified form. By withdrawing A from the group, you do not simply leave B plus C, and the rest in the same pattern of relationships, but in some degree you modify the whole relational system. In human group studies this dynamic view was most cogently expressed by Lewin (1935), but its implications for the study of social learning in the nonhuman primate group do not seem to have been properly formulated. Again and again, in observing a group of monkeys, whether in the field or in captivity, one is impressed by the more or less continual awareness of each member of the behavior, the distance, and the situation of the others. This is inevitable when the group is the natural and more or less constant unit.

If we extend the argument we may distinguish two main characteristics in the dynamic interaction processes, both of which contribute to the total pattern of group organization, the group organization itself being a dependent part in the greater organizational whole of the population. The coordination of the activities of the group and the general conformity of the habits of its members appear to be derived from very simple social learning processes. These we have described as *following, social facilitation,* and *observational learning* (Hall and Goswell 1964). The infant follows the actions of its mother, to some extent and for an early period "modeling" its actions and the directions they take or the objects to which they relate upon her behavior. Any animal's behavior in the group may be set going or facilitated by the perceived example of another member of the group. Any animal may observe the behavior of another, with or without awareness of its reference, and later may behave in the same manner in the same type of situation. These three processes, which intergrade one with another, constitute a more or less continuous interactional system through which survival habits of feeding, avoidance, dispersal, or aggregation are acquired. "New" habits are in this manner spread throughout the group. Although, in terms of the experimentalist's "trials," nothing could appear simpler than the learning processes here involved, in terms of the group the cumulative effect of repeated adjustments in this manner may be considerable. It is, as we know, easy to transform the meaning of some object from negative (avoided) or neutral (ignored) to strongly positive merely by the observing animal's seeing the positive behavior of another animal with whom it has a friendly relationship.

A second characteristic is that of the more active processes of adjustment going on within the social structure of the group. There are two examples we can use to illustrate it. Where rank relationships within the group can be clearly defined (as they can be in macaques) in terms of behavioral interactions, social distances, and so on, it has been suggested that the social behavior of the high-ranking mother favorably affects the chances that her offspring (or at least the males) will attain high rank in the group (Kawai 1958, as quoted by Imanishi 1960; Koford 1963b). Allowing for the likelihood that there are genetic predispositional factors involved, this means that the confidence or security, or simply the example of successful assertiveness afforded by the mother's behavior, serve to build up similar habits in the social behavior of the young. These habits then supposedly transfer to the interactions with agemates, and the "successful" young male remains in the group rather than going to the periphery or into isolation. The converse situation must also presumably produce the opposite effect.

Although there is no doubt that such cases do occur, and indeed would be expected from what we now know about social learning, it has to be appreciated once again that these social systems are anything but static and that the great advantage of observing these systems in action in the natural state is the complexity of the information they provide and the number of alternative hypotheses they may suggest. If we consider an extreme case, that of the one-adult male patas group, we do find in our laboratory group that the ranks of the adult females are over a period very clearly defined. The female is extremely possessive and protective toward her young. The behavior of the highest ranking female was such that when her son reached the age of about two years the pair of them were regularly in the ascendant over the adult male in feeding tests and positioning in the living room. The young male would even threaten the adult male without provoking retaliation. Immediately the mother was removed from the room or was confined in a small cage within the room; the social situation of her son changed radically, and the son accepted the change without its being necessary for others in the group to impose it forcibly upon him. He now takes his place in low rank. The son of the second ranking female in the group, although he is a year younger than the other, when his mother is present is in the ascendant over the first young male. At all times the one-year-old male, whether his mother has been in high rank or not, has been exceedingly bold, and the relationship between the two mothers (although entirely clear in terms of rank) is friendly and without tension, each having more or less equal access to the adult male.

If we now consider the much greater complexity of the wild patas group we have to suggest how it is that all males (with the exception of one) are eliminated from the group when they reach sexual maturity. There are several alternatives, any or all of which may operate according to the dynamic situation of the group and the population, which includes isolate adult males, and, once, a group of bachelors (Hall 1965a). Only four of the many possible combinations of events are shown below:

The following are some alternative interaction patterns whereby changes of ingroup adult males may occur in a wild patas monkey population:

$$
\begin{array}{ll}
\text{In-group adult } \male & = A. \\
\text{Adult } \female\ \female & = 1, 2, \text{ and } 3. \\
\text{Their sons} & = B, C, \text{ and } D. \\
\text{Extragroup adult } \male\ \male & = X \text{ and } Y.
\end{array}
$$

1.   *A* attempts to drive out *B*.
        1, 2, and 3 combine in attacking *A*.
            Then either
        B remains with the ♀ ♀; *A* leaves.
            Or
        B leaves the group; *A* remains.
            Or
        The group splits; some ♀ ♀ going with *A*, some with *B*.

2.   Sons *B* and *C* reach sexual maturity at the same time.
        A attempts to drive out both.
        1, 2, and 3 combine in attacking *A*.
            Then either
        B remains; *C* and *A* are left.
            Or
        A remains; *B* and *C* leave.
            Or
        The group splits.

3.   ♀ 1 loses her ascendancy to ♀ 3 (younger).
        Her son, *B*, is still subadult.
            Then
        Sons *B* and *D* reach sexual maturity.
        A attempts to drive out *B* and *D*.
        3 and other females attack *A*.
            Then either
        B and *A* leave; *D* remains.
            Or
        A remains; *B* and *D* leave.
            Or
        The group splits.

4.   *A* is killed by a predator.
        B, *C*, and *D* still subadult.
            Then either
        X or *Y* join group.
            Or
        Group splits, *X* joining one part, *Y* the other.

When these combinations of interactions are examined it is at once apparent that the ascendancy of a particular male depends upon a great many factors, and that the rank of the mother is likely to be only one of them. It must be pointed out that we know of the ranking behavior of the patas females only in the very restricted situation of the laboratory group. Where the spacing between individuals is as great as it is in the wild group, and where competition for food is very rare, it is possible that female ranks are ill defined and that the high rank of females does not bestow any particular advantage on their young. The mother is scarcely in a position to further the social ascendancy of her son after he has grown beyond the juvenile stage, for he is rarely close to her. It is therefore possible that by their interactions among themselves the young males achieve social learning that is relevant to their social position on reaching sexual maturity. We can scarcely doubt, from what we have seen in the laboratory, that the reaching of sexual maturity is an exceedingly stressful situation both for the young male and for the whole group.

A further problem in social learning arises out of the special role that the adult male of the wild patas group has in watching for predators and for other patas. This male is usually very remote from the rest of the group, and behaves in a manner that is quite distinctive and unlike the other members of the group. This behavior must surely be learned, not simply by unguided trial-and-error, but from the young patas having observed the behavior of his predecessor in the group. If this is so, then the learning is acquired (without any obvious reinforcement at the time of its acquisition) when he is a juvenile or a subadult. The learning is latent because it does not come into effective performance until the young male becomes the successor to the adult male already in the group. The only clue we have to this process is from our observations of the behavior of the young male in our laboratory group. He is obviously attracted to the adult male, approaching close to him, but he is also obviously fearful and ready to flee at the first sign of antagonism from the adult. He is also, in a nervous way, ready to be aggressive. His behavior is thus clearly and repeatedly ambivalent, attraction, fear, and antagonism all expressing themselves in his gestures and movements. In human psychological terms we could suggest that there is a tendency for the young male to identify with the adult male. This would imply that the young one learns to match the behavior of the adult, the reinforcement for such learning arising from the arousal of quite intense conflicting motivation.

It has seemed useful to elaborate these examples as illustrations of the dynamic processes of learning and motivation that may have to be taken into account in the actual group situation. The major point requiring emphasis is that conflict of tendencies, sometimes in only mild degree, is not uncommon in the interactions of monkeys such as baboons, macaques, and patas, and that such conflicts may serve repeatedly to reinforce discriminatory social learning. The adaptability of a species, to use a term that comprehends all possible aspects of social learning, consists not only of the conformity patterns achieved in the routine activities of daily life, but also of the manner in which a stability in the social organization is acquired, maintained, and restored when social or ecological change occurs.

SOCIAL EXPERIMENT

If, in our field studies, we discern such social complexity and so many possible factors that interplay in social learning, we may wonder how far we can succeed in elucidating some of these problems by experimenting on captive animals. An alternative is, as has been mentioned, to try systematic experiments in the field. The difficulties of so doing are considerable, partly because wild groups are not available in many areas in such conditions that they can be observed for long periods at very close range, and partly because there is far too little control of the relevant factors. We wish to discover precisely how these animals learn with reference to one another, not merely how a group responds, for example, to an unusual situation induced by the investigator. The latter is an informal aid to the task of description, not a method of further analysis. So far no sustained experimental program has been undertaken with the primary objective of linking with the field data on a monkey species, with the exception of our own small beginnings on the patas monkey. There are, consequently, evident limitations in the experimental social learning studies that are available, both in terms of species (mainly rhesus macaques) and of situations used. We shall consider a few examples of such studies.

Observational Learning

The problems of method, the experimental limitations, and some of the difficulties involved in selecting relevant social factors for study have been reviewed elsewhere (Hall 1963a). As Harlow (1959a:43) pointed out in discussing a discrimination problem-solving solution that was achieved as a consequence of the observer monkey (O) perceiving cues given by a model (M): "It is more than possible that the conditions existing in nature have been more favorable in terms of suitable social environment in this regard for producing imitative learning than have the conditions in the laboratory. Mother monkeys, for example, may have far more insight into the importance of these variables than do experimental psychologists."

One of the standard methods of experimenting is to use two cages or compartments. In one, M tackles the problem, achieving solution and obtaining food reward. At the same time, O observes M's behavior through a screen. After M has been removed or the screen has been obscured, O is given the opportunity to attempt the same problem. The correct performance by O may be significantly quicker or subject to less error consequent upon his having observed M. The location of the correct cue (that is, the cue in response to which M has been rewarded) has been indicated by M; without M's example, O would have had to find it by trial and error. This type of result, on the one hand, appears to be a straight-forward confirmation of what we can see occurring daily in the interactions of young animals in our patas monkey group (Hall and Goswell 1964) or in wild groups of macaques or baboons. It appears to add nothing to our knowledge of social learning, except by isolating the animals from any other influences or distractions. On the other hand, a negative result (one in which O does not benefit from M's performance) can usually be attributed to some factor such as O's inattention, and cannot be explained as an inability to learn by observing.

It is possible to achieve results that are in one sense more meaningful by adopting a much less rigid experimental schedule. Negative or neutral response by O to an unfamiliar food object can be transformed into positive responses after O has watched M's positive responses, even though O's behavior continues to indicate some aversion (Hall and Goswell 1964).

In avoidance learning, Mirsky, Miller, and Murphy (1958) have demonstrated "communication of affect" from the startle and avoidance behavior of M to another monkey observing this behavior. O responds to the perceived "mood" of M after O has been conditioned to associate such a "mood" with unpleasant stimuli, such as shock. Our own results on the patas confirm this finding, and show that a previously positive response to a box by two young animals can change to a strong avoidance of that box after seeing their mothers reacting by startle to it. The startle was induced in the mothers by our having a live snake under a perspex cover that was visible to them when the lid of the box was lifted; it was not visible to the young patas, in a cage nearby. Similar results have been obtained when a positive response by M (such as opening the box lid) has resulted in shock, with O watching from an adjacent compartment. The point these results seem to make, although they are only preliminary, is that the young animal can learn to respond with avoidance to an object simply as a consequence of perceiving that avoidance in another monkey and without itself having experienced any unpleasant stimulation. This "vicarious reinforcement" seems indeed to be the probable process whereby the young in the wild group acquire their avoidance habits, and, to this extent, the findings substantiate what could only be set up as an hypothesis from field observation.

Expressions of mood or intention, positive or negative, may well serve as very important social indicators to the young monkey, and these expressions, rather than direct experience or primary reward or punishment (food or injury), may provide the reinforcement of discriminatory learning to approach or to avoid particular classes of objects. Whether such reinforcement can be classed as secondary (since the responses are derived originally from primary reinforcers of the usual kind) is still not clear.

Interanimal Conditioning

It is reasonable to suppose that the relative rank of two animals of a group, maintained assertively by one so that the other accepts a subordinate role, may be subject to change. Any experimental effect (for example, lobotomy, drugging, or induction of "neurosis") that substantially alters the behavior of the "dominant" animal may lead to such change. Murphy, Miller, and Mirsky (1955) have described the effect of punishment by shock of the dominant of a pair of rhesus monkeys associated with its seeing the subordinate at the far side of a glass screen. The shock could be terminated by the animal's pressing a lever, which results in the disappearance of the subordinate. Dominance was assessed before and after the experimental trials by the usual food tests. After the shock trials the subordinate in some pairings would take proportionately more food items than the dominant, reversing the pretest food relationship.

The question that at once arises from such findings is to what extent this change of rank, as thus determined, is specific only to the food-test situation or is generalized to other situations in which the two animals are placed. Smith and Hale (1959) found that there was no generalization from the same type of experiment on domestic fowls, and Hansen and Mason (1962) confirmed this on rhesus monkeys. We have also confirmed this lack of generalization with one pairing of adult female patas; although the postshock trials on the food test showed a marked change in favor of the subordinate female, as soon as the two animals were returned to the group in which they had been living before the experiment the ranks were immediately reasserted as before. Although it is perfectly possible that severe and prolonged shocking of the dominant animal in the visual presence of the subordinate would produce a change in rank that would carry over into the group situation, it is equally possible that the same amount of punishment would produce such a change without the view of the subordinate ever having been associated with the punishment. In other words, the shocked dominant would temporarily be reduced to a state of "neurotic" disturbance.

These experiments seem to suffer from two defects. First, they fail to consider in what ways rank relationships are really determined in the group, and the complexity and variability in such rankings. The food test is very far from being an adequate indicator of *social* rank. Second, they take no cognizance of the fact that the discriminatory learning processes that are likely to affect rank are not necessarily brought about by actual fighting or punishment, at any rate in wild groups, but through persistent interactions of a much subtler character. The data on the lack of generalization achieve at least the negative result of demonstrating that the other experiments were meaningless in reference to the group setting.

Studies of "experimental neurosis" per se have likewise failed to reveal much, if anything, that is relevant to the social environment of monkeys. The clash of competing or conflicting stimuli is usually brought about in the standard classical or operant conditioning situation. When the "normal" behavior of the monkey is thus disrupted it can usually be restored by a period away from the experimental situation, particularly if the monkey is returned to the group in which it has been living. Russian experimenters now appreciate the fact that the most lasting physiological disturbances in these animals can be produced where "natural" conditioned stimuli are involved, particularly those associated with the gregarious tendency of the animals (Miminoshvili, Magakian, and Kokaia 1960). If we are interested in using these animals to provide analogues of human neurosis, which is primarily a social condition, or to study the effects of sustained conflict or even mild conflict, it would indeed be appropriate to arrange for more relevant experimental variations.

INTEGRATION OF NATURALISTIC
AND EXPERIMENTAL SOCIAL LEARNING

With the exception of the series of studies from the Wisconsin laboratories, it would probably be true to say that experiments on monkeys relating to any aspect of social learning have contributed very little beyond the simplest confirmation or negation of hypotheses derived from field study. The experimental and the field situations are too far apart and bear too little relationship to one another. A more hopeful prospect lies in the opportunity to compare the social learning processes of groups of the same species kept in captivity under different environmental conditions. Meanwhile, we may at least attempt to see whether any integration is at present possible between the principles that seem to operate in the social organization of wild groups and the principles of learning derived from laboratory experiment.

Mowrer (1960) appears to have been the first experimental psychologist to attempt explicitly to apply his "theory" of learning to the task of explaining social behavior of primates. He examines in particular Carpenter's (1952) generalizations. The key statement from Carpenter concerns

> . . . the formulation of an important principle for understanding the dynamics of group integration: Namely, *that the drives, tensions or needs of one individual which are satisfied by activities of another individual or individuals of a grouping modify the previous adjustments between or among the individuals.* This may be termed the principle of reciprocal interaction (1952:243).

In any group the equilibrium in the relationships of the individuals composing it is a resultant of centripetal and centrifugal forces that, respectively, attract and repel. The former

. . . attract the animals to each other, regulate their interactions and cause the society to persist as an identifiable group [whereas] . . . competition, conflicts, and various forms of aggression . . . operate . . . to differentiate the statuses of individuals, to create intra-group stresses and to cause some individuals to leave the society (1952:244).

According to Mowrer this analysis

. . . lends itself admirably to restatement in terms of revised two-factor learning theory. In fact, it is, in itself, almost such a statement. To begin with, there are the primary rewards and punishments which are mediated by group living. These are, on the one hand, food (especially for the young), shelter and mutual protection, and sexual gratification and, on the other hand, the aggressions which arise as a result of unregulated competition for the available rewards. To the extent that an individual experiences rewards in the group, various forms of stimulation associated with group life take on secondary reinforcement and tend to hold the individual in the group. To the extent that an individual experiences punishment in the group, the associated stimuli take on the capacity to arouse fear and tend to *repel*, or "expel," the individual. Hence the image of "centripetal" and "centrifugal" forces acting upon each individual in the group which thus determine, for the group as a whole, its stability or instability (1960:384).

Mowrer goes on to point out that each individual in a group

. . . learns to react selectively, discriminatingly toward other members of the group, while at the same time establishing attitudes of friendship or enmity on the part of others toward him. . . . Hence, in a stable, successful social group the total learning associated with life in the group is relatively great (1960:385).

Much of what we have discussed earlier in this chapter, notably field studies data, seems, at least superficially, to fit Mowrer's formulation. "Secondary reinforcement" is so general a conception that it enables a learning theory to extend itself indefinitely so as to include learning that occurs in or results from situations in which no primary reward or punishment is evident. Thus the basic "trophallactic" situation (Schneirla 1946) of infant dependence on the mother, the "rewards" of which for the infant in security, comfort, food-getting, and the like are obvious, is considered extensible by some authors to account for the persistent social learning seen in the young as it grows older and loses close dependence on the mother. As drive-reduction and drive-induction are supposedly involved in reinforcement, the secondary reinforcers are derivatives of situations in which, for example, the young animal actually has been rewarded or punished.

Although Mowrer's is a useful attempt to cope with the complex social phenomena with which all field workers are concerned, it is doubtful if it is adequate to explain all that is probably to be included in the social learning of group-living primates. Further, the value of such a theory can be assessed mainly in terms of the hypotheses it suggests that ought to be testable in experiment, and so far, it does not appear to have led to anything new in this respect. This may be because the experimenters working along the Mowrer line have not been personally acquainted with the diversity of social behavior in a monkey group, and hence have been unable to formulate hypotheses from direct observation. Being essentially a drive-reduction (induction) theory, it focuses on problems of social learning in which the motivational factors causing approach or avoidance, hope or fear, are evident. It also probably suggests a framework in which the differential reinforcing effects of conflicting tendencies in social disciminatory learning can be set out for experimental testing.

Berlyne (1960:10) is likewise concerned with the role of conflict in learning, pointing out that conflict of several varieties, other than the more drastic forms described in the psychoanalytical literature, ". . . is an inescapable accompaniment of the existence of all higher animals, because of the endless diversity of the stimuli that act on them and of the responses that they have the ability to perform." In the monkey group there is a more or less continual interplay of friendly, sexual, aggressive, and fearful impulses, which require that they be balanced among the individuals. There is little doubt that these, rather than actual experiences of punishment and reward, serve to maintain the social framework, and, when the balance within the group is disturbed, permit relearning and modification by other means than fighting. Indeed, the significance of social conflict in social learning studies is something that might very well be investigated experimentally in ways more meaningful than those of the interanimal conditioning work so far attempted. The balancing of social inhibitions and excitations in a wild baboon group affords a remarkable illustration of effective social learning based on conflict.

Berlyne (1960) is also much concerned with the problem of exploratory behavior. Such behavior, in a wild monkey group, is scarcely ever carried out by an individual entirely without the awareness or the company of others; it is thus reasonable to suppose that this is simply an important propensity of the species which allows of some degree of flexibility, of opportunity to acquire new habits. It is serviceable to the group, and the learning is reinforced mainly by the friendly, attentive social context in which it occurs.

In summary, the fieldworker may begin with the assumption that all "normal" primate learning is, in essence, social. Where the individual primate is in temporary isolation learning a task without reference to any other member of its species, the learning is not normal. The primary reinforcement for all normal primate learning may come from its social context, the group in which the animal is born and nurtured. Even the sensorimotor activities of observing, manipulating, and exploring that are indicative of individual independence receive some facilitation, some inhibition, and some direction from the group setting. Conflict of tendencies (which varies with age and sex, with group situation, and with species' ecology and social organization) is likely to determine much of the discriminatory social learning with only occasional reference to such primary incentives as food or actual punishment. This paper presents some of the problems of integrating field and experimental studies, rather than attempting to answer them. Learning theory has been in one way too sophisticated, and in another way too limited to provide a useful basis from which to set up social learning experiments. Field studies may be analyzed with the purpose of providing experimentally testable hypotheses and may eventually contribute much to a more generally valid theory of social learning.

chapter eight

AGGRESSIVE BEHAVIOR IN OLD WORLD MONKEYS AND APES[1]

S. L. Washburn and D. A. Hamburg

In this essay we hope to use the problem of aggression as an example of evolutionary perspective as a scientific tool. We are interested in the way the behavior of the nonhuman primates may be useful in increasing our understanding of man. Study of the forces and situations that produced man is one way of attempting to understand human nature, of seeing ourselves in evolutionary perspective. Because the use of man's closest relatives in the laboratory is the nearest approach to experimenting on man himself, the use of primates as laboratory animals is increasing rapidly. As in so much of science, there is no boundary between the pure and the applied, and the answers to our questions lie in the domain of many different disciplines. Since no one could possibly master all the relevant information, we wish to clarify our objectives before proceeding to the main topic of the paper. In discussions at the conference a series of general questions emerged, many of which we have encountered before — since the beginning of our collaboration at the Center for Advanced Study in the Behavioral Sciences in 1957. We propose to deal with these recurrent, general questions first.

Field studies in a comparative evolutionary framework do not permit the manipulation of variables and the precision of measurement that are possible in laboratory experiments. They are, however, likely to give new perspectives which suggest fruitful directions for experiments and for deeper analysis of problems of living organisms. Field studies do not replace other methods of investigation, nor do studies on animals substitute for the direct study of man. For example, Prechtl (1965) has pointed out that much of the

[1] This paper is part of a program on primate behavior supported by the United States Public Health Service (Grant No. MH 08623) and aided by a Research Professorship in the Miller Institute for Basic Research in Science at the University of California at Berkeley. We particularly want to thank Dr. Jack Barchas, Dr. Phyllis C. Jay, Dr. Donald Lunde, and Mrs. Jane B. Lancaster for advice and help in the preparation of this paper.

behavior of the human newborn is understandable only in phylogenetic perspective. This does not mean that comparison is the only way to approach the study of behavior in the newborn but only that inclusion of some comparison enriches the understanding of what is being observed. In advocating the use of the evolutionary approach, we are not suggesting that it is the only method or that it is always relevant. We are suggesting, however, that the comparative, evolutionary study of behavior offers many insights that are unlikely to come from other sources. We believe that the combination of the evolutionary perspective with experimental science provides a powerful approach to biological problems including those of behavior, and we echo Simpson's 1964 statement "100 years without Darwin are enough."

A central problem in the study of the evolution of behavior is that contemporary monkeys and apes are not the equivalents of human ancestors. To what extent their behaviors may be used as indicators of the behaviors of long-extinct fossil forms is debatable. Obviously this is not a simple question, and the possibility of reconstruction differs with each particular behavior. Each answer is a matter of probability, and there is great variation in the adequacy of evidence bearing on different questions. For example, in all the nonhuman primates females have brief, clearly limited periods of estrus. Since this state is universal and highly adaptive, it is virtually certain that this condition was present in our ancestors. Although the time of the loss of estrus cannot be dated, the comparison calls attention to the significance of the loss. The physiology of human females is quite different from that of any other primate, and many specifically human customs and problems are directly related to continuing receptivity in the human female. The human family is based upon a female physiology different from that of any other primate, and the physiology may well be the result of selection for stable male-female relations. The close similarity in much reproductive physiology permitted the use of the rhesus monkey as the most important laboratory animal in working out the human female's reproductive cycle (Corner 1942), but the behavioral approach emphasizes the significance of the differences. Our understanding benefits from seeing the nature of both the similarity and the difference. The problems of reconstructing behavior are discussed more fully in Washburn and Jay (1965).[2]

---

[2] After this paper was in nearly final form we received *Adaptation et Agressivité* (1965), edited by Kourilsky, Soulairac, and Grapin, and *On Aggression* (1966), by Konrad Lorenz. *Conflict in Society* (de Reuck and Knight 1966) also appeared, although we had access to part of that data in prepublication. We find ourselves in general agreement with these new sources, and only wish that we had had them a year ago. We have found J. Altman's *Organic Foundations of Animal Behavior* (1966) exceedingly valuable, and it too would have been very useful if it had been available earlier. None of these references would have led us to different conclusions, and we hope that the point of view presented in the introduction to this paper may help to bridge the gap between the thinking of the students of animal behavior and the social scientists.

Estrus has been used as an example of a problem of the reconstruction of behavior when there is no direct fossil evidence. The same kind of logical reconstruction from indirect evidence can be applied to many other behaviors. We have indirect evidence, for example, that our remote ancestors probably matured more rapidly than we do, lived in small areas, were very aggressive, and hunted little, if any. Let us consider this first statement: If the infant is to cling to the mother, as does the infant monkey and ape, then the infant must be born able to cling. A human infant born with this ability would need far too large a brain at the time of birth, and the adaptation of the mother's holding the immature infant may be looked on as an adaptation to the evolution of the brain. Monkeys are born with more highly developed nervous and motor capabilities than those of apes at birth. A gorilla mother must help her baby for approximately six weeks before it can cling entirely unaided (Schaller 1963). In *Homo erectus* (Java man, Pekin man, and comparable forms) the brain was approximately twice the size of a gorilla, and the infant's behavior must have been much more like that of modern man than like that of an ape.

The implication of the clinging of the nonhuman primate infant can be generalized that biology guarantees the infant monkey or ape a far greater chance of appropriate treatment than it does the human infant. So far as early experience is important in determining later performance, man should show by far the greatest variability of all primates. The study of comparative behavior calls attention to the uniqueness of the human condition, the importance of early events, and of the human mother's need to acquire complex skills of child care.

In addition to the reconstruction of behavior, another problem that has arisen repeatedly in discussions of the comparison of behavior is the extent to which behaviors in different groups of animals are comparable. The issue is whether behaviors labeled, for example, "aggression" or "play" are comparable enough so that comparisons are useful, or whether these are simply subjective human words used to symbolize collection of noncomparable behaviors. There are four important considerations.

The first is that the words are least likely to be misleading when animals, whose behaviors are being compared, are closely related. For example, we think there would be little disagreement as to which behaviors should be called "play" or "aggression" in Old World monkeys and apes, but the problem is very different if the behaviors of monkeys and birds are to be compared.

The second point is that in the observation of the behavior we are not limited to the view of a single animal or to one occasion. An observer sees repeated interactions of animals in the social group or groups. The judgment that an action is threatening, for example, is based on the repeated specifiable response of the other animals. The response to biting in play and in aggression is very different, and the difference is unmistakable in the response of the bitten animals, although it would often be hard for the human observer to perceive the difference in the actual bite.

The third point is that in classifying and labeling behavior we are not limited to the external view of the actions. For example, Delgado (1963b) has shown that stimulation of certain brain areas elicits threat behavior in rhesus monkeys. Although present evidence is limited, we think it likely that the same brain areas are concerned with rage in man. If it can be shown that the behaviors that are classified together are mediated by comparable structures, then the classification is much more likely to be useful. This is best exemplified by the category "sexual," in which the interrelations of behaviors, nervous system, hormones, and experience have been thoroughly investigated (Beach 1965). This takes us back to our first point: One reason that labeling categories of behavior is less likely to be misleading when the animals are closely related is that the internal biological mechanisms on which these actions are based are more likely to be similar.

The fourth point is that the human observer is most likely to detect relevant cues when observing the behavior of animals closely related to himself. We do not underestimate the formidable problems of observation and the need for experimental clarification of what is being observed in the behavior of monkeys and apes, but at least the special senses and central nervous system of the observer are highly similar to those of the animals being observed. Even within the primates the interpretation of observations becomes much more difficult with the prosimians. Since the sense of smell is important in prosimians and the animals have special tactile hairs that we lack, in many situations there is no assurance that the human observer has access to information of importance to the animal being observed. The human can see a dog in the act of smelling, but simple observation gives no knowledge of the information received by the dog. Viewed in this way, the human observer is more likely to be able to see and record the actions that are important in the analysis of behavior when studying monkeys and apes than he is when watching any other animal.

This does not mean that observations are necessarily easy or correct, but it does mean that man is potentially able to see, describe, and interpret actions made by a close relative, especially in a mode that approximates the judgments the animals themselves are making. The more distantly the animal is related to man, the less the human perception of the situation is likely to correspond to that of the animal's. For example, a human observer can learn the facial expressions, gestures, and sounds that express threat in a species of monkey, with some assurance that the signals are seen in the same way a monkey sees them. It is possible to play back their recorded sound to monkeys and most sounds without gestures produce no response. Some monkey gestures can be learned by man, and the human face is sufficiently similar that a monkey interprets certain human expressions as threats and responds with appropriate actions. Without experiment a human observer would

Fig. 8–1.
Baboon threatening through car window in response to human mimicking of baboon threat. (*Photograph by Sherwood Washburn*)

hardly have expected that a male turkey would display to the isolated head of a female (Schein and Hale 1959; reprinted in McGill 1965). Even when a bird and a mammal are responding to the same sense, vision, in this case, the internal organization of perception is so different that comparison can be made only on the basis of experiments. In monkey, ape, and man the internal organizations are highly similar, and the use of a word such as aggression carries far more comparable meaning than when the same word is used to describe a category of behavior in vertebrates in general. We are not belittling the importance of the study of behavior in a wide variety of animals, but we are emphasizing that the problems of comparison in such studies are more complex than they are in these comparing a very few of man's closest relatives.

We are writing on aggression in certain monkeys and apes because we think that there is enough information to make such an essay useful. The word "aggression" only indicates the area of our interest and our use of it does not mean that we think the area is fully understood or precisely defined. Accurate definition often comes at the end of research; initially, definitions serve only to clarify the general nature of the subject to be explored. For the purpose of opening the area for exploration and discussion we find that the definition of Carthy and Ebling (1964:1) is useful. They state that "An animal acts aggressively when it inflicts, attempts to inflict, or threatens to inflict damage on another animal. The act is accompanied by recognizable behavioral symptoms and definable physiological changes." Carthy and

Ebling recognize the displacement of aggression against self or an inanimate object, but rule out predation as a form of aggression. For our analysis, it is not useful to accept this limitation; if one is concerned with aggressive behavior in man, the degree to which human carnivorous and predatory activity is related to human aggressiveness should be kept open for investigation and not ruled out by definition.

In discussing aggression it is important to consider both the individual actor and the social system in which he is participating. In the social systems of monkeys and apes evolution has produced a close correlation between the nature of the social system and the nature of the actors in the system. Societies of gibbons, langurs, and macaques represent different sociobiological adaptations, and, as will be developed later, the form and function of aggressive behaviors are different in these groups (Jay, this volume). Through selection, evolution has produced a fit between social system and the biology of the actors in the system. Aggression is between individuals or very small groups, and individual animals must be able to make the appropriate decisions and fight or flee. Rapoport (1966), in particular, has argued that war between modern nations has nothing to do with the aggressiveness of individuals, but rather is a question of culture, of human institutions. Although agreeing that it is very important to understand the cultural factors in war, we think that it is important to understand the human actors too. It is still individuals who make decisions, and it is our belief that the limitations and peculiarities of human biology play an important part in these decisions. We will return to this issue later; at this point we only want to emphasize that it is necessary, particularly in the case of man, to think both of the social system and the actors. It was not long ago that human war was carried out on a person-to-person basis, and our present customs go back to those times.

Finally, we return to a point of view which we have discussed elsewhere (Washburn and Hamburg 1965a), but which needs emphasis and clarification particularly in the context of aggression. The result of evolution is that behaviors that have been adaptive in the past history of the species, which led to reproductive success, are easy to learn and hard to extinguish. As Hinde and Tinbergen have put it: "This exemplifies a principle of great importance: many of the differences between species do not lie in the first instance in stereotyped behavior sequences but consist in the possession of a propensity to learn" (1958:255, reprinted in McGill 1965). It is particularly important to consider ease of learning, or the propensity to learn, when we are discussing monkeys and apes. These forms mature slowly and there is strong reason to suppose that the main function of this period of protected youth is to allow learning and hence adaptation to a wide variety of local situations.

There is a feedback relation between structure and function starting in the early embryo. Structure sets limits and gives opportunities. Apes cannot learn to talk because they lack the neurological base. Man can easily learn to be aggressive because the biological base is present, is always used to some degree, and is frequently reinforced by individual success and major social reward. The biological nature of man, now far more amenable to scientific analysis than ever before, is thus relevant to aggressive behavior in ways that include learning and social interaction.

THE BIOLOGICAL BASIS OF AGGRESSION

Collias (1944) in a major review of aggressive behavior in vertebrates concluded that the function of the behavior was control of food and reproduction through the control of territory and the maintenance of hierarchy and that males were responsible for most of the aggressive behavior. Scott (1958 and 1962) supported these main conclusions, and emphasized the importance of learning in the development of aggressive behavior. Breeds of dogs differ greatly in aggressiveness, but the expression of these differences is greatly modified by learning and the social situation. Recently, Wynne-Edwards (1965) has stressed the role of social behavior in control of territory, hierarchy, and reproduction as mechanisms of species dispersion and population control. In general, the biological studies indicate that aggression is one of the principal adaptive mechanisms, that it has been of major importance in the evolution of the vertebrates (Lorenz 1966).

We think that these major conclusions will apply to the primates, although adequate data are available on only a few forms and these will be discussed later. Order within most primate groups is maintained by a hierarchy, which depends ultimately primarily on the power of the males. Groups are separated by habit and conflict. Aggressive individuals are essential actors in the social system and competition between groups is necessary for species dispersal and control of local populations. In view of the wide distribution of these behaviors and their fundamental importance to the evolutionary process, it is not surprising to discover that the biological basis for aggressive behavior is similar in a wide variety of vertebrates. As presently understood, the essential structures are in the phylogenetically oldest parts of the brain, and the principal male sex hormone testosterone is significant for aggression. In general, motivational-emotional patterns essential to survival and reproductive success in mammals find their main structural base in the older parts of the brain, particularly in hypothalamus and limbic system. In the higher primates this mammalian "common denominator" is linked to a remarkable development of newer parts of the brain. These are mainly concerned with increasing storage (learning), more complex discrimination, and motor skills; they function in a complex feedback relationship with the older parts (Noback and Moskowitz 1963; MacLean 1963). This interrelation of the older and newer parts of the brain is especially important to remember when monkeys and apes are considered because the expression of the brain-hormone-behavior paradigm may be greatly modified by learning in a social environment. A simple release of a complicated emotion-motor pattern in aggression is not to be expected in such species.

The recognition of a biological factor in aggressive behavior long antedates modern science; stock breeders certainly recognized differences in individuals, in breeds, and in the effects of castration. Comparison of the behaviors of ox and bull shows that a part of the difference in aggressive behavior between males and females is due to the sex hormones. Testosterone makes it easier to stimulate animals to fight, and aggressive behavior in males tends to be more frequent, more intense, and of longer duration than in females. Experimental studies fully support the conclusions of Collias' (1944) survey of vertebrate behavior, and are in agreement with the field studies of primate behavior. It would be extremely interesting to trap a male baboon from a troop that had been carefully studied, castrate him, and release him back into the same troop. The effects of the operation could then be studied relative to hierarchy, predators, and participation in troop life in general.

However, the relation of the hormone to the behavior is not simple, for androgen stimulates protein anabolism in many animals (Nalbandov 1964), and this is one of the factors accounting for the greater size of males. Skeletal muscles of castrated males and females grow less rapidly than those of intact animals, and androgen administration increases the number and thickness of fibers. The sex hormones affect the growth of the brain, and appear to act in an inductive way to organize certain circuits into male and female patterns (Harris 1964; Levine and Mullins 1966). This effect is comparable to that on the undifferentiated genital tract. There is then an interplay, a feedback, among hormones, structure, and behavior, and the nature of this relation changes during development. Testosterone is first a factor in influencing the structure of the brain (especially the hypothalamus) and of the genitals. Later it is a factor in influencing both muscle size and the more aggressive play of males. Finally, it is important in influencing both aggressive and sexual behavior in adults. We stress the complex interplay between hormones, structures, and behavior throughout the life of the individual. It can be seen at once that the concept of a system of coadapted genes is so important because such a developmental and functional pattern depends on the interaction of many different biological entities; variation in any of these may affect the final result. It is probably the primary methodological difficulty in the science of behavior genetics that even apparently simple behaviors are often built from very complex biological bases.

Two groups of experiments yield what seems to be particularly important information regarding factors influencing male aggression. Harris and Levine have studied the sexual behavior patterns in female rats that had received early androgen treatment; treatment of newborn rats with testos-

terone results in abolition of estrous behavior combined with an exaggera-
tion of male patterns, particularly in the aggressive sphere. Young, Goy, and
Phoenix (1964) and Goy (personal communication) gave testosterone to
pregnant rhesus monkeys during approximately the second quarter of
gestation. In addition to producing pseudohermaphroditic females, the
behavior of the prenatally treated females was modified in the male direction.
Rosenblum (1961) has shown that there are marked sexual differences in
the play of infant rhesus monkeys. The masculinized females were allowed
to play for 20 minutes per day, five days a week, from the age of two months;
this continued for more than two years (at the time of writing). They threatened, initiated play, and engaged in rough play more than did the controls.
Like the males studied by Rosenblum, the masculinized females withdrew
less often than did untreated females from initiations, threats, and approaches of other animals. They also showed a greater tendency to mounting; evidently there is a general tendency toward male behavioral repertoire.
Treatment changed the whole brain-hormone-behavior complex, and the
results of the prenatal treatment persisted into the third year of life.

However, it should be stressed that the expression of the pattern
depended on social learning. The infant monkeys were allowed to play, and
Harlow and Harlow (1965) have shown that gross behavioral deficits in
behavior result from early social isolation. It is also important to remember
that these testosterone-treated monkeys were protected. In a free-ranging
troop of rhesus, subject to human harassment, predation, intertroop conflict,
and intratroop aggression (Jay, personal communication; Southwick, Beg,
and Siddiqi 1965), such an animal would probably be punished for inappropriate behavior. The experimentally masculinized female does not have
the large canine teeth, jaw muscles, or body size to be successfully aggressive against males under natural conditions. A similar point is well illustrated in an experiment by Delgado (1963a and b). A monkey in whose
brain an electrode had been implanted so that threat behavior could be
stimulated at the experimenter's will, was put in a cage with four other
monkeys. In the test where the experimental animal was dominant over the
other four, stimulation led him to threaten and immediately attack. But when
the experimental animal was subordinate to all four of the other monkeys,
stimulation led to his being attacked and cowering. Even when there is a
restricted biological base for a behavior, the expression of the behavior will
be affected by socio-environmental factors.

This brings us to a closer examination of the brain in relation to aggressive behavior. Perhaps, the most thorough work has been done on cats
and this is summarized by Brown and Hunsperger (1963). Their experiments
show several areas in which electrical stimulation will elicit threat and escape
behavior. They have been able to elicit such behavior by stimulation in
portions of the midbrain, hypothalamus, and amygdala. Although, just as in

monkeys, threat can be elicited from only a very small part of the brain, it is not a single or simple area. The threat behaviors are multiple and it is especially interesting that escape is closely related to threat. Brown and Hunsperger relate the anatomical facts to the behavioral fact that following threat an animal may either attack or escape. Certainly this is frequently seen in monkeys; whether threat ends in attack or flight depends on the participating animal's appraisal of the situation. In the laboratory, the direction and intensity of the attack resulting from stimulation of the brain depends on what is available for attack, and may be changed by offering the cat a dummy or a real rat.

In man the same parts of the brain are believed to be involved in rage reactions. Obviously, the same kind of detailed stimulation cannot be undertaken on man, but clinical evidence including neurological studies suggests that the limbic system and hypothalamus are very important in mediation of emotional experiences, positive and negative, including anger.

AGGRESSION IN FREE-RANGING APES AND MONKEYS

Just as the biological basis of aggression that we have been discussing can only be seen and analyzed in the laboratory, so the functions and frequencies of aggressive behaviors can only be determined by field studies. The field studies have recently been reviewed by Hall (1964) and Washburn (1966), and here we will call attention only to a few of the major points of interest from an evolutionary point of view.

Conflict between different species is infrequent, even when the species are competing for the same food. Places where the general situation can be most easily observed are at water holes in the large African game reserves. Particularly at the end of the dry season, hundreds of animals of many species may be seen in close proximity in South Africa, Rhodesia, or Tanzania. The Ngorongoro crater affords magnificent views over vast numbers of animals, and from this vantage point it becomes clear that the human notion of "wild," that is, that animals normally flee, is the result of human hunting. In Amboseli it is not uncommon to see various combinations of baboons, vervet monkeys, warthogs, impala, gazelle of two species, zebra, wildebeest (gnu), giraffe, elephant, and rhinoceros around one water hole. Even carnivores, when they are not hunting, attract surprisingly little attention. When elephants walk through a troop of baboons the monkeys move out of the way in a leisurely manner at the last second, and the same indifference was observed when impala males were fighting among the baboons or when a rhinoceros ran through the troop. On one occasion two baboons chased a giraffe, but, except where hunting carnivores are concerned, interspecies aggression is rare. Most animals under most conditions do not show interest in animals of other species, even when eating the same food — warthogs and baboons frequently eat side by side. The whole notion of escape distance is predicted on the presence of a hunter.

Although the general situation seems to be great tolerance for other species (Hall 1964), there are exceptions. Gibbons usually drive monkeys from fruit trees (Ellefson 1968). Goodall has shown remarkable motion pictures of baboons and chimpanzees in aggressive encounters. Baboons have been seen trying to drive vervets (*Cercopithecus aethiops*) from fruit trees in which the baboons had been feeding. There is some deliberate hunting by monkeys and apes. DeVore (personal communication) and Struhsaker (1965) have seen baboons catch and eat vervets. Goodall (1965) records chimpanzees' hunting and eating red colobus monkeys. Nestling birds and eggs are probably eaten by most monkeys, but the majority of interspecific encounters among the primates appear to be neutral, causing little or no reaction among the species.

Monkeys and apes certainly take aggressive action against predators, and this has been particularly well described by Struhsaker (1965) for vervets. Vervet alarm calls distinguish among snakes, ground predators, and birds, and the monkeys respond with different appropriate actions. Baboons have been seen to chase cheetahs and dogs. Monkeys and apes make agonistic displays against predators, including man, and these behaviors have been reviewed by Hall (1964, this volume) and Washburn (1966). The amount of this agonistic behavior leads us to think that predation and interspecies conflict may have been underestimated in the field studies so far available. The problem is that although the primates may have become conditioned to the observer's presence, he is likely to disturb the predator. A fuller picture of interspecies conflict requires field studies of a nonprimate species involved in conflict with primates.

Relations among groups of the same species range from avoidance to agonistic display and actual fighting. In marked contrast to the normally neutral relations with other species, animals of the same species evoke interest and action. This can be seen when strange animals of the same species are artificially introduced (Gartlan and Brain 1968; Kummer 1968b; Hall 1964; Washburn 1966), on the occasions when an animal changes troops, and when troops meet. We think these behaviors suggest that intertroop conflict is an important mechanism for species spacing. The spacing represents a part of the adjustments of the species to the local food supply. The quantity of food is a very important factor in determining the density of primate populations. It has been shown in both Japan and a small island off Puerto Rico where rhesus monkeys were introduced that population expands at a rate of more than 15 percent per year if food is supplied ad libitum (Koford 1966). Intertroop aggression either leads to one group's having the resources of an area at its exclusive disposal, or at least creates a situation in which one group is much more likely to obtain the food in one area. The clearest description of extreme territorial defense is Ellefson's account

Fig. 8-2. Typical scene at water hole at Amboseli Game Reserve. Baboons, warthogs, and zebras mingle and feed without interaction. (*Photograph by Sherwood Washburn*)

(1968) of gibbons. The relation of food supply to population size is considered by Hall (this volume). A very clear case of the relation of food supply to territorial defense is given by Gartlan and Brain (1968): Where food was abundant and there was a high density of vervets, the monkeys showed territorial marking and defense. These behaviors were absent in an area of poor food supply and low population density. From this and other examples (rhesus, langurs) it is clear that one cannot describe a primate species as "territorial" in the same sense the word has been used for species of birds. In monkeys and apes the behavior of a part of a species will depend both on biology (perhaps best shown by the gibbon) and on the local conditions. The intertroop fighting of city rhesus monkeys appears to depend both on high density and on the great overlap of living areas that is a product of the city environment (Southwick, Beg, and Siddiqi 1965). The intertroop conflict of langurs described by Yoshiba (1968) also occurred in an area in which the population is estimated at possibly more than 300 langurs per square mile.

It is our belief that intertroop aggression in primates has been greatly underestimated. No field study has yet been undertaken with this problem as a focus, and no effort has been made to study situations in which conflict is likely to be frequent. More important, the groups of a species are normally spaced well apart, and the observer sees the long-term results of aggression and avoidance, not the events causing it. (In this regard, as in so many others, gibbons are exceptional.) A further complication is that the groups of monkeys which are likely to meet have seen each other before. The relations among groups has been established in previous encounters, and one is exceedingly unlikely to see strange troops meet or some major event change the relative strength of the troops. Carpenter (1964), particularly, has called attention to the importance of sounds in species spacing, and, in species in which this mechanism is important, group avoidance does not even require that the animals see each other. The importance of both sounds and gestures in intertroop relations is discussed by Marler (1968). Lorenz (1964, 1966) has stressed the ritualistic nature of the vast majority of aggressive encounters.

In evaluating the amount of intertroop aggression in Old World monkeys and apes, it is important to keep in mind that the data have increased very rapidly. In Scott's (1962) review on aggression in animals the only major sources of information on primates were Carpenter's studies of the howling monkey (1934) and of the gibbon (1940). In Hall's (this volume) review of aggression in primate society the data are chiefly from publications in 1962 or later. We stress the frequency and importance of aggressive behavior more than Hall does, in part because of our greater emphasis on the biological importance of aggression in species spacing, but more importantly because there is much more information available in recent accounts. Aggression in langurs is described by Ripley (1965) and Yoshiba (1968). Many more observations of aggressive encounters in rhesus, including intertroop fighting in forest troops, are now available (Jay, personal communication). Shirek-Ellefson (personal communication) has observed a complex pattern of intertroop fighting in *Macaca irus*. Ellefson (1968) has given a much more complete account of intertroop encounters, including actual fighting, in gibbons. For vervets, Gartlan and Brain (1968) and Struhsaker (1965) have provided descriptions of intertroop encounters and of the settings that increase their frequency.

AGGRESSION WITHIN THE LOCAL GROUP

Conflict between individuals within the local group or aggregation is far more frequent than intergroup or interspecies conflicts. It is impossible to watch monkeys and apes for any long period without seeing conflict over food or in interpersonal relations. Scott (1962) has emphasized the importance of learning in the development of aggressive behaviors and Hall (this volume) has shown that most learning in monkeys takes place in a group and is appropriate to the group's social structure, individual biology, and ecology. In the societies of nonhuman primates aggression is constantly rewarded. In baboons (DeVore and Hall 1965; Hall and DeVore, this volume) the most dominant male can do what he wants (within the limits of the traditions of the troop), and he takes precedence in social situations. As

DeVore first emphasized, the dominant male is attractive to the other members of the troop. When he sits in the shade, others come to him to sit beside him and groom him. When the troop moves, it is the behavior of the dominant males in the center of the troop that ultimately decides the direction the troop will follow. The whole social structure of the troop rewards the dominant animal, or animals, and when a dominant animal is sick or injured and loses position the change can be seen in the behavior of the other animals. No longer is precedence granted to him for social position, grooming, food, sex, or leadership. Thus, monkeys not only have the biological basis for aggressive behavior, but also use the equipment frequently, and success is highly rewarded.

There are marked species differences in aggressive behavior and in the dominance hierarchies that result from it. Baboons and macaques are probably the most aggressive of the monkeys, but even here there are species differences. *M. radiata* is far less aggressive than *M. mulatta* (Simonds 1965). The behavior of *Papio hamadryas* is certainly different from that of other baboons (Kummer 1968*b*). But *interindividual conflict is important in all species described so far.* Even in chimpanzees with their very open social organization (Goodall 1965; Reynolds and Reynolds 1965) some males are dominant. Goodall had to make elaborate arrangements to prevent a few large males from taking all the food when bananas were provided.

The position of the individual animal relative to other animals in the group is learned, and this process starts with the mother and her support of her infant in aggressive encounters (Yamada 1963; Sade 1965, 1966). Sons of dominant females are more likely to be dominant. The passing of the infant langur from one female to another may be one of the factors in the lack of development of clearly defined dominance hierarchies in this species (Jay, this volume).

Since the animals in a local group know one another, the dominance order is understood, is normally maintained by threat, and usually serves to preserve a relatively peaceful situation. For example, a small group of crab-eating macaques (*M. irus*) kept in a runway (16 by 75 feet) at Berkeley was dominated by one male. For more than two years there had not been a single serious bite by any member of the group. When the dominant animal was removed, no change occurred for two weeks; the social habits continued. Then the formerly number two animal asserted his power, and four adult animals received deep canine bites. (These bites are quite different from incisor nipping, which hurts the other animal but does not do serious damage. Incisor nipping is the normal mode of biting when an animal gives mild punishment.) Two infants were killed in the encounters. This incident clearly shows the role of dominance in preventing fighting. It also shows another characteristic of dominance behavior in macaques. In the runway all animals had access to ample food; they had comfortable social position including opportunities for grooming; and the dominant animal, although he copulated

Fig. 8–3.
Female baboon grooming wound
that was the result of a fight for
dominance. (*Photograph   by
Sherwood Washburn*)

more than the others, did not prevent them from access to females. Being
dominant appears to be its own reward—to be highly satisfying and to be
sought, regardless of whether it is accompanied by advantage in food, sex, or
grooming. In the long run, position guarantees reward, but in the short run,
position itself is the reward, as this monkey's actions suggest; satisfaction
apparently comes from others being unable to challenge effectively, as well
as from more tangible rewards.

Evolution of Conflict

The aggressive behaviors that are the basis of dominance within the
group, that are a factor in spacing groups, and that may result in some predation are rooted in the biology of the species and are learned in the social
group. As noted earlier, the biological roots of these behaviors are complex,
and the individual animal, which carries out the threat or other aggressive
action, must have the necessary structure, physiology, and temperament.
For example, males tend to be more aggressive than females, and this difference depends on testosterone and is altered by castration of the male or
prenatal treatment of the female. The aggressive actions are practiced and
brought to a high level of skill in play. Then, as the male monkey becomes
fully adult, the canine teeth erupt. Notice that the whole practiced, skillful,
aggressive complex is present before the canine teeth erupt. The really

dangerous weapon is not present until the male monkey is a fully adult, experienced member of the social group. As the canine teeth erupt, the temporal muscles more than double in size, and the male changes from a roughly playing juvenile to an adult that can inflict a very serious wound, even death, with a single bite.

All the parts of this aggressive complex evolve, and this is best shown by the differences between species. The differences between baboons and patas monkeys give an example of very different ways of adapting to savanna life (Hall, this volume; DeVore and Hall 1965; Hall and DeVore, this volume). Differences between *Cercopithecus aethiops* and *C. mitis* are noted by Gartlan and Brain (1968). Since selection is for reproductive success, it is clear that there must be a balance between all the different structural and physiological factors that make aggressive actions adaptive; although the biological elements seem remarkably similar in primates, the pattern and degree of development may be very different in various species. It is no accident that the differences between male and female monkeys are in body size, tooth form, neck muscles, hormones, brain, play patterns, and adult behavior, and this whole pattern of sexual differentiation may result in sex difference that is extreme (as in baboons and macaques) or very minor (as in *Presbytis rubicunda* or *Cercopithecus nictitans*). But as these species have evolved, the process has been slow enough so that selection has modified the whole complex of the adapting aggressive behaviors and their biological base. In man, however, the whole technical-social scene has changed so rapidly that human biological evolution has had no opportunity to keep pace. Throughout most of human history societies have depended on young adult males to hunt, to fight, and to maintain the social order with violence. Even when the individual was cooperating, his social role could be executed only by extremely aggressive action that was learned in play, was socially approved, and was personally gratifying.

In the remainder of this paper we wish to consider human aggression, and the problems created by the nature of man.

As Lorenz (1964, 1966) has stressed, most conflict between animals is ritualized. Gestures and sounds convey threats and greatly reduce the amount of actual fighting. This is certainly true for the primates, and many structures are understandable only as the basis for displays. Dramatic structures of this kind, such as the laryngeal sac of the siamang gibbon, have long been recognized, but many less noticeable (from a human point of view) should be included—for example, the pads of connective tissue on the head of the male gorilla or those of the male orangutan's cheeks. Motions of the ears, scalp, eyelids, are important in gesture. The posture, or the position of the tail, may signal social status. Hair, particularly on the shoulders and neck, erects, signaling aggressive intent, and the manes of many male primates probably are to be interpreted as structural adaptations for agonistic display. Man lacks the kind of structures that the other primates use in threat and agonistic display. Although the structures used in display may differ to some extent from species to species, it is remarkable that man has none—no erecting hair, colored skin, callosities, or dramatic actions of ears or scalp. The kinds of gesture that communicate threat in the nonhuman primates have been shifted to the hand (freed by bipedalism and made important by tools) and to language. The evolution of language as a more efficient method of social communication, including the communication of threat, changed the pressures on a wide variety of other structures that must have functioned in agonistic display, unless it is postulated that our ancestors were unique among mammals and lacked all such adaptations. For example, only about one-half of the behavioral items that Brown and Hunsperger (1963) list as indicating agnostic behavior are anatomically possible in man. It is particularly the kind of structures that signal threat at a distance that have been lost. But even the structures that serve in close, face-to-face social communication may have been simplified. Human facial muscles have been described as more complex than those of the apes, making more elaborate expressions possible, but this is surely a misreading of the anatomical evidence and there is no evidence that the facial muscles of a chimpanzee are less complicated than those of man. Certainly the chimpanzee's mouth is more mobile and expressive, and a much wider variety of mouth expressions are possible in an ape than in man.

If we read the evidence correctly, in man language replaces the agonistic displays of nonhuman primates, and it opens the way to the existence of a social system in which aggressive behavior is not constantly rewarded. As noted earlier, in the societies of monkeys and apes dominance is the key to social order. Even if the dominance system of a group is not a rigid one, individuals in protecting young, gaining access to food, sex, grooming, or social position often threaten, and the threat—or, rarely, actual aggression—is rewarded with the acquisition of the desired goal. Agonistic behavior is an essential element in the day-to-day behavior of monkeys and apes, and language removes the necessity of rewarding this kind of aggressive behavior.

Just as the changed selection that came with tools led to increase in the parts of the brain controlling manual skills and to reduction in the whole tooth-fighting complex, so the origin of language led to changes in parts of the brain (Lancaster 1968b) and to a reduction or loss of most structures concerned with displays. In this sense the human body is in part a product of language and of the complex social life that language made possible. Similarly, the emotions of man have evolved in a way that permits him to participate in complex social life (Hamburg 1963). We think it is probable that individuals with uncontrollable rage reactions were killed and that, over many thousands of years, there was selection for temperaments compatible with moderately complex social situations. This process may have been somewhat like the early stages of domestication that involved the removal of socially impossible individuals, rather than the breeding of animals according to any plan. It is a fact that the human adrenals are relatively small compared to those of nonhuman primates and in this way man differs from the ape as domestic rats do from wild rats.

The expression of the emotions in man is more complex than in nonhuman primates, and, although emphasizing the continuity of the biological nature of aggressive behavior, we do not forget the remarkable differences. Compared with the ape or monkey, all the association areas of the human brain have undergone a three-fold increase in size. These are the areas particularly concerned with the ability to remember, to plan, and to inhibit inappropriate action. The increase in these areas is probably the result of

Fig. 8–4. Lioness threatening baboons who have escaped into a tree at Nairobi Park. (*Photograph by Sherwood Washburn*)

new selection pressures that came with the evolution of more complex forms of social life, and is probably highly related to the evolution of language which made the new ways of life possible. Taken together the new parts of the association areas and the parts of the brain making language possible might be thought of as the "social brain" — the parts of the brain that (from an evolutionary point of view) evolved in response to social pressures and the parts that today mediate appropriate social action. This concept is consistent with the fact that degeneration in these parts leads to senile dementia, the inability of some old people to continue normal social life — to remember, to plan, and to keep actions appropriate to time and place. However, the social world in which the human brain and emotions evolved was very different from the present one.

Throughout most of human history (at least 600,000 years, if by "man" we mean the genus *Homo,* large-brained creatures who made complex tools, hunted big animals, and at least some of whom used fire), our ancestors lived in small groups, and (as evidenced by the ethnographic literature, archeology, and the behavior of the nonhuman primates) males were expected to hunt and to fight, and to find these activities pleasurable. Freeman (1964) has given us an anthropological perspective on aggression; the record of war, torture, and planned destruction is exceedingly impressive. Most of the behaviors are so repugnant to our present beliefs and values that people do not want to consider them; in spite of the vast number of courses offered in the modern university, usually there is none on war, and aggression is treated only incidentally in a few courses. As ordinarily taught, history is expurgated, and the historian considers the treaties that were never kept rather than the actual experiences of war.

The situation relative to human aggression can be briefly stated under three headings. First, man has been a predator for a long time and his nature is such that he easily learns to enjoy killing other animals. Hunting is still considered a sport, and millions of dollars are spent annually to provide birds, mammals, and fish to be killed for the amusement of sportsmen. In many cultures animals are killed for the amusement of human observers (in bullfighting, cockfighting, bear baiting, and so forth). Second, man easily learns to enjoy torturing and killing other human beings. Whether one considers the Roman arena, public tortures and executions, or the sport of boxing, it is clear that humans have developed means to enjoy the sight of others being subjected to punishment. Third, war has been regarded as glorious and, whether one considers recent data from tribes in New Guinea or the behavior of the most civilized nations, until very recently war was a normal instrument of national policy and there was no revulsion from the events of victorious warfare, no matter how destructive. Aggression between man and animals, between man and man, and between groups of men has been encouraged by custom, learned in play, and rewarded by society.

Man's nature evolved under those conditions, and many men still seek personal dominance and national territory through aggression.

The consequence of this evolutionary history is that large-scale human destruction may appear at any time social controls break down; recent examples are Nazi Germany, Algeria, the Congo, Vietnam. Further, it must be remembered that the customs governing our lives evolved in the era when killing animals for fun, the brutal torture of human beings, and war were opposed by few. It is not only our bodies that are primitive, but also our customs, which are not adapted to the crowded and technical world that is dominated by a fantastic acceleration of scientific knowledge. Traditional customs nurtured aggression and frequently continue to do so.

The view that man is aggressive because of his evolutionary past, because of his biological nature, seems pessimistic to some, but we agree with Freeman that if aggression is to be controlled in a way compatible with survival and the realities of the new world of science, "it is only by facing the realities of man's nature and of our extraordinary history as a genus that we shall be able to evolve methods likely, in some measure, to succeed" (1964:116).

The situation might be compared to that of a bank. It is desirable to have employees who are honest people who will abide by the bank's customs. But no bank would rely solely on the honesty of its employees. The best auditing and accounting devices are used to make it virtually impossible for the human element to disrupt the functions of the institution. But on the international scene no comparable institutions for accounting and auditing exist, and reliance is still placed on the judgment of leaders and the customs of states. But these states have used war as a normal instrument of policy, their customs have glorified war, and all history shows that nothing in the human leader will necessarily restrain him from war if he sees success as probable. There is a fundamental difficulty in the fact that contemporary human groups are led by primates whose evolutionary history dictates, through both biological and social transmission, a strong dominance orientation. Attempts to build interindividual relations, or international relations, on the wishful basis that people will not be aggressive is as futile as it would be to try to build the institution of banking with no auditing on the basis that all employees will be honest.

In summary, in Old World monkeys and apes aggression is an essential adaptive mechanism. It is an important factor in determining interindividual relations; it is frequent; and successful aggression is highly rewarded. It is a major factor in intergroup relations, and the importance of aggression as a species-spacing mechanism means that aggression is most frequent between groups of the same species. Both within groups and between groups aggression is an integral part of dominance, feeding, and reproduction. The biological basis of aggressive behaviors is complex, including parts of the brain, hormones, muscles-teeth-jaws, and structures of display; successful aggression has been a major factor in primate evolution.

Man inherits the biological base, modified by the great development of the social brain and language. Aggression may be increased by early experience, play, and the rewards of the social system. The individual's aggressive actions are determined by biology and experience. But an aggressive species living by prescientific customs in a scientifically advanced world will pay a tremendous price in interindividual conflict and international war.

chapter nine
SOCIAL COMMUNICATION
IN SOME OLD WORLD MONKEYS
AND GIBBONS[1]

Judith Shirek-Ellefson

INTRODUCTION

As the previous chapters have illustrated, all the free-ranging Old World monkeys and apes that have been observed organize themselves into social groups. There is wide variation in group composition, which includes: a mated pair and offspring, as is found in gibbons (*Hylobates lar*); one-male groups, as in hamadryas; males, several females and young of all ages, as in bonnet macaques (*Macaca radiata*). There is also variation in how species distribute themselves in space, some species occupying exclusive territories (gibbons), some occupying overlapping home ranges (bonnet macaques), and troops of some species doing either one or the other depending on the local ecological conditions (vervets, *Cercopithecus aethiops*). Species and troops within species also vary in the relative permanence and stability of social groupings, the temporary, open social groups of the chimpanzee (*Pan troglodytes*) representing one extreme, and the enduring, relatively closed social groups of anubis baboons (*Papio anubis*) representing the other. Social organization in Old World monkeys and apes varies so much that what were once thought to be rigid patterns for a species are turning out to be variations on a general species pattern on further investigation.

In this tremendous amount of variation and flexibility, the communication repertories of all the Old World monkeys and apes are remarkably similar. The same patterns occur over and over again in different species living in quite different ecological settings and with quite different social organizations. This similarity is remarkable for at least two reasons:

1. The variation in social organization in the Old World monkeys and apes represents a variety of evolutionary solutions to the problems presented by a variety of ecological settings and is the result, in many species, of millions of years of separate evolutionary history.

[1] The fieldwork on *Macaca irus* reported in this paper was supported by Public Health Service Fellowship 5-F1-MH-21, 726-02 and Grant MH-11,273,-01.

2. Social communication functions to mediate the interpersonal relationships that are the basis of social organization. It is surprising that the patterns have remained structurally similar in spite of mediating a diversity of interpersonal relationships in a wide variety of social settings. How is it that patterns of communication can remain so stable when the results of the communication — social organization — are so varied? The answer lies in the fact that the structural properties of a pattern of communication are not all that makes up an act of communication. The social context provides many additional stimuli that play a vital role in any act of communication. Thus the informational content of a pattern can change without the structure of the pattern itself changing. When a change in context can change the meaning or the affect of a signal pattern, then changes in structure are not necessary.

The term communication can include a wide variety of behaviors. Any behavior of one or more individuals that affects the behavior of any other individuals can be considered an act of communication (see Altmann 1967b). Behaviors the primary function of which is something other than communication can thus be included under this definition. For example, a feeding animal by his act of feeding may cause other animals to approach and investigate. The behavior of the other animals has been affected by his behavior, but the primary function of his behavior was to feed himself. A discussion of communication should include such incidentally communicative behavior as well as the communication that takes place between troops of the same species and between groups of animals of different species. Alarm calls, for example, often alert many animals outside the immediate social group of the sender. In general, however, communication occurs most frequently between members of the same social group. The gibbons are the major exception to this statement, for their communication repertoire includes many signal patterns that are used in territorial conflicts between neighboring groups (Ellefson 1967, 1968).

This chapter will deal with only those patterns of social communication that have communication as their primary function, occurring between members of the same social group. These signal patterns function to mediate the interpersonal relationships that are the basis of group living. By expressing motivational states, the patterns of communication permit troop members to adjust their behavior with regard to the intent and personality of other troop members.

THE STRUCTURE OF SIGNAL PATTERNS
OF OLD WORLD MONKEYS AND APES

The structure of the signal patterns of the Old World monkeys and apes is well suited to the exchange of subtle motivational information at close range. The ears, scalp, eyes, eyelids, eyebrows, mouth, lips, tongue,

Fig. 9–1. Mild grimace: subadult male crab-eating macaque (*Macaca irus*).

torso, arms, and legs, and the facial, head, and body hair can all be incorporated into the visual signal pattern. In some species particular facial features are emphasized by contrasting pigmentation. For example, in many species the eyelids are unpigmented and thus appear as white areas above the eyes when the eyelids are lowered and/or the brows raised. With all these elements, each capable of many different positions and each potentially varying independently, the possible number of visual signal patterns is great.

Clusters of expressive elements (facial or body features combined with particular postures) recur, and thus permit identification and naming of signal patterns, such as the grimace, the threat, and the lip smack. There are two important features of these recurring clusters. One is that they do vary considerably, particularly with respect to intensity. In a grimace, for example, the lips can be slightly retracted exposing a little of the teeth, or they can be retracted a great deal exposing all the anterior teeth, or they can be at almost any intermediate position (Figs. 9–1 and 9–2). This variation, often referred to as grading, apparently reflects variation in the intensity and kinds of emotion underlying the facial expression or body posture. The other important characteristic of these recurrent clusters is that many are morphologically similar to other clusters and in fact clusters merge into one another. Thus it is easy for the animal to make a transition from one expression or body posture to another morphologically similar one, and slight shifts in motivational state can be easily expressed.

In crab-eating macaques (*M. irus*) all the facial expressions but one (open-mouth threat) merge into at least one other facial expression. There is a continuum of facial expressions expressing a continuum of motivational states (Fig. 9–3). The scream threat, a submissive threat, merges with the grimace, a submissive gesture. The grimace merges with teeth chattering, a submissive gesture that sometimes leads to friendly social contact, with which lip smacking is associated. Lip smacking merges with the pucker face, a friendly gesture that results in social contacts between individuals who are tense when in each other's presence. The pucker face merges with the white-pout threat, a threat that occurs between individuals whose dominance relationship is unstable. Thus an individual animal can easily change, for example, from a submissive threat to submission, or vice versa. It can also easily change from submission to an expression that results in friendly contact, and vice versa. Individuals often do make rapid changes from one expression to another as the social context changes. This makes for a very

Fig. 9–2. Intense grimace: adult male crab-eating macaque (*Macaca irus*).

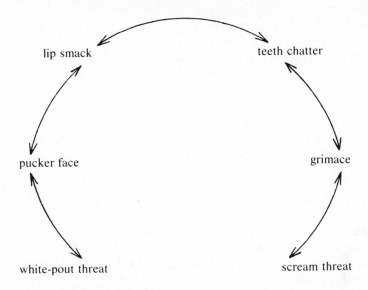

lip smack   teeth chatter

pucker face   grimace

white-pout threat   scream threat

Fig. 9–3. Morphological continuum of facial expressions used in communication by crab-eating macaques (*Macaca irus*).

flexible, adaptable communication system that is capable of handling subtle changes in the immediate social context.

Although expressions can merge into one another, it does not follow that they are ambiguous. An animal does not usually give an expression that is halfway between a grimace and a scream threat as the initial response to a situation. It is in the social context that the animal's mood changes and that its expression changes to reflect internal change. Thus the social context gives added information on which the receiver bases an interpretation.

Visual communication is the most frequently used channel of communication in the Old World monkeys and apes, but auditory communication and tactile communication are also important. Vocalizations often occur in combination with facial expressions, especially when motivation is high. Like visual communication patterns, many vocal patterns are graded. In Rowell's (1962) description of rhesus (*Macaca mulatta*) vocalizations, she reports not only that individual patterns are graded (in intensity), but also that some patterns merge with others, making a continuum of vocalizations. This enables the expression of subtle changes in motivation in the same way that grading and merging do in visual patterns. Not all vocal signals are graded and merging, and different species vary in the number of graded signals they use. Vervet (*C. aethiops*) vocalizations, for example, are primarily discrete and do not merge into one another (Struhsaker 1967a).

The vocal repertories of most species will probably prove to contain both graded and discrete signal patterns, for there are advantages in each. A discrete, nongraded, stereotyped pattern, whether vocal or visual, is unambiguous and thus easy to perceive and interpret. Patterns that are graded and merge into other patterns place a greater burden of interpretation on the receiver of the signal, but they can express the variation in the intensity of the underlying motivation and the shifts in motivational state that are necessary in the context of an ongoing social group. Like visual patterns, the vocal signal pattern itself is not the only stimulus presented to the receiver, and much of the ambiguity generated by grading and merging is cleared up by the characteristics of the social setting in which the pattern occurs.

Patterns of tactile communication, such as grooming, embracing, slapping, and grabbing, have not been studied in the detail that the visual and vocal patterns have. It is thus not clear whether the patterns are graded and whether they merge into one another. Tactile communication is often the end result of an interaction that began with a visual and a vocal overture. Patterns of tactile communication reinforce the social bonds expressed by other forms of communication.

The importance of olfactory communication in monkeys and apes is virtually unknown. There are suggestions, however, that the animals gain much information through the sense of smell. For example, bonnet macaque females (*M. radiata*) do not show sexual swelling when they are in estrus, as do many other macaque females; Simonds (1963) noted that the males routinely sniffed at the perianal region of the females and suggests that perhaps males are informed of the sexual receptivity of the female by her odor. The vervets that occupy exclusive territories go through marking behaviors typical of other mammalian species that mark their territorial boundaries with special scent glands (Gartlan and Brain 1968). Both crab-eating macaques (personal observation) and vervets (Struhsaker 1967*b*) apparently gain information about available food by sniffing each other's mouth while feeding.

In summary, the patterns of social communication in the Old World monkeys and apes are well suited to the expression of complex motivational information. The exchange of this information is necessary to maintain the interpersonal relationships at the basis of social living. Variation in the intensity of underlying motivation is expressed by the grading of elements within a single pattern. Changing motivation can be expressed easily because many patterns morphologically merge into other patterns.

The study of communication has focused so far primarily on the signal sent. Observers have been impressed by the lack of stereotypy as described above. The responses to the variations in signal patterns have not been studied in such detail, and thus it is not altogether clear how much of the variation reported is actually significant and responded to by the receiver, and how much is an artifact of the observers' descriptive methods.

ON AND SOCIAL CONTEXT

...unication systems of the Old World monkeys and apes within social groupings, the functions of the communication should be discussed within their social contexts. Social organization in the Old World monkeys and apes, and therefore the patterns of communication of each species, and even troop, should be discussed with reference to its own social organization.

Here only the common denominators of social context can be discussed. A major problem is that the data come from groups in various social circumstances and thus it is sometimes difficult to make comparisons. The members of a free-ranging social group have usually grown up in that group and know each other quite well. In contrast, animals that have been in laboratory colonies for only a short time do not know very much about one another. Subtleties of communication can occur between individuals who know each other intimately that simply cannot occur between individuals who do not know each other very well. No one has yet explored the question of how long it takes for animals initially strange to one another to become as familiar with each other as individuals who have grown up together in a group that has a continuity over time.

For some species, such as gibbons and hamadryas baboons, in which the adult free-ranging troop is composed of individuals that did not grow up together, comparability of laboratory and field situations can be achieved. Kummer and Kurt (1965) report few differences in social behavior between the hamadryas baboons (*P. hamadryas*) they observed in a zoo colony and those free-ranging troops they observed in Ethiopia. In the zoo colony nine patterns occurred regularly that were not seen in the free-ranging troops. Only two behaviors were seen in the wild and not in the zoo. It is possible to relate these to differences in the setting of the zoo and the wild. For example, the protected threat occurred in the zoo but not in the wild. In a protected threat, one animal threatens another and presents to a more dominant third animal at the same time. The threatener thus insures that he will not be attacked by the more dominant animal to whom he is presenting, and also insures that the animal he is threatening will be unable to present to the more dominant animal. This behavior does not occur in the free-ranging hamadryas troops because the one-male groups sit close enough to each other that if a protected threat occurred it would force one group member into the midst of another one-male group, thus precipitating a fight. It can occur in the zoo because there is only one one-male group in an enclosure.

For other species it is not at all clear how laboratory colonies differ from naturally occurring troops. It can be argued that after one generation captive colonies are comparable to naturally occurring troops because there is a generation that has grown up together, but this has not been demonstrated.

Two colonies of crab-eating macaques, 11 animals imported from Southeast Asia, were established at the Stanford Primate facility, Stanford, California in the Spring of 1967. There is no available information as to whether they came from the same or different troops. I observed the two groups for approximately 150 hours over a 16-month period. Neither of the laboratory groups had a full complement of age-sex classes. Fewer patterns of communication occurred between the individuals in the laboratory colonies than occurred between the individuals of a free-ranging troop I observed for 550 hours in the Singapore Botanic Gardens in 1965. In the larger of the captive groups all the visual signal patterns did occur, but two of the facial expressions that occurred daily in the Singapore troop occurred only once in the captive colony, when individual animals were being removed or returned to the colony and all the animals appeared to be under considerable stress. Many factors may be responsible for the differences, including the fewer number of hours of observation and the discrepancy in age-sex distribution.

An argument against the possible achievement of comparability of laboratory and free-ranging groups rests on the evidence that learning is important in the acquisition of subtleties and appropriateness of communication (Mason 1960, 1961a, 1961b). If the adults of laboratory groups rarely use certain expressions, it is quite possible that the young will never learn to use them appropriately. The problem needs further research in view of the large number of laboratory colonies used in studies on social behavior.

An important point is that the structure of the patterns of communication remains stable even under these widely varying conditions. The number of patterns present varies, social context varies, responses vary, frequency varies, but the basic structure of the patterns remains the same regardless of where the animals are. The laboratory is a good place to get detailed descriptions of the structure of Old World monkey and ape communication systems; the *form* or structure of a call does not yield much information on social organization and the function of communication.

Two basic functions, but not the only functions, of communication in the Old World monkeys and apes are to bring individuals together and keep them together to participate in such bond-maintaining activities as grooming, hugging, huddling, and play, and to keep apart individuals that would participate in such bond-disintegrating behaviors as fighting if they came into contact (Marler 1968). Although this appears to be an oversimplification of the complexities of primate communication, its very simplicity reveals details of complexity otherwise overlooked. There are several patterns that bring animals together and several patterns that keep animals apart; the immediate question is why. Are different signal patterns used to call animals together for different activities? Do the different signal patterns relate to different individuals or combinations of individuals? Are the contexts different? What is different in the situations that generates different emotional states requiring a different message be sent? Some signal patterns can both bring individuals together and keep them apart; what are the variables that indicate to the receiving animal that he should approach or retreat? Different species with different social organizations may use the same expression differently; how is this difference related to social organization? Approach and avoidance are basic tasks of communication. The complexities of social life in the Old World monkeys and apes are reflected in the elaborations in this basically simple system.

The following discussion emphasizes visual communication patterns. The brief descriptions do not do justice to the detailed differences that exist in the structure of the visual patterns; for a more detailed description see van Hooff (1967). I am most familiar with the visual communication patterns used by crab-eating macaques, and thus they form the basis for the following discussion. The information on social context for species other than crab-eating macaques is taken from van Hooff (1967) unless otherwise indicated. Where my terminology and van Hooff's do not correspond, the terminology used by van Hooff will be found in parentheses after mine. Statements as to underlying motivation are my interpretations based primarily on what happened prior to and following the pattern of communication in question.

The facial expressions and accompanying body postures in Old World monkeys and apes that result in friendly physical contact such as grooming, huddling, hugging, play, or just sitting together are lip smacking, teeth chattering, grimacing (silent bared teeth face), pucker face (pout face), and play face. Lip smacking, teeth chattering, and grimacing can result in avoidance as well as in approach, depending on the social context in which they occur, and these three facial expressions merge into one another. Lip smacking consists of rapid opening and closing of the mouth and lips, such that when the lips close they make an audible smacking sound. In teeth chattering, the lips are retracted, exposing the teeth, and only the mouth is rapidly opened and closed. In the grimace the mouth is closed and the lips and lip corners are retracted so that the teeth are exposed in a white band.

The position of the ears, eyelids, and eyebrows in these three expressions varies considerably both with regard to the intensity of the expression and to species differences. The eyes are directed toward the receiver. Body posture is often slightly tense and the approach to the receiver is cautious. The morphological continuum probably represents a motivational continuum—social attraction being high and fear low at one end (lip smacking); fear high and social attraction low at the other (grimace).

Lip smacking is associated with grooming, often occurring during a grooming bout while the groomer brings pieces of dirt and skin to its mouth. It also occurs as a prelude to grooming as the initiator of the grooming bout approaches the animal it intends to groom and lip smacks. If the animal being approached either lip smacks or presents a part of its body to be groomed, the initiator begins to groom and often continues to lip smack. When positional readjustments take place during a grooming bout, the groomer usually lip smacks rapidly until both parties are settled again. In crab-eating macaques, lip smacking occurred as a prelude to grooming and huddling most frequently between animals who spent many hours a day in close contact.

Teeth chattering often alternates with lip smacking, presumably reflecting increasing and decreasing fear. In most species teeth chattering does not function to bring animals together, but occurs only as part of the response to a threat and in that context increases distance (see below). However, in some macaque species (*M. sylvana* and *M. speciosa*) teeth chattering does function to bring animals together, primarily to engage in mutual hugging. High-pitched squealing may accompany teeth chattering in these sequences.

Grimacing results in approach and peaceful contact only in some social contexts. In crab-eating macaques, if a dominant animal approaches a subordinate and in so doing grimaces, the subordinate animal responds by remaining in the same place while the dominant approaches, after which they engage in behaviors such as grooming, huddling, hugging, and the like. No vocalizations occur with grimacing in this context.

The pucker face is characterized by eyes directed toward the receiver, ears and eyebrows retracted and eyelids lowered so that unpigmented eyelid skin is exposed. The mouth is only slightly opened, and the lips are protruded to varying degrees. In the most extreme form of the pattern the lips are protruded and pressed together, and the lip corners are drawn forward so that the lips are wrinkled. The mildest form of a pucker face in crab-eating macaques is simply exposure of the unpigmented eyelids. The expression is sometimes accompanied by a soft cooing vocalization. The body tends to move forward. My observations on crab-eating macaques revealed many more contexts in which pucker faces oocur, particularly among adults, than are reported by van Hooff for any of the species he observed. This may be owing to the differences mentioned above between laboratory colonies and free-ranging groups, or it may be that use of the expression in a variety of contexts is peculiar to crab-eating macaques. In most Old World monkeys and apes pucker faces are given by infants when they become separated from their mothers, and in this context the expression is accompanied by cooing, which may become screeching if the separation is prolonged. Adults and juveniles also give pucker faces when they become separated from the rest of the troop. In crab-eating macaques, pucker faces also occur when individuals that are either very distant in rank or very close in rank approach one another with the intention of making friendly physical contact. Juvenile males pucker-face as they approach adult males to groom them, to feed near them, to sit with them, and, occasionally, to play with them. Adult males also pucker-face to one another as they approach prior to greeting, hugging, grooming, or feeding. Pucker faces apparently express a strong motivation to make social and physical contact combined with the fear of being in close proximity.

Play faces and play postures vary considerably among species. The one thing they all have in common is that they are relaxed. Although play faces often display an open mouth that resembles the position of the mouth in an open-mouth threat, the relaxed (normal) position of the eyes and ears is quite unlike a threatening expression. At times the white eyelids are exposed in combination with the open mouth, resulting in an expression quite unlike an open-mouth threat. Body posture is relaxed and locomotion is often erratic and inefficient, and the animal makes many body and limb movements to cover very little ground. Play faces, as the name implies, lead to play, and they often occur during a play bout as well as prior to it. In crab-eating macaques a pucker face, when accompanied with a playful body posture, can also lead to play. A soft vocalization called "girning" by Rowell and Hinde (1962) often accompanies play faces and play bouts.

Soft coos and grunts occur during troop movement and when the troop is resting or feeding. These vocalizations apparently function to keep troop members in vocal contact, thus preventing individuals from becoming separated from the main body of the troop.

The signal patterns that result in an increase in the distance between the sender and the receiver are the various kinds of threats, grimaces (frowning bared-teeth scream face and silent bared-teeth face), and teeth chattering. Every species of Old World monkey and ape has at least two distinct threat patterns. One is some form of open-mouth threat, which consists of an opened mouth with the lips either covering the teeth or retracted so that the teeth are exposed. The position of the lips varies between species, with most species covering the teeth. The eyes stare at the opponent, the brows are either lowered or raised, ears either forward or pulled back, depending again on the species (Fig. 9–4). The accompanying vocalization is variously termed the bark, growl, or roar. In its mildest form this threat consists of a consistent stare. Open-mouth threats are given by the dominant member of an interacting pair. A grimace is the usual response to an open-mouth threat. If the motivation of the threatener is strong enough, it may give chase, catch, and bite the animal it is threatening.

The other common threat consists of a combination of the open-mouth threat and the submissive grimace. In crab-eating macaques it consists of a scream vocalization, a grimace with the mouth open and the lips retracted so that the teeth and sometimes even the gums are exposed, and a forward body posture (Fig. 9–5). The eyes, rather than being evasive as they often are in submissive grimaces, stare directly at the opponent. In all species this threat is given by the subordinate member of the interacting pair. In crab-eating macaques, scream threats are responded to by retreat, accompanied by an open-mouth threat. The scream-threatener never physically attacks the animal it is threatening. Quite often other troop members come to the aid of the scream-threatener and depending on their rank in relation to the animal being threatened, they either scream-threaten or give an open-mouth threat.

In crab-eating macaques, in which not all troop members have stable dominance-subordinate relationships, a third threat occurs between such troop members. This threat is morphologically similar to the pucker face and in fact merges into it. In the white-pout threat (Fig. 9–6) the white eyelids are very prominently exposed, the lips are puckered and slightly parted, the hair on the shoulders is erect, and a soft roar can sometimes be heard. Since the white-pout threat merges with the pucker face, it probably represents a motivational continuum with it. White-pout threats appear to be motivated by social attraction as well as aggression. Often the response to this expression is a grimace, but sometimes the animal being threatened returns the threat. If this happens, the interaction can last as long as 15 minutes, and many other troop members can become involved as they line

Fig. 9–4. Open-mouth threat: an adult male crab-eating macaque (*Macaca irus*) directs an open-mouth threat to a young visitor to the Singapore Botanic Gardens.

Fig. 9–5.   Scream threat: an adult female macaque (*Macaca irus*) scream threatens
a subadult male.

up behind either one combatant or the other and white-pout threaten. It may
be that dominance rankings are settled by these long interactions, but my
observations were too brief for me to see any results of that kind. This is the
only well-documented example of a threat that occurs between troop members where dominance-subordinate relationships are unstable. Simonds
(1963, 1965) reports that in bonnet macaques a jaw threat occurs that is different from either of the other two threats he observed. Simonds cannot
correlate the occurrence of the jaw threat with dominance, which suggests
that the context of the jaw threat may be similar to that of white-pout threats.

Threats are made for various reasons and it is difficult to give the common denominators of the situations in which they occur. Often threats occur
over desired objects, such as food, a preferred sitting place, or an estrus
female. They are also precipitated by the behavior of one individual toward
another and often in these situations it is difficult for the observer to understand the exact cause of the threat. Perhaps some troop members have long-
standing antagonisms, which erupt into threatening behavior over small
incidents that would go unnoticed between other pairs of individuals.

Grimacing and teeth chattering can increase the distance between the

Fig. 9–6. White-pout threat: four juvenile male crab-eating macaques (*Macaca irus*) white-pout threaten an adult female.

sender and the receiver when they are given in response to a threat or to the approach of a dominant animal. At high intensities of fear, grimaces are often accompanied by screeching and defecating. These signals increase the distance between the sender and receiver by the withdrawal of the threatening animal.

It is clear that the same signal pattern can either increase or decrease the distance between the sender and the receiver depending on the social context in which the pattern occurs. Thus, not only does the structural form of the pattern contain information on which the receiver bases his response, but so also does all variety of contextual variables, such as the identity of the sender, the identity of other individuals in the vicinity, what happened just prior to the sending of the particular pattern, what priority objects are nearby, plus numerous other possible factors. The informational content of a signal pattern can change without the form of the pattern itself changing, because contextual changes result in informational changes. The structural characteristics of Old World monkey and ape communication systems enable the transmission of complex, subtle motivational information. The incorporation of context into each pattern further increases the complexity and subtlety of the information passed between the sender and the receiver by increasing the number of variables to which the receiver responds.

This discussion is only a brief summary of social communication and social context. It describes only a portion of the total communication that occurs in a social group of Old World monkeys. The variety of social contexts and social relationships that exists in such a group leads to much more complexity and variation in occurrence than I have been able to suggest here. The descriptions of facial expressions and body postures do not do justice to the variation in detail that occurs within a species and between species. A few hours spent observing a group of monkeys or apes at a local zoo will give the reader a more realistic picture of the variety, complexity, and subtlety of their communication.

SUMMARY

This review chapter on communication in Old World monkeys and apes has focused on a few of those visual patterns that occur within the social group. The communication systems of the Old World monkeys and apes are well suited to the exchange of complex motivational information. Many signal patterns are graded, enabling the expression of variation in emotional intensity. Many signal patterns are morphologically similar to other signal patterns, and thus they merge easily into one another, enabling the expression of changes in emotional state. Basically all patterns either bring together and keep together individuals to engage in such bond-maintaining activities as grooming, playing, hugging, huddling, and sitting, or they keep apart individuals that would engage in disruptive behaviors such as fighting.

Social relationships and social contexts within a troop of monkeys or apes are varied and complex, and there are several patterns to express the differences in emotional state generated by differences in relationships and context. Some signal patterns can either bring animals together or keep them apart depending on the social context in which they occur. The receiver of a signal pattern gains information not only from the structure of the pattern itself, but also from the social context in which it occurs. The inclusion of context permits the structure of Old World monkey and ape communication repertoires to remain stable with regard to form and number while social organization has become varied. Changes in social organization can occur without necessitating changes in the structural form of the pattern.

Differences in social organization should be reflected in parameters such as frequency, age-sex distribution, and response. More data on these variables are needed to gain a real understanding of the relationship between the evolution of social organization and social communication.

chapter ten

THE DEVELOPMENT OF MOTOR SKILLS AND SOCIAL RELATIONSHIPS AMONG PRIMATES THROUGH PLAY[1]

Phyllis Jay Dolhinow and Naomi Bishop

IN SEARCH OF PLAY

Introduction

Important as play is to mammals, remarkably little attention has been given to its development, forms, and effects. Although play has been a topic of interest and research for decades, early theories of play were based on anecdotes and speculation rather than on controlled observation and laboratory experimentation (Beach 1945). This paper will concentrate on the play of free-ranging nonhuman primates and in particular on the Old World monkeys and the apes.

Only a small number of the living nonhuman primates have been observed in the field, and only a few of the species studied have been observed in different habitats. Play has been recorded as an incidental activity in studies that emphasized other aspects of social behavior. The data are limited and the vocabularies used by different observers are sometimes difficult to equate. Observers vary tremendously in the selectiveness of what

[1] This paper is part of a program on primate behavior supported by the United States Public Health Service Grant No. 8623, Analysis of Primate Behavior. Mrs. Bishop was supported by the National Institute of Health (National Institute of General Medical Sciences), Training Grant No. GM-1224. We wish to thank Dr. S. L. Washburn for reading the manuscript and for the helpful suggestions he gave us; Anne Brower for her editorial assistance; and Lynda Muckenfuss for bibliographic assistance. Many of the comments in this paper are based on the authors' observations of nonhuman primates in the field and in captivity. Mrs. Dolhinow's major field studies have been on langurs and rhesus monkeys in North India; she has also observed baboons and chimpanzees in East Africa and primates in Ceylon and South India. *Macaca irus* have been observed at the Berkeley Behavior Station by both authors. We assume full responsibility for generalizations that are made in this paper, at the same time acknowledging our debts to many researchers who have not been quoted directly.

The full version of this paper appeared in the *Minnesota Symposia in Child Psychology*, vol. 4, pp. 141–198, John P. Hill, ed., University of Minnesota Press, Minneapolis, © Copyright 1970 University of Minnesota.

they record and in their manner of recording it. Most studies concentrating on play have been undertaken in captive colonies or in laboratories, and the problems of comparing results of the field with those of the laboratory are exceedingly difficult.

Since Beach's article in 1945 animal behaviorists have become increasingly interested in play. A comprehensive review of their recent work, primarily with laboratory and captive animals, is summarized by Loizos (1967). Welker (1961) has compiled a partial list of studies that include descriptions of animal play and exploration within the classes Mammalia, Aves, and Pisces. According to his survey, play has been described in 54 genera of Mammalia, in 10 of Aves, and in none of Pisces. Play is primarily limited to mammals and probably does not exist in invertebrates, fish, amphibians, or reptiles.

Within the class Mammalia, there are great differences in variety and complexity of play patterns and great variability in capacity for learning. An examination of the many different kinds of mammalian play demonstrates the relevance of the activity and of the specificity of patterns of play to the life of the animal. Many special skills and behaviors important in the life of the individual are developed and practiced in playful activity long before they are used in adult life.

The social behavior of monkeys and apes is far more complex than that of any other mammal except man, and the variability and versatility of their play is due, in part at least, to the increased complexity of the brain including areas involved in memory and learning. Humans have developed the most complex of all play phenomena with the addition of human language and abstract thinking afforded by an even larger and far more complex brain. Human language and the ability to name objects is a unique evolutionary development late in Hominid history and its effects on behavior have been extraordinarily great (Lancaster 1967; Geschwind 1964).

Play in primates is highly variable, has many effects on the individual and on the group, and is exceedingly difficult to describe accurately. Early attempts at definition and explanation were confounded by these attributes. Beach recognized that no single hypothesis would apply successfully to all species and called for "a definition of play . . . based upon a number of predominating characteristics which combine to set it off from non-playful behavior" (1945:539). One feature of play appears to be universal among mammals; it is an activity usually characteristic of young animals. Although the equation of youth and play is usually a valid generalization, adults do play, albeit much less frequently than do young.

A tendency exists to regard whatever the infant or juvenile primate does awkwardly as playful behavior but this is not true; not all activities performed by young are play, however imperfect and clumsy they may appear. Other clues help distinguish between play and the imperfect execution of a nonplayful or so-called "serious" behavior. Various gestures and signals have developed to indicate the playful nature of a movement. Since many play patterns among primates have a potentially agonistic basis and involve quite strenuous physical contact, gestures that signal the intention of play are important. Humans also have their play signals, such as the wink and smile that when combined with a smart slap on the back, for example, prevent the person slapped from taking offense or becoming aggressive in return. Among primates when two conflicting signals are given simultaneously, the play signal usually predominates. Altmann has called this communication about communication, "metacommunication" (1962).

Many qualities associated with play will be discussed in the following sections. Among them repetition has often been cited as a major characteristic of play, and it has been suggested that repetition brings the animal back to something familiar and thus reassuring. In play, usual sequences of motor patterns are often not completed and there are frequent reversals of accustomed roles. For example, in a chase sequence A will chase B, and, before A catches B, B will turn and chase A. It appears that the real pleasure is in the chase and that it is relatively unimportant who is doing the chasing or whether catching ever takes place. Play is a very tenuous activity in the sense that it only occurs in an atmosphere of familiarity, emotional reassurance, and lack of tension or danger. Its course is also unpredictable in that it easily changes into other nonplay behaviors.

Three variables that affect play behavior are ecology, biology, and social context. These are interrelated and interdependent, and we find that the play of a group of animals in a particular location is apt to be slightly different from that of groups of the same species in different habitats. It is evident that the order Primates has utilized a great many of the possible variations in these three contexts and that play patterns reflect these differences in adaptation.

Summary

From the field studies of other Old World primates we can offer some generalizations. For the newborn monkey the social world is quite small and for weeks may include only its mother; for the infant chimpanzee this period lasts for months. The world soon enlarges to include more of the environment, the area immediately around the mother. Early movements include a great deal of exploration, and the infant tests his own body and the inanimate world. If siblings are a part of this immediately surrounding environment, then they too are investigated, felt, smelled, and tasted. In a few weeks, or as soon as the young primate is no longer restricted by its mother and is able to move around, he becomes interested in peers as play partners. In this manner he becomes acquainted with the individuals that he will live with for the rest of his life. Once social play begins, there is a marked preference for it over solitary play or play with inanimate objects, although among chimpanzees play with objects is also very important and continues to be frequent throughout the years of immaturity.

Group composition may vary with individual troops of monkeys. Normally, however, when there is a wide cross-section of ages in a troop there will be several play groups consisting of animals of similar age and size. The existence of a birth season produces this effect, since each year most of the infants in species living in certain habitats are born within a few consecutive months.

The play group is an important focus of the attention of young primates, and in it they spend hours a day for the years of their immaturity. It is a context for learning social and physical skills and as such is an important factor in development.

PLAY

Introduction

Some information on activities that have been called play has been gathered for most species of nonhuman primates observed in the field. Our aim in this section is not to categorize each action or pattern of behavior as play or not play, nor is it our intention to arrive at a typology of play. Rather, we will ask what general sorts of events called play are meaningful in the life of the individual primate and how may these activities be understood in terms of their effects on the individual and the group. Although many aspects of playful behavior have meaning for group life, this does not necessarily indicate that the specific play patterns were selected during evolution for those purposes. On the contrary, play is viewed here as a powerful set of behaviors, adaptive to the extent that they facilitate the individual's integration into his troop and otherwise affect his eventual reproductive success. These relationships are suggested rather than demonstrated, since demonstration would require experimental evidence that is not available.

Experimental work is needed on all aspects of play behavior and its social consequences if we are to understand the processes of learning that take place in play. Hall stated emphatically that the only kind of learning that is normal is that which takes place in a full and natural social group (Hall, this volume). The richness of the social group can only be appreciated in the field, and the same is true for the complexity of social relationships that structure the life of the individual. Unfortunately, in the field one sees the results of learning, the results of play, rather than the processes by which behavior is modified and by which learning occurs. In addition to careful quantification of behavior in the field (Altmann 1965, 1968a, 1968b), laboratory and colony research are needed. Field experimentation will also clarify some of these questions even though behavior in the field is the result of so many variables that it is exceedingly difficult to control or to measure their effects.

Discussing primate play often creates special problems for humans. At one time or another everyone has experienced the ease of projecting human feelings into monkey or ape behavior. The closer we get to man phylogenetically, the more cautious we must be in commenting on the motivational state of the animal, even when it is well known as an individual. In fieldwork, when the animals are studied for a long time, they become very well known as individuals, and after hundreds of hours an observer can become very skilled in understanding the reactions of individuals and in predicting their behavior. It is very tempting to speculate on what is happening in play in terms of what it might mean in the feelings of the player. Since we are all primates, and very closely related to some species, it is likely that we do experience many of the same emotions in response to similar stimuli.

A good portion of this predictive ability and knowledge of the animals as individual personalities is extremely difficult to document. But with more rigorous detailed descriptions of actions and reactions in the context of the troop and with information on an individual's physiological reactions obtained by telemetry, it should be possible to assess the amount and kinds of reactions of an animal to a social interaction. At present it is impossible to measure the involvement of an animal that merely observes or is near an interaction and that shows no visible, outward signs of reaction. Nevertheless, many social experiences must be obtained by the individual animal through observing other animals, even though the observing animal is not directly involved in the activity.

We will refer in this paper to the pleasurable aspects of play. Our assumption that play is pleasurable is based on observation of nonhuman primate play in several different areas of the world, as well as in captivity. The behaviors we have observed appear similar in many ways to the motor and behavioral expressions that we associate with pleasure in humans. The monkey cannot of course tell us his feelings, it cannot give us an introspective report, but that a monkey will engage in this activity whenever it can, for long hours at a time, and sometimes despite threats by adults, indicates that the motivations for play are strong. One of these motivations, we believe, is that it is pleasurable to the animal. There is repetition of play behavior because such behavior gives pleasure to the individual and this repetition leads to increasing competence in motor behavior and social interactions that are important in the life of the individual and hence of adaptive value to the species.

The Social Context of Play

The major contribution of field studies has been to portray primate behavior in its full social context, a structure of relationships that is extremely sensitive and responsive to events within the confines of the troop and to the ecological environment. Differences in responses of individual animals constitute an additional source of variation. Play takes place within the social context of the troop and is influenced in its frequency and forms by many of the same factors that affect the behavior of the troop.

Play can occur at any time of the day, when the troop is quiet or when it is moving. The daily routine of the troop is such that there are times during the day when the troop rests or remains quietly in one place, and it is during these times that most play occurs. These periods are in the early morning before the group moves away from sleeping trees, during midday when it is warmest, and again in the evening when the group is getting ready to settle down for the night.

At the first sign of alarm if a predator threatens, or when fighting breaks out in the troop, play is one of the first activities to cease. Tension and fighting are almost wholly antithetical to play. If a troop is passing through a relatively unfamiliar portion of its range, or if something new and strange appears, play ceases. Mason has said that "extreme departures from the familiar tend to depress play, whereas moderate degrees of novelty tend to enhance it" (1965:529). In general, play is more frequent in situations that lack pressure, whether physiological, psychological, or environmental. Play normally occurs in an atmosphere of familiarity, emotional reassurance, and lack of tension or danger.

The play of individual infants is further influenced by the mother's behavior. If she is involved in fighting, her infant is unlikely to continue to play; it may become very agitated and often will try to run to her and cling even though she is interacting aggressively with other monkeys. When the mother signals to her infant that she wants it to come to her and cling, the infant usually responds immediately, regardless of what it was doing, even though it had not perceived any source of threat or danger. In this way the events of the troop need not affect the infant directly but may do so through the mother.

Describing Play

Play is possibly the most difficult of all primate activities to describe. Its rapid starting and stopping and its myriad variations on many themes defy detailed description except by film analysis. When there are four, five, six or more players, as there frequently are in many species, the observer is presented with a formidable task if he is to transfer what he sees onto paper. Often there is a rapid alternation of play with nonplay and individuals may join a group, drop out or slow down, and then enter into the activity once again. This may be repeated constantly, especially when the animal is tired after having played for some time. The moods of the players may also change, producing variations in the kinds of activities that form play.

The observer's problem of description and analysis is compounded by the rapidity with which everything happens in play, but this is only part of the problem. The play participant is receiving much more information than is the uninvolved observer. Subtle glances, tensing of muscles, firmness of grasp, nipping, are but a few of the cues that are unavailable to the human who watches. The players are feeling, smelling, tasting, and experiencing far more than any observer can measure by looking. Here again, the use of telemetry to obtain information about an individual's internal physiological changes in response to observing, hearing, or participating in a social interaction will enable us to analyze the effects of play more accurately.

No list of actions can convey the feeling of play even though almost every field study that has been presented in the literature includes a list of play activities typical for that particular species. A comparison of lists quickly shows that there are many common elements regardless of whether the species spends more time in the trees or on the ground during the day. Trees are often chosen as the focus of action since trees present a third dimension of pathways for chasing and a set of footholds and handholds. Species that spend a great deal of time in tall grass may develop play patterns of hiding and freezing and then chasing in the grass.

Certain motor patterns appear regularly in the play behavior of each species. A group of crab-eating macaques (*Macaca irus*) in the course of 10 minutes of group play may spin, push, pull (hair, limbs, fingers, ears, tails, etc.), squint, drag feet, rub with hands, scratch with the nails, run, jump, chase, charge, fall down, swing by the feet, mouth, lean — and the list could be lengthened. Such a list would also include mounting, presenting, and making readily recognizable gestures of threat and submission. When gestures of threat appear in play or if one animal "feels like" fighting, tension may develop that stops the play or turns it into aggression. Wrestling may turn into aggression if one animal bites a little too hard. It is not always clear to the observer what constitutes biting too hard, since at one time a juvenile may bite the fur and skin on the back of a slightly smaller monkey and literally lift the smaller one off his feet without any outcry. At another time what appears a much less severe bite will produce loud cries of pain.

To describe or attempt to measure the intensity of play or the involvement of the players is possible in only a very imprecise way. The observer cannot use the frequency of cries as an index of pain; cries may reflect the mother's closeness and her willingness to back up her infant more than it reflects how much the infant is hurt. Social context can affect play in ways that are not always obvious to the observer.

To think of play as inefficient performance of specific gestures or motor patterns may be misleading. Very young infants' poor coordination makes it impossible for them to perform many of the movements of an adult. Play may appear inefficient because of its repetition of items and its tendency at times to display only bits and pieces of longer motor sequences. Repetition enhances and facilitates skill, and the piecing together in different ways of patterns of moving and of gestures is one way of learning all the possible combinations. This may well be an important part of becoming thoroughly familiar with the patterns of behavior and movement that will be essential when the young monkey or ape is grown.

A signal that helps us identify play is the "play face." It is commonly used by Old World primates to denote to others that play is intended and not harm or aggression. It is a useful gesture and is adopted by both adult and young alike to facilitate starting and continuing play. The play face is combined with movements that are also typical of soliciting play. The animal may swing its head from side to side and move in an uncoordinated manner. Among humans the child or adult may also signal play with a facial gesture and combine this with smiling or laughing that tends to set the stage for play and to reduce the likelihood of the following gestures or words being taken seriously.

In describing play and associated gestures, there is always a problem that the units we choose to record may not in fact be the most meaningful ones. We tend to assume that what we recognize and label as play, such as chasing and wrestling, are the important aspects of the activity. It is possible that other parts of the activity or that qualities of the way animals interact or handle one another in play are aspects very important to the development and expression of the players. The amount and kinds of physical contact may be as important as the motor patterns of play, and are the most difficult dimensions to detail and quantify. This may be true of other social activities as well. Consider, for example, grooming. Crucial aspects of grooming are the degree of physical closeness and the quality of handling that takes place during the activity. The act may have variable importance among individuals socially as well as physically.

Species-Specific Patterns of Play

Behavior patterns that are specific to one species or to troops within a single species must be analyzed individually. Several of those have already been discussed, such as the termiting among chimpanzees. The manipulatory play of the chimpanzee predisposes the young for termiting, but the actual behavior is not practiced in the context of peer-group play. There are other behavior patterns that are practiced in play; for example, the jumping bounce of the male patas monkeys.

Ethologists have traditionally been interested in species-specific patterns of behavior and they have looked at play from this point of view. One of their major efforts has been to describe behavior in components of fixed action patterns (Barlow 1968) and the stimuli that release them. The resulting analyses of play have been very descriptive with no emphasis on contextual information. Poole's study of aggressive play in polecats is an excellent example of this type of analysis. He compares the fixed action patterns of both aggression and aggressive play, thereby isolating the elements that signify play (Poole 1966). The ethological approach, while useful in the description of play, does not direct itself to determining the effects of play.

A few attempts to correlate play with other activities of the troop have taken us a step beyond the purely motor description of the ethologists. Ellefson in his study of gibbons (1967, 1968) suggests that the lack of play among young gibbons is due to the absence of possible playmates and to the disparities in their size, strength, and coordination; these conditions, in turn, are due to age differences among them, the ultimate result of their isolated "family" social structure. This hypothesis could and should be tested in an experimental laboratory situation by testing different age combinations of primates for differences in both amount and kinds of play and the subsequent effects on later adult behavior. In this way one could determine whether the kind of play experience itself was the deciding factor in the lack of play among gibbons, as Ellefson suspects, or whether it is something particular to gibbons as a species with their special social structure.

Play as Practice

In general, the practice gained by playing develops two kinds of skills: physical skills and interpersonal or social skills, and both are necessary for troop life. Other skills such as maternal behavior are also necessary but are practiced and learned in other contexts. As a result of play, there are benefits to the individual that are immediate and there are longer range benefits to the social group. The carryover of competence in locomotion and social skills into adult life is important, and it is likely that selection has taken place in the evolution of primates to favor those that play a substantial amount during their immature years.

The usefulness of a play pattern to an individual is seldom apparent if only play is observed. Consider, for example, young monkeys dashing at high speed in trees, chasing one another, making what appear to be hazardous jumps, running with alarming speed out onto the smaller branches, and falling to the ground. The observer may well wonder why this happens when it seems so risky. If the observer also sees escape behavior in a serious fight among adults, then it is possible to see how important the practice in youth may be. The ability to escape from an attacking adult or from an attacking predator may mean the difference between serious damage or death, and safety. As is true with so many patterns in play, it is necessary to see the animals in other situations using the locomotor skills and social relationships used in play to appreciate the eventual value of the play action.

It has been stated that "it is not necessary to play in order to practice, there is no reason why the animal should not just practice" (Loizos 1966:5). But, does the animal or child "just practice"? And, would it "just practice" enough for skillful use? Again we are faced with assessing motivation. There is no reason why it should be necessary to play in order to practice, and there is ample evidence that monkeys and apes—particularly apes—will practice skills outside of play. However, play's importance is that it insures sufficient practice for competence plus, and the plus can be thought of as the margin of skill that may make possible a higher dominance status, escape from harm, or greater ease in getting along in the social group. For whatever reasons the young play, in doing so they gain many hours of experience in the countless repetitions of behaviors that will be used when they are older. It is our suggestion that this is one of the most important aspects of play.

It has also been asked if there are not more economical ways of obtaining information than in play. If economy is measured only by the immediate output of energy and time and by the immediately observable and measurable effects of play, the answer might be yes, there must be more economical ways of getting information than in play. But once again, when the long-term effects of play are considered, then it is not at all clear that play is uneconomical, but rather may be very efficient.

Some things are not learned or practiced in play; for example, the young primate does not learn about predators in play. He does, however, probably learn and practice many of the rules of dominance; if learning which animals are stronger involves some pain, the young primate may learn the rules rapidly. It is impossible to tell just how fast certain behaviors or relationships are learned because there is so much repetition of them in play. Repetition may be needed to learn all features of those patterns, and the most efficient and economical way of insuring repetition may be in play.

Harlow (1959b) has cited curiosity as one of the determinants of play among rhesus monkeys. Field and laboratory observations suggest that play and curiosity appear only if the infant or juvenile monkey is secure enough to be willing to explore. Hutt (1966), describing human children's investigations of new objects, reports that children first investigate an object and then the object is incorporated into their play. Any further learning is incidental. If during play a new property of the object is discovered, the child responds with more investigative behavior rather than with further play.

Hutt states that play, with its high degree of redundancy, can actually prevent learning. This may be true for the human child. In observations of the nonhuman primate, however, increments of manipulatory skill learned in play, combined with smell and taste investigations of an object, are difficult for an observer to measure and the total impression may be that the animal has ceased to learn *new* things. We suggest that the definition of learning should not be restricted to the initial acquaintanceship with an object or an event, or its properties. It is not at all clear how much repetition a child, a monkey, or an ape needs in learning.

Hutt reports an experiment in which a chimpanzee failed to use a stick to solve a problem because he played with it. This conclusion may be based on the human's failure to understand the chimpanzee way of behaving and limitations. The learning natural to chimpanzees was not goal oriented as the human observer in the experiment wished it to be. If the chimpanzee was immature, playful handling of the stick could be expected in addition to motions that might have solved the problem at hand. The chimpanzee's limitations according to a human definition of what a primate can do with a tool, are due to its biology, brain, and experience, and the limits these factors place on its behavior.

Loizos (1967) has noted that play is one way of achieving relief from boredom. If the primates being observed are living in typical captive conditions this is very likely true. For an animal in a sterile, bare, uninteresting environment of a few square feet, with few if any other animals, what to do during the long hours of daytime becomes a problem. The solution is usually endless repetitions of highly stereotyped locomotor patterns, and manipulation of the cage and walls. The observer of free-ranging primates in a natural context seldom if ever gains the impression that the primates are bored. The primate is surrounded by a varied universe of objects to manipulate, to move around on and in, and there are other animals, old and young, to play with. If the young monkey or ape does become bored, it is only for the length of time it takes to spot something or someone interesting to play or interact with.

Play and Exploration

Knowledge of the environment is an indirect result of playing. At first the very young infant is exceedingly cautious in its approaches to anything apart from its mother. It leaves her with some hesitation and moves rather slowly, deliberately, and awkwardly, since it is not well coordinated in its movements. It may tumble or trip on a small twig and then dash back the few steps to its mother's arms. But, with repeated investigation and gradually more freedom, even though there are also more tumbles and stumbles, the infant gains familiarity with the surrounding objects and appears less frightened when it falls or gets a few feet from the mother. It is after this initial phase of gaining confidence that most monkeys will begin to leave the mother repeatedly and the behavior then appears more playful. Often the infant romps around and over the very objects that made it stumble and will repeat the movements over and over again.

When the infant is older and regularly leaves its mother to join in social play with other young, play tends to take place in areas that are very familiar.

Exploration of unfamiliar areas for the first or even second or third time is not an obvious factor in play, but after the animal has been in a place several times, then play will add the number of repetitions of experience with the area that are needed for thorough familiarity and skill in moving in that place.

For the first months of life the infant monkey generally does not go very far from the mother, but after weaning, play is no longer tied to the proximity of the mother. At this point it is more important to stay within, or within sight of, the social group. This means that the radius of potential play sites is much greater than it was earlier in the infant's life. However, play groups, even those of large juveniles, seldom move far from the group and whenever contact is lost, if only for a few minutes, among many species the young animals tend to stop playing and give location calls that quickly reunite them with the troop.

Learning Locomotor and Manipulative Skills in Play

The first tumbles of a very young infant are purposeful and there appears to be some anxiety associated with the falling that stimulates the infant to alarm and the mother to hold it or pick it up quickly. She comforts it and the infant is off again to move around her. At a later date in development the same infants may be deliberately selecting a play site that involves jumping or falling a long distance down to the ground. Rhesus macaques have been filmed dropping repeatedly some 30 feet from trees into a shallow pond of water. The lure of climbing the branch to get repeated falls raises the speculative question whether the falling might not be in part the excitement of mastering the previous helplessness. There is some evidence that objects that initially elicit caution will be manipulated longer than objects that provoke little or no fear (Loizos 1967:198), and this might be generalized to places such as trees.

There are great differences between monkeys and chimpanzees in the manipulation of objects. Handling and playing with objects is an important feature of chimpanzee life and play. Chimpanzees do not, however, practice hitting with a stick or throwing stones, and without this practice skillful performance is not possible. This is clear in contrast to the performance that man is capable of after training and practice encouraged and rewarded by his social group.

Dominance and Sexual Behaviors in Play

Sexual and dominance behavior develop and are practiced in play. It is in play that mounting and presenting first appear among langurs (Jay 1965b), baboons (DeVore 1965b), rhesus (Lindburg 1967), and other Old World primates. Although the motor patterns in these behaviors are similar to those in the adult's repertoire, the motivations for mounting and presenting appear to be as much related to dominance as to sex. It is unwise to assume that similar motor patterns appearing in different contexts and at different times in development necessarily share a common set of stimuli or of underlying motivations.

Playing animals are involved in a great deal of physical contact and continuous social interaction, and the ability to control one's own behavior and the behavior of others becomes very important. With the amount of physical contact in play, the chasing, hitting, wrestling, and mounting, the individual soon learns the differences in size, strength, reaction times, and tolerances of each of the other players. The total experience makes ranking almost inevitable.

In older infants and juveniles there is a marked increase in the frequency of the transition of play to aggression. Play often produces situations in which one may hurt another and the result is aggression that causes play to stop. Probably more dominance interactions among juveniles arise from play than from any other type of activity. Among langurs only a few dominance fights among infants and juveniles occurred in a nonplay interaction (Jay 1965b).

Among those species in which communication of submission, dominance, and appeasement is part of behavior, these and other gestures are often seen in play, although the precise situations in which they will be used later in life are not reproduced in play. The motor patterns sometimes fragmented and sometimes out of sequence, are observed over and over again in play long before they are used in social interactions among troop members in nonplay circumstances.

The Mother's Influence on Play

As mentioned earlier, the personality of the mother monkey or ape may affect the infant's contact with other troop members both directly and indirectly. The effect may be restricting or may encourage wider contact. On the one hand, if the mother is very subordinate and is constantly tense when in close contact with other adults, she may stay away from most of the other troop members and deliberately restrict the movement of her offspring so that it will not bring her into contact with the others in the group (Jay 1963a, 1963b). If, on the other hand, the mother is a very dominant and socially active female, her infant may come into frequent contact with other young monkeys. A mother that is quick to threaten young who solicit her infant to play reduces the total amount of time that her infant will spend in play (Jay 1963a).

Van Lawick-Goodall (1968b) has described a chimpanzee in the Gombe Reserve whose infant tended to display stereotyped motions such as rocking and pirouetting. The mother usually avoided social interactions with other chimpanzees consequently reducing the opportunity for her infant to interact with peer groups. There may have been a factor of boredom in this infant's behavior, since young chimpanzees do not usually play alone most of the time.

The effect of the mother's status in the group on her infant has been discussed earlier, especially in connection with chimpanzees in the Gombe Reserve. Studies of the rhesus monkeys on Cayo Santiago reveal important effects of the status of the mother on the activities of her offspring (Sade 1965; Wilson 1968). Here too the young of dominant females can afford to take liberties against other monkeys when the mother will back them up against reprisals. In these examples the offspring of more dominant females tend to have higher dominance status when grown than the young of less dominant females.

Experimental Investigation of Play

Experimental research is needed to test almost all the suggestions in the preceding portions of this paper. In many instances the field workers themselves can manipulate some of the variables in a free-ranging situation. To date very little experimental work on play has been attempted in the field and the results of those under way are not yet available. The laboratory, however, has been the locus for many investigations of primate behavior, some of which have related directly to play. Unfortunately, the design of a great many laboratory experiments is such that the stress and trauma of the investigative techniques create a situation in which play patterns may not be normal. Mason (1960, 1961a, 1961b) has discussed some of the social restrictions of captivity.

Harlow, in his studies of social deprivation among rhesus monkeys, has uncovered some extremely valuable information about play and the importance of peers (Harlow 1959b, 1962, 1964; Harlow and Harlow 1962; Harlow, Harlow and Hansen 1963; Rosenblum and Harlow 1963). His research indicates that monkeys raised with peers but without a mother grew up to be "normal" adults, whereas monkeys raised with only a mother showed normal aggression, little play, and no sexual behavior when adults. Fieldwork has indicated that behaviors such as sexual or maternal behaviors are learned and developed in a very complex manner as the result of the interaction of biological factors, social situation, and the basic behavioral flexibility of primates. Obviously then, the presence or absence of play alone will not account for the development of a behavior as complex as sex. This does not mean that peers are more important than a mother in the development of "normal" behavior. It is not clear from these experiments that contacts with peers as opposed to contacts with individuals of other age levels are necessary for normal development. This could be tested by raising one infant with a full complement of adult animals, and then comparing it to infants raised only with mothers and infants raised only with peers. The problem remains whether it is peer contact, or the act of playing that results in normal behavior, and this would be very difficult to test experimentally. Another problem studied in the laboratory centers around the concept of normality. The concept is limited by the parameters of cage behavior and it is unlikely that the "normals" with their limited experiences could survive in the wild.

Subtle variations in behavior are not emphasized in the publications of Harlow's work. It would be interesting to know in what ways the juvenile and adult behavior of an infant raised with a full complement of adult animals but no young animals, differed from the juvenile and adult behavior of an infant raised with both adults and peers. Presumably, according to Harlow's standards, both situations would produce animals that would be termed "normal" adults. Harlow and Harlow (1966) does hold reservations about his earlier statements concerning the role of peers and mothers, and states that he now feels contact with both mothers and peers is the best arrangement for infant development.

Harlow's studies raise an interesting question relevant to the allegations of field observers that the adaptive importance of play is shown by the amount of time spent playing. Harlow's infants were allowed only 20 minutes of play per day and yet they developed into "normal" responsive adults. It should be noted, however, that Harlow's rhesus had a good deal less to learn that a rhesus living in a north Indian forest, and it may be that much more time is needed to learn and master the knowledge essential for the complex environment of the free-ranging rhesus monkeys.

Adult Play

Play is not a notable feature of adult life; although the adults of some species play, their play is neither frequent nor prolonged. Adult langur monkeys rarely play, and when they do, the play usually involves an adult male playing briefly with a large juvenile (Jay 1963a). In general adult female baboons do not play and adult male baboons seldom play except with large male infants and juveniles. Two adult males in one troop in Nairobe National Park were observed to spend a great deal of time in association with, and on occasion played with, two large male infants. On the basis of observations of these pairs over a period of only one week this appeared to be a persistent association in which the young male stayed close by the adult and was at times carried by the adult male. Adult female chimpanzees play with their infants frequently and at times also with their adolescent offspring. Infant and juvenile chimpanzees play with siblings and other unrelated young, but the adolescent male chimpanzee plays only with his siblings and not with the offspring of other females (van Lawick-Goodall 1968*b*).

It is not at all clear why adults play as little as they do. Part of the explanation must relate to the potential danger to the would-be player in the running approach, the physical contact of play, and the difficulties of indicating precisely to the play partner that the intention of the stronger is only to play. Display of a play face is not always sufficient to convey this message.

Physical contact and personal space are very important among adult primates and especially among adult males. When adult males are tense, they are surrounded by what has been called a personal or social space. This is an area into which other animals may not enter without signaling their intentions. The more dominant the male in the center of the space and the tenser or more aggressive he is, the larger the space surrounding him. Tolerance for physical contact even when the animals are relaxed varies among different species from those that seek passive contact to those that tend to avoid it.

Probably some of the most important reasons adults do not play more often are those that are the most difficult to describe and to substantiate. Action or activity without an observable reason or goal is simply much less frequent among postadolescent primates than among younger ones. Humans often contrast the unbounded energy of the very young to the activity level of adults. Adults may be active in sports and in games where there are rules and reasons for the action and rewards for the activity, but they do not normally expend great amounts of energy in the same ways that children do. Instead we expend most of our daily activity in pursuits related to work and to planned events or hobbies.

Much of the exploratory and learning activity of the playing infant or juvenile has been completed by the time it reaches maturity. The land is known, social skills are mastered, and motor coordination has long reached its peak. In an article summarizing work done by the research teams of the Japan Monkey Centre and Kyoto University, Tsumori (1967) reports that among Japanese macaques (*M. fuscatta*) exploratory play learning virtually ceases at maturity.

Itani (1958) observed the introduction of caramels as a food for one troop of Japanese macaques at Takasakiyama and found that infants and juveniles less than three years old quickly accepted the caramels. In general, the younger the monkeys, the more willing they were to eat new things and his data suggested that they were less adaptable to new situations after they reached three years of age. Other items were introduced in different troops. Some of the new items required different ways of peeling, washing, or eating so different skills were needed in eating them. Potatoes were first washed in sea water by a female in one group in 1953, and by 1962 every member of the troop except very young infants, one-year-old infants, and adults of more than twelve years of age were washing the potatoes. Tsumori (1967:209) writes that the "Lively and playful, two and three year old Japanese monkeys are highly curious and flexible in their behavior, possessing high imitative ability and adaptability compared with the older ones. But when more complicated tests are given the monkeys, an accumulation of experiences and understanding abilities are required." Other tests were made relative to other food objects and the success rate varied with the ages of the animals. In general the older the animal the lower the success rate. In one series every animal over 12 years of age failed on the first trials.

DIMENSIONS OF VARIABILITY IN SOCIAL ORGANIZATION

All primates are faced with the same problems of living; getting food and water, avoiding predators, locating safe sleeping places, reproducing and raising young, communicating with one another, to mention but a few of them. Although there is substantial similarity in the manner in which these problems are solved, there is also variability. Complications of the problems may be unique to one troop or to troops living in a specific type of habitat. The amount and kinds of variability that exist were not anticipated by students of primates until comparative studies of the same species were undertaken in different habitats.

Differences in social systems correspond with taxonomic groupings and with different ecological habitats. In addition to variations that characterize different genera and different species within a genus, there may be significant differences in the composition, sizes, and behavior of troops within a single species (Gartlan and Brain 1968; Frisch 1968; Jay 1968a; Yoshiba 1968). Infants growing up even in adjacent troops may need to acquire different knowledge.

Crook and Gartlan (1965) have summarized recently collected data on primate social systems and have divided them into five grades. Within each grade there are "close correlations between habitat and diet preferences, diurnal activity rhythms, group size, type of reproductive unit, sexual dimorphism and population dispersion patterns indicating co-adaptations of these features to aspects of biome types forming major selection pressures. The five grades represent levels of adaptation to forests, forest fringes, rich savanna, and arid environments" (Crook 1970b).

The majority of species fall in the grade in which there are several adult males with adult females and young of all ages. The northern Indian langurs (Jay 1963a, b), vervet monkeys (Gartlan 1966; Gartlan & Brain 1968; Strusaker 1967b, c), and Japanese macaques (Imanishi 1957b, 1963; Imanishi and Altmann 1965) are a few of many examples. Other types of organization and composition have been illustrated above.

THE ENVIRONMENT AND BEHAVIOR.   Patterns of locomotion appropriate to the environment are practiced in play. The patas monkey studied in Uganda by Hall (1965a) is an excellent example of this. Patas show a constellation of behavior patterns, including play, that are adapted to the ecological environment. In Uganda the patas live in flat open areas covered with tall grass. Play consists mainly of running, chasing, and a special game of bouncing up and down in the grass and off trees. Since there are few and scattered trees to provide safety from predators, adults must be able to run very swiftly. If a predator approaches and threatens, the adult male patas acts as a decoy and runs noisily away from the rest of the group, which, instead of running, freeze in the grass. During the male's flight he also bounds off trees and bounces up and down in the tall grass, thereby making himself very conspicuous. Hall (1965a) has suggested that the perfection of locomotion abilities in young patas is particularly adaptive because they are exposed to day predators and the young must be very swift if they actually are chased.

Hamadryas baboons (Kummer 1967, 1968a, 1968b; Kummer and Kurt 1963, 1965) live in the extremely arid ecology of the highlands of Ethiopia, where food is very scarce and widely scattered on the rugged countryside. Although they sleep together at night in large herds of several hundred animals, during the daytime these monkeys live in small units of one adult male, his adult females, and their offspring. Males leave the group in which they were born to join all-male groups when they become subadults. Young males spend time daily playing together in groups composed almost exclusively of males. In contrast to the pattern for the male, the young female remains in the group of her birth and either is taken as a consort eventually by the male of the group or is adopted when she is about one year old by a male from another small group. Sometimes this male is a subadult, but often he is a fully mature adult male. The male cares for her until she is reproductively mature at which time they will mate. For the young female this means that after she is about one year old she has little or no experience in play with other young baboons.

The effects of this lack of play or peer contact for the young female hamadryas are not known. Her lack of play is further enforced by the position of the female in the hamadryas troop. Females are the exclusive property of the male that adopts them; thus she will always remain with other females and will not be free to leave even for a few hours.

Biologically Based Variations in Behavior

The biology of the monkey or ape not only sets limits and provides guidelines for development of play patterns, but also affects nearly all forms that its play will take. What an animal learns, as well as how, and what use it makes of it depend both on biology and experience.

One important variable among the nonhuman primates is the rate of maturation of the young. Compared to the normal length of infant dependency of a macaque or baboon, from $1\frac{1}{2}$ tp 2 years, the period of dependency for the chimpanzee is at least double, and an estimate of 3 times that length of time would probably be more accurate.

Endocrine factors important in behavior are evident in primate play. In all sexually dimorphic species young males play more roughly and soon grow larger and stronger than young females so even an untrained observer can observe sexual differences in play patterns in the first few weeks of an infant's play. These differences are pronounced in the juvenile and subadult stages, when the animals are near physical maturity. Infant and juvenile females that have been treated while in utero with male hormone threatened, initiated play, and engaged in rough play more often than the normal control females in the experiment (Young, Goy, and Phoenix 1964).

In summary, there are major limitations and influencing factors on the range of possible behaviors for any primate species. Ecology, biology, and social structure are all interrelated and act together to shape forms of social behavior. Some species are capable of a wide range of behavioral adaptations to very different habitats whereas other species exist only in quite restricted types of habitat and are not noted for variability in patterns of social behavior. Individuals do the adapting and it is the experiences each individual has during growth that most likely will affect most of its subsequent actions.

HUMAN AND NONHUMAN PRIMATE PLAY
ANTHROPOLOGICAL STUDIES OF PLAY

All human activities and institutions are the result of the interaction between basic human biology and the environment including culture. Man has inherited a great deal from his primate ancestors; in addition, with his larger and more complex brain, man has uniquely developed language. The use of language makes it possible to name objects, symbolize, plan, and to have a conception of the future. Through language man has developed his own variations on activities common to other animals. Human play reflects very clearly the basic biological similarity between man and other primates, but it also reflects the differences that result from the addition of language and its correlate, culture.

Every culture has its own attributes, attitudes, and concerns. Through child training, each individual human child is socialized to become a viable member of his or her society. Cultural values may be expressed in many modes such as song, dance, folktales, religion, and games. Children's games are early expressions of the basic values and issues that concern their culture, just as each species of monkey or ape has different play patterns that the young must learn as preparation for adult life.

The interaction between a culture and a mode of expression, as for example a game, is quite complex and it is necessary to be very wary of causal theories. It is a gross oversimplification to make a one-to-one correlation between any particular aspect of child training with a particular type of game played in that culture. It is more likely that clusters of elements establish a tendency that becomes evident in the expressive modes of a culture. Nevertheless, it is interesting to look at some examples of the way the values of a society are mirrored in childrens' games.

In New Guinea the Tangu engage in a ritual food exchange in which strict equivalence is maintained (Burridge 1957). Equivalence is determined by mutual agreement between trading partners. The Tangu children play a game called "taketak" in which 2 lots of 30 spines of coconut palm fronds are stuck into the ground 5 yards apart. Individual spines within the lot are placed approximately six inches apart. The children have tops that are spun and let loose to try to touch the spines of the opponent's lot. The teams need not be equal, but the number of tops must be equal. The game proceeds, hypothetically, as follows:

1. Team A spins top into lot A and touches 3 spines that are removed.

2. Team B spins top into lot B and touches 2 spines that are removed. Team A then puts *back* into Lot A 2 spines.

3. Team A spins top into lot A and touches 1 spine which is removed (now both teams have 2 spines out.)

4. Team B should not hit any on their next turn, but they must take their turn and their top must land in the middle of their lot. The game ends when both teams have an equivalent number of spines in and out, or, since this rarely occurs, when an end is mutually agreed upon.

The object of the game is equivalence, just as in the food exchange ritual of the adults, and in both cases the outcome or equivalence is decided upon by mutual agreement. There is no winner or loser; the object is to tie.

Sutton-Smith, Roberts, and others have utilized a cross-cultural approach in a number of studies correlating types of games with social structure, with socialization practices, as a reinforcement of other expressive modes, and with class structure within a single society (Roberts, Arth, and Bush 1959; Roberts and Sutton-Smith 1962; Rosenberg and Sutton-Smith 1960; Sutton-Smith and Roberts 1964). These studies are all based on the assumption that children's games contain and reinforce cultural values and attitudes.

The Relevance of Nonhuman Primate Studies to Human Play

Insights into human behavior gained from studies of the nonhuman primates vary with each behavioral topic under consideration. Patterns of behavior that are common to most nonhuman primates are likely to have been characteristic also of the immediate primate ancestors of man and as such may give us some understanding of the basis for human behavior (Washburn 1963*b*, 1968*a*, 1968*b*; Jay 1968*b*). The extent to which these basic patterns have been altered or entirely changed in the course of the last 2 to 3 million years is a subject of great concern to the anthropologist and to many other students of human behavior.

In our efforts to gain comparative information from the behavior of the living nonhuman primates to use in the investigation of the evolution of man's behavior, we must not make the assumption that because monkeys or apes behave in certain ways, man also did, does, or should behave similarly. It is also a mistake to assume that the behavior of primitive cultures is most exemplary for us today. Neither the nonhuman primates nor primitive cultures are necessarily the most appropriate or the most "natural" models for us to use in reconstructing human evolution or in our analysis of human behavior, yet the study of both the nonhuman primates and primitive cultures can be enlightening and should be used when applicable.

Until there is more detailed understanding of the functions and effects of play in different monkeys and apes, the basis of comparison to human play should be from nonhuman primates in general. Chimpanzees are most closely related in evolution to man, as is evident in the similarities of morphology and behavior, but caution must be exercised in assuming that the behavior of chimpanzees represents the behavior of early man. In fact these behaviors may be similar only and then only with respect to certain aspects.

Chimpanzee play is noted for the variety and frequency of object manipulation, but chimpanzees are not unique in this ability; some of the New World monkeys frequently manipulate objects. Objects are exceedingly important also in human play; however, if the distinctive qualities of human play are those connected with language, games, fantasy, and symbolism, then the manipulative skills of chimpanzees are only a small step toward the complexity of man's play. Of all the communication systems of nonhuman primates the chimpanzee's is by far the richest; they are able to reassure, to beg, and to exchange subtleties of mood and intention that are beyond the range of monkey communication.

Nonhuman primates may offer important insights into the biological bases of play behavior. Both human and nonhuman primate play studies indicate that rough-and-tumble play is an activity of young males and that females tend to withdraw from play groups during the juvenile stage when play becomes too aggressive for them. Male-female differences in play among the nonhuman primates strengthen the view of a biological basis of play and raise questions concerning the additional effects of culture and learning on play patterns among human children.

Nonhuman primate studies of play provide a model situation for the investigation of the effects of play on development, both physical and social;

this model can be analyzed in terms of its applicability to humans. We have emphasized the importance of the practice of interpersonal behavior in the peer group play of monkeys and apes. This emphasis should be extended into anthropological studies of human play, and social scientists might well pay more attention to the interpersonal and interactive elements of play in the cultures they study and focus less upon the adult imitative behaviors included in children's games and play.

Play is obviously an important activity for young primates, and its complexity is such that all types of study are needed if it is to be understood. Play is a phenomenon that spans the class of mammals and studies of play in animals other than humans, and particularly in nonhuman primates, are important for the insights they offer for the understanding of human play. To assess this relevance realistically, however, elements that are comparable must be used. The study of play as practice for social interaction and individual adjustment will be more fruitful for comparative purposes than the study of play as learning and practice for specific adult behaviors and specific manipulatory skills.

A Description of Some Aspects
of Monkey, Ape, and Human Play

Human play exhibits many of the characteristics described for monkey and ape play. Games are chosen appropriate to the age and sex of the player. This can be observed on a playground or in a nursery school playroom where girls play with dolls or in small groups and boys heap up on top of one another on the floor. As children mature the dichotomy between male and female activities becomes more pronounced. Whatever the elements of biologically determined sex differences in play, there are also cultural factors that teach and approve certain games exclusively for each sex.

In humans, age-appropriate toys are designed based on the coordination and interest of the child, and nonhuman primates also play at and with certain things in each stage of development. Preschool-aged children build with blocks and girls from 7 to 12 play with jacks; both activities are challenging to the appropriate age group.

Like nonhuman primates, human children are social animals, and play is a social activity. Instead of the "peer play group" of monkeys and apes, we find nursery schools or neighborhood play groups. The human infant goes through the same phases of solitary play, play with his mother, and social play that the young monkey or ape does, and as with other primates, the play group occupies an important role in the socialization of the young child. He learns the use and control of his own body — how to draw or how to climb on jungle gyms ("monkey-bars"). Later he develops his physical skills and endurance with sports such as kickball, tennis, and swimming. These sports, and play in general, have another important function, that of teaching the child how to get along with others and to be a socially acceptable individual. Just as the nonhuman primates learn and practice the appropriate "rules" of adult behavior in the play group, so does the human child. He is actively discouraged from developing a second set of standards for conduct.

The play experience of the human child differs from that of the nonhuman primates in two respects: his ties to his mother aren't severed when the next sibling is born and his peer play group is not permanent. These two factors work together, the first providing a continuity of emotional reassurance and the second promoting a more variegated and diverse play experience — one better suited to the adult human situation in Western societies. At adolescence, when the maternal ties are loosening and the child is reaching adulthood and independent status, the peer play group becomes the focal point and main influence for the child. This happens at the same *stage* of development in both human and nonhuman primates, although in terms of years there is a major difference.

Our emphasis so far in this paper has been on nonhuman primate behavior. The primatologist may apply questions suggested by the study of man to nonhuman primate behavior. If we are searching for comparable data we must ask comparable questions. In framing these questions we may make more explicit the limitations of the nonhuman primates as a source of relevant data, and at the same time make it possible to place the emphasis of our future research on a firmer basis.

The most obvious limitation of nonhuman primates, when compared with man, is their lack of language in the human sense. The nonhuman primates can communicate emotional states and a very few more specific reactions to objects in the environment, but they cannot make statements about the past or the future. Vervets have several alarm calls indicating different kinds of predators (Struhsaker 1967a, 1967b) and other primates often give loud calls when they locate concentrations of food. The latter calls, however, express excitement rather than a statement about the food. When chimpanzees call in this way other apes also gather as a result of hearing them. They may or may not associate the calls with a ripe fig tree, since similar or identical calls are also given in other circumstances not directly related to food.

The time reference of primate sounds is limited to the present. This does not mean that primates do not remember events in the past; their behavior indicates that they do. Whether they anticipate much beyond the next few minutes is open to question, and it would be valuable to consider their actions with this in mind. An example of thinking ahead may be a troop leaving sleeping trees in the morning and heading directly for a fruit tree. It may take them an hour or so to get to it.

The symbolic use of objects, words, and actions that is so important in our lives is also probably unique to man. The extent to which a monkey or an ape has this capacity, if in fact it does have it, would be extremely difficult to test. An ape may pound on a tree when he cannot pound on another chimpanzee that frightened or threatened him, but whether a young ape will destroy a twig in play because he would like to do the same to a new sibling is a matter of conjecture.

We do not know how much a monkey or an ape recalls or thinks ahead but on the basis of the limited areas of their brain associated with memory and planning, it is likely that this capacity is very limited.

A monkey or an ape has aggressive feelings, tensions, conflicts, and ambivalences, and there are ample instances that indicate this. Van Lawick-Goodall (1967:169) cites an example of an adolescent male chimpanzee frustrated by his inability to get some food from adults; as the young male's tension grew he "finally gave vent to his frustration by leaping up, swaying branches, and dragging vegetation." Another male when frustrated beyond his endurance released his frustration against a tree in a display of pounding and shouting; when he had done this he appeared much less tense.

Nonhuman primates are constantly taking into consideration many elements of their social relationships. Often the observer sees a mixture of gestures that makes sense only in the context of the influence of other animals on the actor. A female juvenile began to chase an infant with the obvious intent of punishing the smaller animal. At the approach of the infant's very dominant mother the attacking juvenile immediately changed her entire posture from one of aggression to submission towards the infant. The juvenile presented in submission and lip-smacked to the infant. After such a sequence the juvenile may then redirect her aggression towards another animal in the troop that is clearly subordinate to her. She will make certain that her attack will not provoke another animal to support the one she attacks so that she will succeed in domination and in giving vent to her aggressive feelings.

A juvenile wrestling in the arms of an adult male may appear tense and inhibited in his movements, and finally may utter a squeal of fear and succeed in breaking free and dashing away from the adult male. The adult may have been making a play face toward the young male the entire time they were in contact, but this gesture was not sufficient to avert the juvenile's breaking away. What often surprises the observer is the juvenile's immediate return to the adult male. The pattern may be repeated again and again; each time the juvenile flees, only to return for more rough play. The ambivalence of the juvenile is clear and his actions continue to shift from approach to withdrawal. Such conflict situations, with their tensions and anxieties, are present in the behavior of monkeys and apes: it remains to be demonstrated whether or not any of them are resolved in play.

The biological limitations of the nonhuman primates in terms of brain and behavior are evident. On the basis of our limited knowledge at present we do not understand what the monkey or ape experiences emotionally, or what the experiences mean to the animal. We understand to some extent the limitations of monkeys and apes, but we are at a stage in our investigations when we need to observe important periods of development with some new questions in mind. Perhaps we can look for cues that we have not noticed before in our studies.

In particular, it would be very useful to observe the patterns of play prior to, during, and after weaning to see whether the young monkey or ape does in fact express some of its problems in play forms; whether there are observable differences in relationships with peers and older animals that might relate to the trauma of weaning.

Monkeys and apes probably master anxiety in play although this is difficult to describe. Helplessness for the very young infant monkey relates most obviously to locomotion and anxiety about leaving its mother, although it is probable that as it grows it will also experience anxiety related to inter-individual relationships as well. The play group may be a locus for working out aggressions that the animal might otherwise direct toward larger and stronger individuals in the group. The opportunity to play dominant as well as subordinate roles in play may be a part of the attractiveness of play for the young monkey. It does appear, however, that the young monkey or ape will not play during periods of substantial stress; this does not rule out the possibility that less severe tensions and anxieties resulting from normal daily life will be expressed in play-group activities.

Clearly, this is highly speculative, but if we are led by this speculation to view play a little more closely, we may discover new qualities and effects of play for the nonhuman primate.

SUMMARY

Play is a major category of nonhuman primate behavior. This is demonstrated by the amount of time the animals devote to this activity (often hours every day), by the energy spent playing, and by the complexity of the behavior. According to the theory of natural selection this degree of expenditure of biological resources must be adaptive; it must serve important biological functions for the individual.

An evolutionary approach to primate play stresses the importance of play for the order and particularly for man. The amount and complexity of play has increased greatly in the evolution of man. The length of the period of immaturity, in which most play occurs, is more than doubled between monkey and ape, (*Cercopithecus aethiops* and chimpanzee), and it at least doubles again between ape and man. The protracted period of youth increases the quantity of play during the lifetime of the individual. It also carries with it a much lengthened period of infant dependency; the young must be protected and looked after for a much longer time and there is a greater delay in attaining maturity and full participation in adult life. From a biological point of view, play is not, as it is often described in our culture, "only play." It is a major category of adaptive behavior that must be understood if we are to understand primate behavior.

Perhaps in part because of the values of our culture, play has been the subject of relatively little research. Traditionally it has not been viewed as a major research problem. Not one field study has made the understanding of play a major goal, so in this paper we have had to piece together information from many reports primarily devoted to other topics. This is particularly unfortunate because if play is preparation for adult life, it must be evaluated in the setting of the natural life of the species. As we have outlined, this natural setting varies markedly, and play is a behavior that can be greatly distorted by captivity.

chapter eleven

EVOLUTION OF PRIMATE BEHAVIOR

S. L. Washburn and R. S. Harding

The study of evolution provides an intellectual background, a setting, for the understanding of the contemporary forms of life. Knowledge of evolution should help in the selection of animals for experimental purposes, in the formulation of comparative problems, and in interpretation. No one thinks that evolutionary investigations should dominate biology as they did in the nineteenth century, but we believe that they should be a useful part of biological science.

THEORY

The present theory of evolution has been clearly stated by numerous authorities, especially Huxley (1964), Simpson (1949), Roe and Simpson (1958), Dobzhansky (1962), and Mayr (1963). The synthetic theory, which was formulated in the 1930s and accepted in the 1940s, has been greatly strengthened by the discovery of the nature of the genetic code and the new techniques which the discovery of DNA has made possible. The synthetic theory may be briefly stated as follows: Evolution is the result of changes in the gene frequencies in populations. Behaviors leading to reproductive success are favored by natural selection, and the genetic bases for these successful behaviors are incorporated into the gene pool of the population. There is a feedback relation between behavior and its biological base, so that behavior is both a cause of changing gene frequencies and a consequence of changing biology. Evolution is adaptation over time, and there are no trends except those that result from continuing selection.

The most important conclusions that can be drawn from evolutionary studies stem from this basic theory, rather than from descriptive studies.

Reprinted with permission from Schmitt, Francis O., (ed.), *The Neurosciences: Second Study Program.* New York: The Rockefeller University Press, 1970, pp. 39–47.

For example, many persons have believed that large parts of the human brain (such as the frontal lobes or corpus callosum) had little functional importance, and surgical destruction of these and other parts of the brain has been defended on this basis. But, whatever particular theory of human origin one may believe, the brain has become larger over time, and, if selection has been for more and more frontal lobes, these parts must have important adaptive functions. Evolutionary gain or loss clearly indicates functional importance, even if the precise nature of the function is still debated by scientists. For example, the threefold to fourfold increase in the size of the human cerebellum in the last million years requires explanation in terms of new human adaptations.

Evolutionary theory demands a radical departure from the assumptions and methods of traditional comparative studies. Not only do a frog, a rat, and a man not constitute an evolutionary sequence, but the division of creatures into systems makes it almost impossible to understand adaptation. Obviously, as a temporary research strategy, it is useful to look at the skeleton, for example, but interpreting evolutionary changes in the skeleton necessarily involves other systems. Such an involvement is an inevitable consequence of the fact that evolution is the result of selection for the biological base of successful behaviors of populations. Because the nervous system is a coordinating system, to view its evolution in isolation from the adapting animal of which it is a part is particularly misleading.

The theory of evolution requires that we consider both the fossils and the living forms, and that this information be considered in the light of genetics and experimental biology. Each kind of information is essential and supplements the other, and both are briefly considered before the problems of the primates and human evolution are discussed in more detail.

THE FOSSILS

Science has revealed an incredible variety of life and an evolutionary process that has gone on for some three billion years. From the rocks comes evidence of multitudes of creatures which no longer exist and which have left no close living relatives. The variety of forms that are alive today are only a few end products of a few of the ancient lineages. For example, the 150 million years of dinosaur evolution could not be deduced on the basis of the study of contemporary reptiles, nor could the diversity and complexity of their adaptive radiations be appreciated. Comparative anatomy of the surviving few is a completely inadequate approach to understanding the diversity of life, and it is no accident that many of the scientists who have contributed most to the understanding of evolution have considered the fossil record, anatomy, and behavior (Romer 1966; Simpson 1953; Young 1950; Roe and Simpson 1958).

It is not only the faunas that are remote in time that are no longer represented, but without fossils there would be no evidence of the glypto-donts, giant sloths, camels, or sabertoothed cats of North America of only a few thousand years ago, and there would be no traces at all of the extinct faunas of South America (Simpson 1965). It is clear that, even if a few forms survive with little change (such as the opossum), the succession of faunas can be appreciated only through a study of the fossils. There is no way to reconstruct the past without using evidence from paleontology.

LIVING FORMS

The contemporary forms of life are the representatives of those few lineages that were successful, and all have been evolving. The point is simply that the study of both contemporary forms and the fossils gives a better understanding than studying either one alone (Jay 1968b). For ex-ample, fossil skulls show that the earliest primates had long faces, wide interorbital regions, and small brains (Clark 1962). In the contemporary prosimians these same features can be seen, and part of the primitive struc-ture is directly related to the sense of smell (e.g., turbinate bones, cribriform plate of ethmoid, large olfactory bulb). Study of the living forms provides detailed information and permits reconstruction of the adaptive functions of the parts seen in the fossils. But, in addition, the living forms have a rhinarium, tactile hairs on the nose and elsewhere, scent glands, and a sense of smell that is far more important than it is in monkeys. These features were characteristic of primitive mammals in general, so it is reasonable to reconstruct far more anatomy and behavior than could possibly be deter-mined from the fossil bone alone.

But the contemporary prosimians are highly diversified and have evolved away from the form seen in the fossils. There is a great diversity of locomotor patterns, from the slow-moving lorises to the rapid-jumping galagos and tarsiers. There is no suggestion of anything like an aye-aye or an indri in the ancestry of monkeys. Some have evolved far from propor-tions of the Eocene forms, but some have changed remarkably little. Ex-amination of both the living animals and the fossils shows which have changed least, the nature of some of the changes, and gives a far fuller un-derstanding of both.

No living form is *assumed* to be primitive or to represent an unmodified ancestral stage. No feature of a contemporary creature is *assumed* to be the same as that in ancestors of many millions of years ago. Studying both the living creatures and the fossils of their ancestors shows that the basic adapta-tion of the primates is in the hands and feet, that the sensory adaptation was that of primitive mammals, and that the adaptation of the special senses and brain to arboreal, diurnal life came millions of years after the appearance of other primate features. These general conclusions provide a setting for the understanding of the primate special senses, brain, and biology that is compatible with all that is known of evolution and in no way depends on the great chain of being or an assumption that any particular primate is an unmodified ancestor.

TRENDS

When the adaptive radiations of any group of animals are investigated, trends can usually be discerned. As time passed, animals became larger, brains became larger, and teeth evolved. Such sequences were once attributed to orthogenesis, an inbuilt evolutionary momentum. Now it is believed that they are the result of selection, which, in the main, has been in one direction. Where data are abundant, the trends seem much less regular and much more complicated than was once thought, and, because the trend is the result of selection, there may be reversals. The existence of long-term adaptive trends brings some order into evolutionary comparisons and makes it easier to compare contemporary forms.

For example, the brains of the contemporary prosimians are much larger than those of their early Eocene ancestors. Comparisons with insectivores and rodents show that there has been substantial adaptation to arboreal life, and much of this may have taken place after the separation of the lines leading to contemporary prosimians and monkeys. For this reason, differences in the intelligence of *Lemur catta* and a macaque, for example, may be less than between *Notharctus* and a macaque. Knowledge of the trends makes it possible to compensate for the biases which are introduced when the living forms are compared without reference to ancestral ones.

New World and Old World monkeys are probably descended independently from some group of prosimians (Omomyidae?). The apparent similarities are the result of parallel evolution, which should help in the interpretation of adaptive trends. Whether in the New World or the Old World, brains associated with stereoscopic color vision and reduction in the sense of smell (and related systems) have evolved to a greater degree than have prosimian brains. The New World monkey is important in research on the central nervous system mainly because its nervous system affords us the opportunity to discover the selection pressures that resulted in this parallel evolution. No Old World monkeys have prehensile tails, but in New World monkeys there is every variation from long, primitive, balancing tails to prehensile grasping tails, complete with hairless skin that is comparable with that of a hand and is associated with a large area of cortex.

This series offers a model of the evolution of the interrelations of ability and brain. The issue is not ancestry, but the understanding that comes from the differential adaptation in a series of related forms.

In summary, the existence of evolutionary trends, adaptive radiations, and diversified behaviors offers the opportunity to interpret the process of evolution, to plan experiments, and to understand biological differences or similarities.

TIME

Most of the research in comparative anatomy was done prior to the development of time scales based on the use of radioactive elements. Viewing contemporary forms as relatively unchanged is much more credible if the times involved are short, as they were thought to be when the methods of comparative anatomy were taking shape. For example, the age of the world was supposed to be something on the order of 30 to 40 million years in 1900 (Sollas 1905). This is approximately half of the time in which the mammals have been dominant, a quarter of the age of the reptiles, and less than one percent of present estimates of the age of the earth. When Lord Kelvin concluded that organic evolution was impossible because there was not enough time, he was entirely correct—it was only the estimates of time which he used that was wrong.

When Keith wrote that the ancestors of man and apes must have separated in the Oligocene, he thought that the time of the separation was approximately a million years ago. Middle Oligocene is now regarded as some 30 million years ago. Consider the difference of a separation between man and monkey a million years ago with subsequent evolution dominated by orthogenesis, and a separation 30 million years ago with evolution the result of selection. Old World monkeys have been evolving, adapting, and radiating for at least 30 million years after their lineage separated from that leading to man. And on the human side our ancestors were evolving, adapting, radiating for 30 million years; therefore when men and macaques are compared, it must be remembered that there are at least 60 million years of evolutionary events that separate the two.

It is useful, therefore, to look at the adaptations that have evolved in each group in light of this long biological separation. On the monkey side, locomotor patterns have remained much the same. On the ape-human side, quadrupeds evolved into brachiators, brachiators into knuckle-walkers, and knuckle-walkers into bipeds. On the monkey side, animals remained primarily vegetarian, mostly arboreal, and did not use objects. On the ape-human side, animals began to hunt, lived on the ground, and used objects. There is every indication that the contemporary apes have evolved farther from early Miocene ancestors than the monkeys have from theirs, and man has advanced structurally and behaviorally beyond the apes.

ADAPTATION AND COMPARATIVE BEHAVIOR

The problems which arise in comparing behaviors that result from adaptation over long periods of time may be illustrated by considering three of the most commonly compared animals: pigeons, rats, and men. Birds have had a distinct evolutionary history for at least 150 million years, which means that at least 300 million years of evolutionary events separate pigeon and rat. There are some parallel adaptations, such as a high metabolic rate, but, for the most part, adaptations to flight have been quite different from mammalian adaptations. The resulting behaviors, on the other hand, may seem similar at first glance. Two experiments typify the kind of behavioral adaptation in birds that appears understandable in terms of mammalian behavior, but is not. First, the courtship display of the male turkey before the female turkey seems quite comprehensible to the human observer. Yet Schein and Hale (1959) have shown that the male turkey will go through a complete display when confronted with nothing more than the stuffed head of a female; in other words, most of what the male turkey sees does not influence his behavior at all. Or, to take another example, a hen turkey shows what appears to be normal maternal behavior toward her chicks, and protects them vigorously. Yet a deafened turkey will kill her own chicks (Schleidt, Schleidt, and Magg 1960). Conversely, a turkey hen with normal hearing will attack a polecat or any other small predator, but will not attack a stuffed polecat that has been made to chirp like a turkey chick. In other words, the chirp of the turkey chick functions to keep the hen from attacking, and she will attack any small, moving object that cannot be heard to chirp. Clearly, then, avian behavior achieves the same ends as does comparable mammalian behavior, but it rests on a fundamentally different biological basis. The adaptation to flight has dominated the evolution of the brain of the bird, just as it has that of the skeleton and many other parts of the body. Complex behavior is highly adaptive, but flight puts very severe restrictions on the size of the brain. The avian solution is complex stereotyped behavior (dependence on vision and hearing) which is controlled by a highly simplified relation to the environment and a minimum of learning.

Now consider the social rat. When a strange rat approaches the nest, it is attacked (Barnett 1963). The rats appear to recognize the members of their own group, but, if the stranger is rubbed with litter from the floor of the nest box, it is not attacked. And if a member of the group is removed and cleaned, it is then attacked. Recognition is by the sense of smell, and the rat makes a distinction that neither the pigeon nor man can make.

Obviously, pigeons and rats have different abilities which enable them to adapt to different ways of life. From the point of view of evolution, to ask which is more intelligent is meaningless. Pigeons cannot live like rats, nor can rats live like pigeons. Both represent successful ways of life, and the differences cannot be measured by any simple psychological test. Because man is visual and equates intelligence and learning, man presents pigeons and rats with learning sets based on visual clues, but rat intelligence is based on the sense of smell, touch, and proximity, rather than on vision and distance. It would be easy to devise tests in which the senses of rats were favored; the intelligence of pigeons and men would then be rated as low.

In summary, tests may be given to pigeons, rats, and men, and their performance on these tests may be compared. Performance will depend at least as much on the tests as on "intelligence." One may also be interested in the evolutionary process that has produced the very different adaptations of these highly diverse forms.

The rate of maturation in pigeons and man gives a useful measure of the fundamentally different adaptive nature of their nervous systems. Birds mature in less than one per cent of their lifetime span. Imprinting may take place in a few hours. Only a little practice is necessary for the maturation of skills. In man, on the other hand, critical periods (as in language acquisition) have ranges of many months or years. Adult skills (such as chipping a stone tool, throwing a spear) can take years to acquire. It has been adaptive for man to store a very large amount of information and to be able to learn a wide variety of behaviors. For a human to mature normally, there must be years of social interaction and learning.

Obviously, scientists may be interested in what a pigeon and a man have in common, and some generalizations about learning may apply to the most diverse vertebrates; however, the study of evolution attempts to give some historical understanding of how birds and man have adapted differently.

From an evolutionary point of view, the study of rats is very different from that of birds. The sense of smell and touch, and nocturnal activity, are as important to rats as they are to prosimians. Many of these features are basically mammalian, and many of the early primates had enlarged incisor teeth similar to those of the contemporary rodents. Study of the fossils, contemporary prosimians, and rodents helps in the understanding of an actual stage in human evolution.

HUMAN EVOLUTION

Human evolution is the result of the success of a series of ways of life. Each of these was based on abilities which have changed over time.

The general stages may be discerned in both the fossils and the most closely related living forms. The primates became separated from the insectivores as a result of their adaptation to climbing with grasping hands and feet. The replacement of claws by nails is anatomically exceedingly complicated, and this major locomotor adaptation appears to have taken place only in one group of mammals. The unity of the early primates is shown in features of the skull (Van Valen 1965) and teeth (Szalay 1968). Increase in the size of the brain followed (Hofer 1962); the olfactory bulb remained large and the cerebrum and cerebellum small. Fossil forms were numerous and highly diversified (Russell, Louis, and Savage 1967), and the same is true of the surviving prosimians. The monkeys of both the New World and the Old World are probably descended from one family of prosimians, and the surviving prosimians have been evolving in separate lineages for some 50 million years. The existence of a wide variety of living prosimians and of fossils from many families allows considerable reconstruction of the way of life of our ancestors for some 30 million years, after the primates had separated from other mammals and before the evolution of the monkeys.

During the Oligocene period, some 40 to 30 million years ago, monkeys evolved. Again, with our understanding based on the living forms and the exceedingly fragmentary fossils, the principal changes were the evolution of stereoscopic color vision, together with a great increase in the size of the brain and a loss of the structures associated with the sense of smell. The major changes are clearly reflected in the anatomy of the skull, but the fossil record is so incomplete that there is no way of deciding how far the evolution may have progressed in the latter part of the Eocene. There are only fragmentary skulls from the critical period (Simons 1959, 1967).

As shown by the limb bones, our ancestors were quadrupedal, arboreal forms for some 20 million years before the ape pattern of locomotion evolved. During this period, the general pattern of adaptation appears to have been very similar to that of the contemporary Old World monkeys. In spite of the variety of kinds of monkeys, many structures and behaviors are shared. All are quadrupedal, with many common features of hands, feet, limbs, and trunk. All sleep in a sitting position, which is reflected in the callosities and in the ischia. All are highly social, mature slowly, and bear single offspring. These monkeys have been highly successful, with the density of a single species rising to more than 200 per square mile in favorable localities.

Interest in human evolution leads to an emphasis on the behavior and structure of apes, but the fossil record shows that the monkeys have displaced all the small African apes. Small apes were numerous in the Miocene, but now only the large ones can compete in the African forests, and they are rare and most unsuccessful compared with the monkeys of the *Ceropithecus, Cerocebus,* and *Papio* groups. In Asia, where competition was with monkeys of the Colobinae only, small apes (gibbons) survive.

In short, study of the contemporary Old World monkeys does give many clues to the way our ancestors lived for millions of years – arboreal, visual, social, with learning important. Comparisons with parallel adaptations in New World forms allows controls on many of the interpretations.

APES (PONGIDAE)

Long after Old World monkeys (Cercopithecidae) and apes (Pongidae) could be distinguished on the basis of the dentition, a new way of climbing-feeding evolved in the apes. As shown in the contemporary apes, this pattern involves the independent use of individual arms and legs in climbing, hanging, and feeding. This new arboreal pattern is correlated with the structure of the trunk, the upper limbs, and the arrangement of the thoracic and abdominal viscera. It is clearly reflected in the bones forming the fingers, wrist (Lewis 1969), elbow joint, and shoulder (Washburn 1968*b*). Man shares this structural-functional pattern with the contemporary apes, but in the fossils of some 20 million years ago only the shoulder shows the beginning of this locomotor specialization. Men and apes must have shared a common ancestry long after the kind of adaptation seen in the limb bones of *Proconsul* (Napier and Davis 1959) or *Pliopithecus* (Zapfe 1958). Unfortunately, there are no ape skeletons of Pliocene age.

The gibbons are the most extreme in their adaptations to hanging in the trees, and many lines of evidence suggest that their lineage diverged before that of the great apes and man, but after the locomotor adaptation described above. For example, gibbons sleep sitting and have the monkey kind of ischial callosities, whereas the great apes build nests, sleep in positions very similar to those of man, and have, at most, traces of callosities.

The great apes manipulate objects and use them in aggressive displays. Van Lawick-Goodall (1967, 1968) described chimpanzees throwing branches, grass, and stones in displays, and playing extensively with objects. Leaves are used for cleaning the body, and grass or sticks for obtaining termites. Lancaster (1968*a*) has put these activities into perspective by pointing out that, although they do not seem to be much from a human point of view, chimpanzees put objects to more different kinds of uses than all other mammals combined. Termiting involves a long period of learning for the young chimpanzee, and this activity, which requires much skill and concentration, may last for hours.

Chimpanzees hunt to a limited degree, and as many as four males may cooperate in extended hunting. Reynolds (1966) pointed to numerous features in which chimpanzee social behavior may afford the most useful model for the possible behavior patterns of our ancestors. But it must be stressed that chimpanzees have also evolved since their separation from the human lineage. No claim is made that chimpanzees are unmodified ancestors, but it is asserted that the rich data on chimpanzee behavior give a very different picture from what would be the case if the behavior of contemporary monkeys were used as the only standard. Further, any speculations may be checked against gorilla behavior, as described by Schaller (1963), and the whole understanding of ape behavior will soon be enriched by studies now in progress (Horr on orang; Fossey on gorilla). It should be emphasized that man is most similar to African ground-living, knuckle-walking apes, not to the forms that remained arboreal (Washburn 1968a).

The brain of the apes has been traditionally described as advanced in a human direction over that of monkeys, and the behavioral data certainly support this interpretation. Certainly, slower maturation, object play, object use, cooperative hunting, some food sharing—all suggest a brain-behavior complex that is much more human than that of the monkeys.

MEN (HOMINIDAE)

Isolated teeth show that the human lineage had separated from that leading to the apes by 3.7 million years ago, and the actual separation must have been more than four million years ago (F. C. Howell, personal communication). Many fossils dated on the order of two million years ago show that, by that time, our ancestors (*Australopithecus*) were bipedal. Apparently these creatures used stones and shaped some by knocking off a few chips. They lived in the savanna and hunted at least small game, and also possibly large animals. The teeth of these fossils were remarkably human. The foot was fully human, and the ilium was similar to that of man in most respects and very different from the comparable bone of any other primate.

The brains of these creatures were no larger than those of the contemporary great apes, representative capacities being from 400 to 550 cc. One hand has been well preserved, and the phalanges are halfway between that of a knuckle-walker and that of man. Judged from a terminal phalanx, the thumb was small compared with that of modern man, but large compared with an ape thumb (Napier 1962). The molars erupted on the same delayed pattern as in man and very much more slowly than in the apes (Mann 1968).

It is probable that this small-brained, bipedal stage in human evolution lasted for at least two million years and, possibly, twice as long. By some 600,000 years ago *Homo erectus* had replaced *Australopithecus,* and the brain had doubled in size. The large-brained men are associated with stone tools, which take great manual skill to make. Even when suitable material is supplied and careful instructions and demonstrations are given, it takes months to learn to copy some of the better tools of half a million years ago. These people used fire, killed large animals, and were human in the usual sense of this term. But, even so, there was very little change in the form of the stone tools. The same assemblages have been found over large geographic areas and over many thousands of years. It was not until 40,000 to 30,000 years ago, when men who were, in skeletal characters, virtually indistinguishable from ourselves appeared, that history starts to proceed at a rapid rate.

In the last 30,000 years before agriculture, the kinds of stone tools changed quickly, were locally highly diversified, and many technical advances were made, such as bows, boats, harpoons, ground stone, and so forth (Laughlin 1968). The rate of change and degree of local cultural adaptation stand in marked contrast to earlier conditions in which the same kind of stone tool was made for hundreds of thousands of years.

If this view of human evolution is at all correct, then most of the distinctive evolution of the human brain was *long after* the separation of man and ape and was in response to uniquely human selection pressures. If the brain evolved in a feedback relation with hand skills, social skills, and, finally, language, the explanation of what is unique about the human brain lies in the evolutionary events of the last million years. In this sense, most of the human brain is a product of the success of the human way of life, and it cannot be understood by the study of brains that have not undergone that particular evolutionary history.

The point may be illustrated by a consideration of the hand. Among the contemporary primates, the human thumb is unique in its size, strength, and degree of opposition of the fingers. The hand of *Australopithecus* shows that its human features evolved long after the use of stone tools began. The human hand is an ape hand rebuilt by selection pressures for efficient tool use. Likewise, the large area of human brain related to hand skills is the result of the success of skillful object using. The brain is important, not only in determing the possibility of manual skills, but in making learning possible. Object use is fun for men, and they easily learn a wide variety of manual skills (Hamburg 1963). Chimpanzees throw stones and may hit the intended target, but they do not make piles of stones and then practice throwing. Other chimpanzees do not applaud a hit, nor does an organized society help the individual to see the possibilities and develop his skills to the utmost. (A human brain may easily guide a chimpanzee to a level of performance that lies well beyond the normal behavior of the species.)

What has evolved in man is an ability, and one can see how remarkable this human ability is by an appreciation of the fact that, with all the families of primates equipped with grasping hands and feet, with dozens of genera and hundreds of species, there is only one instance of substantial object use. The study of evolution may be used either to see the common elements in hands and their actions or to see the unique elements and their special histories.

Numerous social skills are also unique to man. There is virtually no planned cooperation among the nonhuman primates. Estrous behavior is universal. Rage may be only slightly controlled. Dominance behavior is common. Sharing of food is almost nonexistent, except to a slight degree among chimpanzees. Put in evolutionary terms, the success of the human kind of social system has been so great that it has exerted a major influence on the evolution of the central nervous system. Man has evolved a brain that can mature normally only in a social setting; the advantage of slow maturation may be primarily that it allows the learning of social systems, including the knowledge and skills that make them possible. It must be remembered that, until very recently, there was little specialization in society, aside from the primary sexual division of labor. Each individual in a social system had to know all the skills, magic, religion, and so on, of the society. Only after the advent of agriculture, and especially after the scientific revolution, were there many individuals who had not mastered most of the skills of society. Division of labor changes the whole relation of the way of life to the evolution of the brain.

At the present time the greatest functional difference between man and the nonhuman primates is language. In the communication systems of the nonhuman primates, the information conveyed is concerned, for the most part, with the emotional states of the actors. Such systems are multimodal (Lancaster 1968b), using gesture, postures, and sound to convey general meanings, but there is no way to express a name. In social situations, animals may convey their intentions (to be aggressive, to submit, to groom, to flee, and so forth), but no specific information is conveyed about the environment. For example, sound may mean "danger" (more likely fear on the part of the sender of the message); on some monkeys different sounds may signal ground predator or avian predator (Struhsaker 1967a), but there is no way to signal leopard or hyena, although this would seem to be of the greatest adaptive value. Language has been so important in human adaptation, and sounds are so common among nonhuman primates and mammals in general, that it is very difficult to understand why such a system of naming evolved only once. It is tempting to see this as originating in consequence of early tool using, which may have forced concentration on objects in a new manner. However the naming adaptation may have started, Lenneberg (1967) has suggested that all modern languages may have had a common origin on the order of 50,000 to 30,000 years ago, or about the time anatomically modern man appeared and the historical record rapidly became complex. Obviously, even if this speculation is correct, simpler antecedent forms of language may have existed. In any case, man is now so constituted that he can learn language, any language, and may learn several. This ability depends directly on an extensive biological base (Masland 1968; Sperry and Gazzaniga 1967) and indirectly creates a pressure for vastly more memory and the possibility of complex planning. Nowhere is the difference between man and ape more apparent. In spite of great efforts, chimpanzees cannot be taught to speak because they lack the biological base for this learning (Kellogg 1968).

The brain has evolved in a feedback relation with the way of life, so the present structure of the brain reflects its evolutionary past. For example, the large area of brain associated with hand skills is an indication of the importance of this behavior in human evolution. Because the brain increased in size by approximately three times in less than a million years, the more that can be known about the new parts of the brain, the more we can understand the events of the last million years. Uncertainties of the fossil record and the gaps in our knowledge of the functions of the brain, however, make this correlation difficult. Penfield and Rasmussen's (1950) description of the function of the cortex corresponds with our view of this period in human evolution. What evolved were the hand skills and the technology they make possible, social skills with increased memory and planning and control, and linguistic skills, which are the basis of human communication. If Cobb (1958) was correct in his view that the newest part of the cerebellum is primarily concerned with learned hand skills, this would support the inferences we have drawn from Penfield and Rasmussen's interpretation of the function of the cortex. The threefold to fourfold increase in the size of the cerebellum in the last million years must be explained by the evolutionary importance of substantial new functions. Eccles et al. (1967) wrote that the cerebellum is important in learned skills and stated that "We have to envisage that the cerebellum plays a major role in the performance of all skilled actions. . . ." (p. 314).

As pointed out at the beginning of this chapter, those structures that have increased over the millennia as a result of selection must have important functions. The more fully the functional differences between man and ape can be understood, the more we can analyze the events that historically separate man and ape. The more complete the fossil record becomes, the more accurately the origin of the more recent behaviors can be dated, but it may well be that in the next decade the increase in understanding of the human brain will do more to illuminate the fossil record than the other way around.

In summary, human evolution must be seen as a succession of ways of life. The nature of each way of life may best be understood by a study of both the fossils and the living primates. Fortunately the wide variety of living primates provides a rich source of information. Unfortunately, the fossils, although numerous, are fragmentary, leaving wide areas for reasonable disagreement.

The ways of life are made possible by abilities, and each ability is made possible by inputs, internal coordination, and skilled performance. Over time, an ability may decrease (as has the sense of smell in the evolution of man or that of birds), or it may increase because of continued selection (as have hand skills in human evolution). Abilities may be determined almost entirely by heredity or be greatly influenced by learning. In the latter case, the biology determines the ease with which the behaviors can be initially learned or subsequently modified.

Evolution (adaptation over time) is a complex process in which behavior is in a feedback relation with the biology that makes it possible. This process has resulted in an almost infinite variety of forms of life, most of which have been long extinct. There is no way that the few extant forms can be arranged in any simple order of complexity or intelligence. Study of the fossils and of the living forms may give us deeper understanding of adaptation, of evolution, and of the processes and settings that have produced the existing varieties of life, including man.

NOTE: There have been many different theories of human origin and primate phylogeny. Recently, numerous studies of albumins, transferrins, hemoglobin, DNA, and chromosomes prove that men (Hominidae) are particularly close to the apes (Pongidae) and much less closely related to Old World monkeys (Cercopithecidae). The recent evidence supports the traditional conclusion that, among the apes, the order of relationship is: African apes, orangutan, gibbons. The times of separation of the various lineages is still under debate, but a relatively recent separation between man and ape seems the most probable. The evidence has been reviewed by Wilson and Sarich (1969) and by Goodman (1968).

chapter twelve

PRIMATE PATTERNS

Phyllis Dolhinow

INTRODUCTION

Patterns are beginning to emerge from the myriad volumes of details as to how primates live, and many of these patterns appear to be characteristic of very different kinds of monkeys and apes. Our knowledge of primate behavior has been accumulating for several decades as the number of studies is increasing rapidly. It is a good time now to step back and look at what is known—to focus on those aspects of life common to most primates and in so doing, to see to what extent primate behavior is variable.

Although much is known about monkey and ape life, it is clear that many parts of the total set of patterns are missing; those that are known form a mosaic, or a three-dimensional puzzle, rather than a clear complete picture. What at one time was conceived of as *the* primate pattern has been transformed into variations on themes. To define with precision what it means to behave like a primate is even more difficult than to include every primate in one simple definition of the mammalian order Primates.

Some aspects of primate social behavior vary within a single species, whereas other patterns are remarkably invariant. For example, although the general organizational patterns of social groups within some species may differ from one habitat to another, the basic behavioral elements, such as gestures of communication, that constitute an individual's repertoire appear not to vary within a species regardless of the differences in habitat.

Only a few of the behaviors in the constellation common to most primates will be discussed in this chapter. No attempt is made here to report all the available information on any one, and the topics chosen (those most intriguing to the writer) are not necessarily the most completely understood or, from some points of view, the most important. The information used in this chapter is drawn from studies of Old World monkeys and apes, and when reference is made to the primates, the reader is asked to recall that the subject is thus limited. Many of the following descriptions are true for New World monkeys, and to a lesser extent for some of the prosimians, but these primates are not so well known as Old World monkeys and apes and for that reason they have been omitted from the discussion.

THE IMPORTANCE OF LIVING IN A GROUP

The great majority of primates live in a rich, complex, and relatively stable social group. It is a small world circumscribed by tradition and daily routine with a minimum of unusual intrusive events from outside its perimeter. But what is this social group and how does an observer know whether the animals he is watching constitute a social group? Ten years ago the answer would have been simple. The group was a number of animals that lived together in an area. Its members might disperse over a short distance during the day, but they were never very far apart and they gathered together again for the night. Now we know that not all species are characterized by living in neat and easily countable stable units that are quite separate from other such units.

Hamadryas baboons are an example of a species for which it is difficult to describe a simple unit comparable to the troop of savanna baboons. During the daytime an adult male hamadryas will lead his small group of adult females and immature monkeys, and in the evening he will return with his small group to the sleeping cliffs where he rejoins hundreds of other hamadryas to sleep (Kummer 1956, 1967, 1968a, 1968b; Kummer and Kurt 1963, 1965). Chimpanzees living in the Gombe National Park enter and leave variable subgroupings that form according to age, sex, reproductive status, and genealogical relationships, so that, in the course of a year, one chimpanzee may have spent weeks or months with different chimpanzees. (van Lawick-Goodall 1967, 1968b). Here the unit of high interaction frequency among a specific group of animals is the local population.

Frequency and duration of interaction, then, are two important measures of a social grouping. These factors are not always easy to measure, as for example if the primates are living in thick forest where visibility is poor. Solitary or peripheral adults may live between groups, and it often takes a long time for the observer to be certain that what has been labeled as a group includes all the regularly participating members.

The basic nature of most primates is an intensely social one and all early experience occurs in a group. The young animal learns by observation and practice in a context of affectional bonds. It is protected, carried, and nourished for weeks or months, and long after it gains a degree of independence from caretaking adults, it continues to learn and develop in relation to other members of the group. There are many ways of learning and many rewards for gaining both motor and social skills.

There are many causes for being social and many channels for expressing sociability. Numerous suggestions have been made as to what attracts and keeps together the members of a group. The list of possible explanatory factors is long and not all may apply to every species. Included in this list are reproduction, kinship ties, the attractiveness of adult males because of the protection they afford, the pleasures of social interactions such as grooming, the social bonds that develop early among the young and extend into adulthood, and the security of well-known companions and daily routine. None of these factors may be conscious, of course, but the net effect seems to render a stable social group very attractive, and primates appear to prefer life with the group to a solitary existence. Sociability surely must be enhanced by the pleasurable experiences of group life, and it has been demonstrated experimentally that a monkey will work simply for the reward of seeing another monkey (Butler 1954). The nature of sociability and the expressions or forms it takes, from specific behavior patterns to social structures, are shaped in part by the structure and complexity of the primate's brain.

Normal life patterns have their genesis, development, and expression only in the social group. Not only the amount of social experience is important, but also the occurrence at the right time during development of the necessary forms of social stimuli from other group members.[1] The newborn is concerned with a tiny portion of its group and the rest becomes significant only very gradually. The young primate soon encounters most of the group, and within months it becomes an active participant in an increasing number of social activities. There is still a fairly long time, however, during which events that occur in and to the group affect the infant more through its mother than directly; later these events will concern the juvenile and its peer group. A stable group provides each young member with a locus for learning, experimentation, and social as well as physical development during the months when the young animal most needs protection, security, and stimulation from group members.

Being social is a major theme of primate life and each monkey or ape must adjust to life with other individuals, even if it means constant subjection to aggression and very low status in the group. Within groups structured in part by relatively rigid adult dominance ranking it is easy to see the re-

---

[1] Complexities of infant development and the social context in which the young primate lives have been investigated and discussed by many researchers including Bernstein and Mason 1962; Hansen 1966; Hinde, Rowell, and Spencer-Booth 1964; Hinde 1965; Seay 1966; Spencer-Booth, Hinde, and Bruce 1965.

wards enjoyed by high-status animals and even by those in the middle of the group's power structure, but why others remain in the group is sometimes less apparent. In fact, a few subadult and adult males of some species do leave their group of origin, either to live alone for a while or to join another group directly. Reasons for leaving a group are undoubtedly complex and in most instances involve far more than status, although this is certainly often one factor.

Although stability and predictability are important aspects of social life, there are adjustments and changes of personnel within the network of group relationships over the years. Divisions and recombinations of group-ings do occur. In times of crises these changes may take place rapidly, but such times, of course, are exceptional. Groups may grow too large or may shrink in size for many reasons, including climatic changes, disease, and predation. In addition, unusual proportions of sexes in successive birth seasons eventually may affect group life.

There are varying degrees of permanence and of definition to the boundaries of primate social groupings. Similarly, the internal structures of groups in different species may vary greatly. Not all species of primates live in groups characterized by classical dominance hierarchies (Bernstein 1966), in which each animal occupies a definable position relative to most or all others in the group. For groups that do have dominance ranking, shifts in dominance may occur, but the overall structure of leadership and power within the group persists even though the individuals that occupy different ranks change. Such changes do not alter the essential organization and struc-tural characteristics of the group unless changes occur frequently over a long period or are in response to external pressures such as changes in the environment. There is far more stability than novelty in primate life and changes that do take place tend to follow predictable lines.

Every group, however, is permeable, and the membership of all groups will vary over time, regardless of species or habitat. For short periods there may be a very small number of exchanged or recruited animals, as among some savanna baboons, or a considerable number, as among forest baboons (Rowell 1966a, 1967a; Ransom, personal communication). Lindburg's (1967) study of rhesus indicates that they tend to change troops most fre-quently during mating seasons. If excluded from their original group, rhesus have been known to move into a langur troop rather than live a solitary life (see Chapter 5).

Advantages other than those mentioned above accrue to living in a group. The location of every group within a well known area of land, a home range, assures that no individual in the group will be endangered in a strange and unknown place. Optimal use of a habitat to assure sufficient food and living space for resident groups is also facilitated by the traditions of spacing and land usage characteristic of each group.

SOCIAL SKILLS AND TROOP KNOWLEDGE

Every primate must master many complex social skills during its brief period of maturation, skills that are necessary if the animal is to become an effective participating member of its group. The group offers opportunity to observe social abilities, and in the group these are encouraged to be developed and to be practiced until through constant repetition these skills gradually become an integral part of the life of the individual. Subtleties of communication, responses to threat and submission, and appropriate patterns of interaction are only a small and obvious part of each primate's total necessary knowledge. The more that is known of how complex behavior development is, the more evident it is that the effects of different factors of attachment and the specificity of social stimulation need to be investigated. Not only the variability of his social experiences, but also the personality or temperament of each individual contributes to the outcome of development and learning.

The young monkey or ape must learn to know every other animal in the group, their personalities, tolerances, and habits, and this knowledge is generalized into an awareness of which animals must be avoided and under what circumstances, as well as which other animals may be approached and when. Less social but equally necessary are the many kinds of information an animal must acquire concerning edible foods, pathways for traveling from one place to another, safe sleeping trees, the daily routine, and which animals and objects in the environment to fear and avoid.

Little, if any, of this knowledge is immutable; instead modification appears to be constant. Each individual must be able to respond to such factors as changes in the personnel of the group, age and status changes of other group members, and alterations in details of the habitat — to name but a few of the possible stimuli that require adjustment in a group member's behavior. In order to maintain working relationships within the group, it is essential that every animal be able to accommodate its behavior over time, to learn constantly, to modify action, and to remember. This does not usually entail learning completely new patterns, but, rather, requires a series of modifications of existing ones. Most accommodations are to other members of the group. Individuals are subject to a long series of changes from the time they are born, through their maturation, and into adulthood.

There are many "rules" for what can and cannot be done in a group, rules that are usually related more to social roles than to the specific individual in a role. These rules will be discussed later in this chapter, but it is important here to recognize them as part of the body of acquired understanding of each individual. Many if not most of the strictures on social activity are learned early in life when the consequences of breaking them are minimal. The rules include some very practical ways to behave, such as not challenging an animal that is much stronger when the results are likely to be punishing to the challenger, not leaving a position of safety and approaching a predator, and similarly avoiding other obviously hazardous actions. Each is an implicit rule only in the sense that abiding by it increases the likelihood of survival or well-being.

Since every social group has a home range, knowledge is specific to each area and group. This may not vary much among some groups, but in others it may vary markedly. The environment of a mountain-living troop of rhesus is almost totally different from that of a troop living in a crowded city bazaar. The forest-living troop may have almost no contact with man and may gather its food from trees and plants. In contrast, the city-dwelling troop has an environ of walls and buildings, where there may not be a single tree or shrub. Food for the city rhesus may be limited to items stolen from man or gathered from man's refuse. Instead of trees, branches, and grassy earth, the city animal is surrounded by mud huts and stone buildings, doors, latches, screens, ladders, pipes, and locks. Even more importantly, the city environment includes humans who are often hostile and intolerant of the larceny and intimidating attacks of the monkeys.

Singh (1969) has compared the reactions of city and rural rhesus in competitive and noncompetitive situations. As anyone would predict who has ever had occasion to compete with an adult male city rhesus over a lunch or right-of-way in a narrow alley, the city monkeys were completely successful in dominating the rural ones in competitive food situations. Singh's results indicated that the city animals were more aggressive in all contexts and were definitely dominant over their country cousins.

SOCIAL TIME

Group behavior is more than the sum of ongoing interactions; it is also the result of past adjustments. This is not to say that there is a social memory residing in any individual, but only that relationships are the results of long periods of accommodation and learning among group members of successive generations. Some older animals may act as leaders, but the very old animal may become peripheral to daily activity and decline to interact. The presence of several generations represents years of interaction and relationships, many of which may be based on genealogical ties with other group members. The role of experience in these accommodations is far from understood, but most wild troops are composed of animals that have adjusted to social changes over time.

Knowledge of the history of adjacent groups would probably be an asset to understanding group relationships in any area. Some groups may be closely related, having split or branched off from a common group. Thus adjacent groups might contain related animals, which would influence group encounters, and might be very significant in the exchange of members.

HOME RANGE

Nonhuman primate social groups, regardless of size or composition, have circumscribed areas of land designated as home ranges within which they live.[2] Some social groups may move over miles, whereas the ranges of others are concentrated into acres.

By following a group for an annual cycle, the observer can plot on paper exactly what routes were taken and how long the group stayed in each place (see Chapter 5). It is soon apparent that groups tend to spend most time in certain areas and in general these are locations where there is a special food, a good water source, or perhaps favorite sleeping trees. Different parts of a range may be used as seasonal changes in vegetation occur. In tropical regions where there is minimal seasonal change in plant life, groups may move about from one place to another in their ranges depending upon the fruiting of single trees or the availability of preferred food plants. Parts of the range that are used a great deal are sometimes referred to as "core areas." Still other areas may be used seldom if ever. Thus the picture of land use over an annual cycle is usually quite uneven.

Often there are other groups of the same species occupying nearby or adjacent areas. A picture of ranges in a large section of land may look like a mosaic, with each neighboring troop having its own area in which it stays most of the time. The same portions of land are often included in the range of several troops and these overlapping areas may be used, though usually at different times, by two, three, four, and conceivably even more groups. This joint use may present problems if something in the area of overlap is desired by several troops at the same time. Normally this does not occur, or if there is some such feature (for example, a water hole), then the groups that habitually use it tend to do so in ways that minimize the chances of groups' meeting. If one group is larger than another, the smaller may hesitate and wait at a distance until the larger is finished drinking; groups may use the area at different times. Groups of some species tolerate being near conspecific groups, and for these species it may be possible for several groups to use such a location at one time.

Species of several genera live together in many areas. For example, some forests in parts of Uganda are inhabited by chimpanzees, baboons, black and white colobus, red colobus, two species of *Cercopithecus*, and mangabeys. Sometimes each species has a preference for a certain level of the forest, or, if the area is open, one kind of monkey may move farther from the trees during the daytime. Often a species has food preferences that

---

[2] Even a local population of chimpanzees, such as the one studied by van Lawick-Goodall (1967, 1968b) in the Gombe National Park, Tanzania, stays primarily within a specific area. The chimpanzees of this area do not move together as a single group but nonetheless each animal usually is to be found within the area.

differ from those of the other kinds of primates in the area so there is a minimum of competition for what may otherwise be a limited quantity of edibles.

The size and shape of a home range may change over time, but in many species it is surprisingly stable from year to year. Changes in group size and composition are usually reflected in the amount of land a group uses or controls. For example, if a group grows in number more land may be occupied, or if an especially large number of aggressive adult males assumes leadership more space may be commandeered from smaller or less dominant neighboring groups. A small group with many highly aggressive adult males may control a larger area of land than an adjacent group with many more members but fewer males, or males of less aggressive temperament. In this, as in many aspects of behavior, one part of social life is related to others. Changes in water sources or food supply can also be a stimulus for changes in the size and shape of the groups' living areas.

Home range sizes vary tremendously among species and are related to the type of terrain and habitat. Savanna baboons may move over 15 square miles (Hall and DeVore, this volume), whereas gibbons will content themselves with a few acres (Ellefson 1967). Of course, savanna baboons and gibbons occupy very different types of habitat, and the adaptations of the animals themselves are very different in the relative amount of time they spend on the ground and in the trees, as well as in their dietary preferences (Napier and Napier 1967).

Hall (1965b) described differences in the home range patterns of baboons, vervets, and patas all living in the same area of Murchison, Uganda. Patas lived in home ranges considerably larger than most baboon home ranges. Patas ranged in a manner similar to the baboons but with some important differences that Hall correlated with differences in adaptations between the two genera. Patas, for example, have a large number of night resting areas and the numbers of a patas group disperse over a wide area to sleep each night. In contrast, both vervets and baboons returned each evening to the same clusters of trees.

Use of home range is related to the way of life of the species. It is important for a baboon troop to stay together night and day because safety lies, in part, in group action. This is in contrast to the very different system of protection from predators characteristic of the patas. The adult male patas acts as a decoy and draws the attention of the predator from the rest of the group. At night the members of a patas group sleep at a distance from each other and should a predator come upon one the others would be safe.

Under some circumstances such as crowding or food shortage a home range may be actively defended and other members or groups of the same species prevented from using the land or the trees or any resource in the range. Such ranges are usually referred to as territories. Exclusive use, a diagnostic feature of a territory, may be achieved in several ways. Daily routine may keep neighboring groups from even coming into contact. The boundaries of the range may be marked in some manner, such as with substances that smell (urine or feces, or in some species glandular secretion). Loud calls may be given periodically to signal where the group is and where the other groups had better not go. Actual fighting with other groups may be engaged in.

Gibbons live in small groups composed of a mated pair and one or two immature animals. A typical day for a gibbon adult male includes several hours of vocal battling at the boundaries of the group's territory (Ellefson 1967, 1968; Carpenter 1940). The males seldom actively struggle physically but a large part of the day may be spent in boisterous boundary arguments. Gibbon ranges are small (a few acres or less), groups are small, and population density may be quite high in an area. Because the gibbon way of locomotion is brachiating, these small apes are not efficient at moving long distances. Rather, they concentrate their lives in small areas and have developed effective social means of assuring that each small range contains only the number of gibbons the area can support.

For the majority of species fighting is rarely resorted to in an attempt to maintain exclusive use of range unless, and this must be emphasized, there is some unusual situation prevailing, such as overcrowding, or rapid change in the habitat because of flood, farming, hunting, or drought. On occasion, several troops may skirmish, but pitched battles in which wounds are inflicted are exceedingly uncommon in a natural habitat. In contrast to groups in a normal or an undisturbed habitat, a city-dwelling group of rhesus monkeys, for example, one living in a crowded bazaar or similar location where there are too many people and too many monkeys for the space, water, and food available, will very likely do anything they can to assure exclusive access to as important parts of the area as possible (Southwick, Beg, and Siddiqi 1965). In this situation a doorway may sometimes become a boundary, and if a strange or nongroup rhesus enters it, he may be subject to instant attack and suffer possibly fatal wounds. This is excellently illustrated in a film entitled "Rhesus Monkeys in India" (C. H. Southwick). There are many indications that most primate species, if not all, are capable of territorial behavior if their living conditions deteriorate sufficiently (Gartlan and Brain 1968). Under such circumstances, groups in species that are not normally characterized by territorial behavior may become exceedingly defensive.

There are advantages to the individual primate in knowing the group's home range very well. Details of the location of food, water, travel routes, safety, and sleeping trees are but a few of the items of information that develop with familiarity. The importance of this knowledge is not apparent on a day-to-day basis, but in an emergency the ability to respond immediately and appropriately with respect to the terrain may save an animal from harm or death. In a well-known range an animal is less likely to encounter hostile conspecifics or to be surprised by predators.

SOCIAL SPACE

Man seldom is conscious of his patterns of social space, but they exist and are adhered to as strictly as are the spacing patterns among nonhuman primates. Observers of human behavior have described in detail many of the regularities in the use of space whenever two or more people come near one another or interact (E. T. Hall 1959, 1966; Hallowell 1955; Goffman 1959, 1961; Sommer 1969; Scheflin 1964). How and where people stand or sit, how closely they approach each other, and how long and in what manner they maintain proximity are all factors that give important information about the interactors once the system is understood. Almost everything humans do together is regulated in part by conventions of social spacing.

Very early in life every child begins to learn the patterns of space that are acceptable to his caretakers and the people around him. He is taught the postures, attitudes, and actions that are appropriate in different kinds of social encounters. As soon as the infant is able to interact with his mother, she encourages some kinds of activity and discourages others. Slowly but inevitably the baby and then the child assumes the patterns characteristic of his or her age, sex, status, and ethnic background as they are interpreted for him by the surrounding members of his family and community.

Human cultures vary tremendously in their explicit and implicit rules of social spacing and of how individuals should behave under different conditions. For the members of some cultures physical contact in any but certain specific situations is unpleasant or even intolerable. Other cultures permit and expect people to touch one another. Often the contact is made in a very stereotyped manner—which signals to the participants that everything is ok. The stereotypy prevents suspicion of ambiguity in motivation for the contact. This reassurance is important where the same gestures are used in different contexts and when a confusion of motivations would cause distress. For example, if a kiss is highly formalized and stereotyped, it will be acceptable between men under certain circumstances, or a hug or pat on the back may be defined as "brotherly" and be quite acceptable under specific conditions.

Popular stereotypes of some cultures, such as the aloof Englishman or the effusive Italian have their basis in the quite different patterns of physical interaction among some social classes of the two groups. This is often accentuated in literature, but a comparison of the almost stylized characters of *The Forsythe Saga*, moving in the ritualized manner of their day, with noisy, boisterous peasants interacting in an Italian marketplace offers striking evidence for the existence of great differences in spacing and social interaction patterns in parts of these two cultures.

"Rules" of personal space usually become obvious only when they are breached and the interactors become aware of the problems that result. At one time or another everyone in our culture has experienced discomfort when, for example, a stranger or a casual acquaintance stands too close during a conversation. When this happens the person who feels uncomfortable invariably shifts his position or focus of attention away from his conversational partner in order to reestablish appropriate social space.

It is easy to test this. Simply start a conversation with a stranger and gradually move closer, until you are standing within 12 inches of him or her. The amount of discomfort can be increased greatly by standing directly in front of the listener and not looking away from his face. The invariable response of the listener is to look away to break visual contact, since in our culture when two people talk they normally do not look directly into each other's eyes for more than a few seconds at a time. Avoiding visual contact is one way of minimizing the discomfort of a violation of social space, as for example in a crowded situation such as in an elevator where it is not possible to move away from someone.

Many compensatory behaviors and gestures are used to maintain individual personal space and separateness under crowded conditions: arms may be held tightly to the body to avoid touching someone with the hands; the gaze may be directed downwards or at some distant object; interest may be directed in a very obvious manner to a book or some other object; and should there be contact, the individuals involved may choose to act as though it had not occurred. In these situations it is possible to maintain a degree of separateness or social distance because no sign of recognition or awareness of proximity is given.

To the untrained observer the members of a primate group may appear randomly distributed, some close to one another and others at a distance. After watching for a few days the observer becomes aware of regularities and soon it is possible to say, for example, that certain adult females generally stay together under certain circumstances, while 2 particular adult males appear unable to get within 10 feet of each other without showing antagonism. As the observer gradually gets a feeling for the space relationships of the group, he can begin to generalize as to the characteristic distances among group members according to activity — whether resting, moving, eating, or settling down for the night.

Obviously, there is more to the use of space among nonhuman primates than is indicated by the size of the home range. The use of interindividual space is exceedingly important in the hour-to-hour behavior and relationships of the group. Whether animals stay within a few yards of one another or spread over more than a mile is related to several factors including the spatial tolerances characteristic of the species and of the habitat in which the group lives. A thick forest makes it very difficult if not impossible to maintain visual communication, whereas in an open grassy area a monkey may be able to see for long distances. If animals must keep in touch with each other by calling, then they must stay within hearing distance, or if they keep in touch by gesture, within sight.

Among North Indian langurs, and other species as well, the way animals space themselves while eating is a good indication of relative dominance with respect to foods (Jay 1965a). When a group of langurs moves into a beanfield full of fresh tender blossoms each animal grabs large handfuls and stuffs as much as possible into its mouth. There is usually a limited amount of bean blossoms so that within a few minutes tension among the eating animals builds. Each looks quickly toward every other animal in sight and then begins jockeying for control of as much of the bean patch as possible. Invariably the adult males move into the center of the patch, and females or younger animals are forced either to the very edge or completely out of the patch. The linear distance among the remaining eating animals may be correlated very precisely with each eater's ability to control food in other situations and also with many other aspects of dominance behavior.

In some species, individuals within a group are spaced according to patterns characteristic of the species. Bonnet macaques of South India (*Macaca radiata*) seek a great deal of passive contact (Kaufman and Rosenblum 1966; Simonds 1965), and when bonnets sit quietly they choose to sit touching each other or huddled in groups with their heads on each other or leaning against one another. The pig-tailed macaque (*Macaca nemestrina*) (Bobbitt, Jensen, and Gordon 1966), on the contrary, very seldom sit touching and prefer to space in such a way that there is no passive contact. Contrasts have been drawn between these two species, characterized respectively as "huddlers" and "spacers" (Jensen, Bobbitt, and Gordon 1966a, 1966b). It is not possible to categorize many species in this way because there are few clear or general spacing tendencies. In most species, distances among individuals and the amount of physical contact sought are results more of social and genealogical relationships than of contact perferences of the species as a whole.

Social distance can be measured in feet and yards of distance between animals and also it can be considered a measure of avoidance. Living space is full of objects, and there are many ways of getting out of sight of others. If a monkey does not wish to interact or respond, the simplest way for it to avoid the issue is to disappear from sight. Probably there are more dominance fights avoided by one animal's slipping quietly away than there are actual fights or chases.

There is differential use of space by group members. Some age-sex categories of animals tend to remain peripheral. Subadult males, for example, tend to stay on the edge of the group, where they are least likely to come in direct contact with dominant adults (DeVore 1962; Lindburg 1967; Jay 1965a). If individuals do not get along well they may avoid one another and remain in different parts of the group.

For groups spending much time on the ground in the open, position in the dispersed group may be important because of predation. Females with infants usually remain in more protected parts of the group, in contrast to large juvenile and subadult males that are frequently located on the edge of the group where they may be more vulnerable to predators.

The orangutan is unusual among the nonhuman primates because the majority of adults live alone (Schaller 1961). A mother may have one or two young with her but most adult males live in isolation. In general, orangutans can be characterized as exceedingly inactive, and the wide spacing among adults assures that there is little contact with each other under ordinary circumstances.

GROUP ORGANIZATION AND THE ENVIRONMENT

The primate observer is concerned with reporting on the observable regularities and relationships in the lives of animals. Kummer has stated, "the field studies to date suggest that the patterns of social *behavior* of a species are related to the taxonomic position of the species. Social *organizations*, however, seem to appear here and there in the Primates without apparent relationship to taxonomy or to patterns of social behavior" (1967:361, *emphasis* his). The more studies we have of groups of a species living in different habitats and under different circumstances the more evidence there is for variability in the total group structures rather than in particular elements of social behavior. Gestures, vocalizations, and other behavior patterns do not vary appreciably among groups within a species. Similar types of organization appear in very different taxa and seem to be more related to ecology and habitat than to the type of monkey.

Several authors, in classifying types of group organization, have indicated various ecological factors that influence group size and composition, home-range, size, and the presence or absence of peripheral adult males (Gartlan and Brain 1968). Recently Crook (1970b) has summarized these efforts. Although classifications of types of primate groups vary according to author and emphasis, each list usually includes the following types:

One-male/multi-female groups: *Papio hamadryas,* Kummer 1956, 1957, 1967, 1968a, 1968b; Kummer and Kurt 1963, 1965; *Theropithecus gelada,* Crook 1966; *Presbytis entellus,* Sugiyama 1964, 1965a, 1965b; *Presbytis johni,* Poirier 1968a, 1968b, 1969; *Cercopithecus sp.,* Gartlan and Brain 1968; *Erythrocebus patas,* Hall 1965a.

Larger groups with several to many adult males in addition to females and young: *Macaca mulatta,* Altmann 1962; Lindburg 1967; Neville 1966; Southwick, Beg and Siddiqi 1961a, 1961b; *Macaca nemistrina,* Bernstein (unpublished); *Macaca radiata,* Simonds 1965; Nolte 1955a, 1955b; *Macaca fuscata,* Itani, Tokuda, Furuya, Kano, and Shin 1963; Itani 1954, 1961, 1963; Tokuda 1961–1962; Frisch 1959; Imanishi 1960, 1964; Miyadi 1965; *Papio cynocephalus,* DeVore and Hall 1965; DeVore 1962, 1963,

1965*b*; *Papio ursinus,* Hall 1962; *Papio anubis,* Rowell 1966*a,* 1967*a*; *Presbytis entellus,* Jay 1963*a,* 1965*a*; Ripley 1965, 1967; *Cercopithecus aethiops,* Struhsaker 1965, 1967*a,* 1967*b,* 1967*c*; Gartlan and Brain 1968; *Mandrillus leucophaeus,* Gartlan (personal communication), Struhsaker; *Gorilla gorilla,* Schaller 1963.

Small groups of a mated pair and offspring: *Hylobates lar,* Ellefson 1967, 1968. Other units with less clearly defined membership, often referred to as "open" groups or subgroups within a local population: *Pan,* van Lawick-Goodall 1967, 1968*b*.

There are variations on these general types of group organization and several types may characterize a single population or a species. For instance, hamadryas one-male units forage independently during the daytime but at night a number of one-male units may come together in one of the few sleeping places available in the area. Kummer (1968*a*) has suggested that these larger units are analogous to the troops of savanna baboons in that the same one-male units generally meet during the evenings and remain together during the night. In contrast, gelada nightime groupings are composed of one-male units not substructured into bands as are those of hamadryas (Crook 1970*b*).

Habitats range from tropical forest to open arid country, with many in between these extremes. Some correlations of group organization with habitat type are obvious, others are not. In an extremely harsh, arid environment like that of the hamadryas, where there is little food and what food there is is widely scattered, social units foraging together are usually composed of one adult male with several females and their young. It would be difficult and perhaps impossible for a large group, with many adults of both sexes as well as young, to survive under conditions of very low food density. Females would have to compete with males for whatever was available and the ensuing conflict would certainly break up the group and cause starvation or injury to the weaker members of the group—the females and the young.

There is no single forest habitat; there are many, and this makes it necessary to be very specific as to the type of forest and the parts of the forest a species inhabits. Monkeys that remain in the upper stories of rain forest have a very different habitat from that of monkeys that live primarily on the ground or on the edges of the same forest. Deciduous forest in zones of seasonal changes in vegetation is different still.

In general, although there are exceptions, the group size of primates living on a savanna tends to be larger than that of primates living mainly in the trees of a forest. Baboon groups living in forest are often as large as savanna groups but these monkeys are not characteristically an arboreal forest species. The generalization of increasing group size in the savanna seems to hold if the contrast between forest and savanna species is restricted to those forest-dwelling species that spend their days in the trees and those species that are on the ground in the open during the day. Again, it is necessary to be quite specific as to the type of forest habitat a species prefers, including the level within the forest in which they spend the most time and where they obtain their food.

Small groups with only one adult male are found in several types of habitat. In addition to those found in very arid open country, such as among gelada and hamadryas, there are similar groups among Uganda patas monkeys (Hall 1965a), which do not live in such harsh surroundings. Crook (1970b) has suggested that the one-male units of gelada and hamadryas may be the result of such factors as scarcity of sleeping trees and a dispersed, scarce food supply. Kummer (1968b) suggests that the hamadryas pattern of social grouping, similar in many respects to that of the gelada, has developed from the phylogenetically older pattern characteristic of savanna baboons as an adaptation to the harsh environment of the hamadryas.

One-male groupings are also found in the New World *Callicebus* monkey (Mason 1965). South Indian langurs frequently live in groups with only one adult male (Sugiyama 1964, 1966a, 1966b; Yoshiba 1967, 1968), although in the area studied by researchers from the Japan Monkey Centre many of the groups did contain more than one adult male. The habitat of Dharwar, where the Japanese research team studied, varies considerably during the seasons of the year, with relatively lush vegetation in monsoon months. Male leadership within a Dharwar group changes often and a new adult male leader may kill all the infants in the group. It is difficult to understand how this frequent change of male leadership and the killing of young permits the survival of the langur population over time or how these behaviors are related to the habitat of the group. Langurs living in similar habitats to the north do not behave in this way.

Average population density in some North Indian langur populations is approximately 12 animals per square mile. At Dharwar in one forest area the number of animals totaled 220 individuals in a square mile (Yoshiba 1968) and in another forest area the density of langurs was as high as 349 (Sugiyama 1964). Yoshiba (1968:224) summarized the langur data: Some of the forest is not used by langurs because it has recently been disturbed by humans, so the real density of the area may be between 220 and 349 langurs per square mile. This density is at least 13 times as high as that at Orcha, and much higher than that at Kaukori (Sugiyama et al. 1965). Crowding may be an important factor in the behavioral differences observed in these areas of India.

Some Tropical forest species live in one-male groups in areas with relatively low food density. In these areas there are few individuals of each plant species and this produces a condition of generally low food productivity. However, population density of arboreal monkeys in some tropical forests may be as high as 300 individuals of several species per square mile. These forests are clearly far more complex habitats than are open or savanna areas, and they allow many specializations according to preferred forest level and to foods within levels.

Human population density did not reach 300 per square mile until long after agricultural means of food production were developed. Before agriculture, man relied on hunting and gathering for his food. The range of a band of humans in the earlier stage probably was of a size averaging ten to twelve square miles for each individual in the band (Lee and DeVore 1968). This is much larger than for any nonhuman primate.

An important part of the environment of a primate group is the potential predators living in the same area. The social group affords a degree of protection against predators that an individual living alone does not enjoy. In open areas, such as savanna or the edge of a savanna, the long list of potential predators includes lion, leopard, cheetah, wild dog, and hyena, although not all these are present in every area. Avian predators such as monkey-eating eagles may be extremely effective upon small or young monkeys, and on the ground snakes pose a threat to monkeys spending much time out of the trees. There is a cumulative protection offered by group life since each member adds its alertness and vigilance, thereby decreasing the likelihood of members of the group being attacked by surprise. Each animal acts for itself but the result of an alarm call by one can serve to alert all.

Observers usually underestimate the effects of predation. Schaller's careful study of leopard and tiger in northern and central India revealed that 27 percent of 22 leopard feces and 6.2 percent of 335 tiger feces contained langur remains, far more than anyone estimated on the basis of observing the monkeys. The toll of predation on monkey groups can be estimated only when the predators are studied, not by observation of the primates alone.

Eagles and other birds that may prey upon primates should be observed and their droppings analyzed to estimate how much effect they have on the monkey populations. Predation must be discussed relative to habitat, both as habitat is related to predator and prey and as it affects response to threat. For example, lions do not hunt in forests whereas leopards do. Monkeys living primarily in the tops of tall trees and monkeys on the ground confront different dangers in the same forest.

Of all the nonhuman primates only the gorilla is able to remain on the ground both during the day to feed and at night to sleep. This ape's large size and great strength make it extremely unlikely that any forest predator would pose a serious threat to its safety. The adult gorilla's weight makes it impossible for swift escape in trees but it does provide effective protection on the ground. Baboon troops living in forests may also remain on the ground for hours each day but they are constantly vigilant and when a predator is sighted the monkeys can climb to safety. Although adult male baboons are large and strong for monkeys, they are no match for many predators.

Savanna baboon troops have been characterized by Hall and DeVore (this volume) as responding to the presence or approach of some predators by organizing in such a way that adult males are between the potential threat and females and young. Adult males may approach a cat while the remaining members of the troop stay behind them in relative safety. Although this may be true of a baboon troop on the open savanna some distance from trees, the reactions of the troop will alter under other circumstances. If the approaching predator is a lion or several lions, baboons do not approach, but instead are more likely to move back to the nearest trees. If the group is already near trees, the response to any predator is likely to be flight to the tops of the trees and not a stand in the open, where they are much more vulnerable. Under these conditions each baboon runs to safety regardless of where adult males are in relation to females and young. Adult male baboons will successfully intimidate cheetah and leopard, but not lion, where there is no question as to the potential winner.

Similarly, langurs will respond differently to a predator depending on where they are and what predator threatens. As soon as an alarm bark is given each langur dashes up into the trees if they are sitting underneath them, or they run as fast as possible to trees if they should be in an open area. A major factor in the difference between langur and baboon responses to predation is that the langur is never far from trees whereas the baboon group may be far enough that if each animal tries to make it to the trees whenever a predator is sighted there is a greater risk of being caught. Once a langur group is in the trees, and if it is a cat that chased them there, the adult males may threaten from a protected vantage point. Long after the langur group is in the trees the males continue to give the alarm barks and vocalizations toward the source of danger. Adult langurs have been observed to return to the ground to rescue an infant that has fallen or been dropped in the flight (personal observation).

Uganda patas monkeys respond to potential danger from man or predator in a unique way. The adult male member of the small patas group gives a very conspicuous kind of display, which Hall noted, "bears a general resemblance to the distraction behavior of some species of birds, and which seems to have no precise equivalent in other monkey species" (1965: 72). The adult male bounces very noisily and visibly on the tree branches, comes to the ground, and then dashes away from the source of danger and from the group. Hall (1965:72) suggests that

> the running far away from the observer and the group, however, may perhaps be explained as, in effect if not in intent, a distraction or a diversion which because the male is such a conspicuous animal, might draw off the attention of day-hunting predators sufficiently for all the group itself to escape.

In general, the response to a predator by a primate group will depend upon the kind of predator and the habitat in which they are at the time of danger. Most species have several ways of reacting to danger rather than a single invariable response.

SOCIAL ROLES

Social anthropologists traditionally have analyzed human societies in terms of social structures or organizations and have defined the social roles that occur within them. These roles and structures are abstractions based on events that occur time and again in certain situations. Role analysis may become a useful tool in understanding nonhuman primate behavior since monkey and ape social life is a web of interaction among group members living together possibly for years and even for generations. The following comments will be based on role analysis as it is conceived in social anthropology and as summarized in an excellent recent article by Benedict (1969: 203):

> Role enables us to abstract certain normative or stereotyped aspects of behavior from the full repertoire of an individual's actions. It "operates in that strategic area where individual *behavior* becomes social *conduct*" (Nadel 1957:20). We move away from the idiosyncratic behaviour of an individual towards a type of behavior which can be recognized as occurring in many individuals. We recognize a role performance because of the situation in which it occurs.
>
> This abstraction of certain characteristic forms of behavior in defined situations is what enables comparison not only from individual to individual within the same culture (for humans), but cross-culturally and possibly cross-specifically. It is essential that we study enough individuals in enough interactions so that we can make this sort of abstraction.

In a summary of major characteristics of social roles, Benedict concludes,

> Anthropological definitions of role have stressed: 1) its interaction aspects; 2) its normative aspect; 3) the importance of setting and cues for role performance; 4) the fact that roles are made up of sets of interrelated characteristics some of which are more crucial than others for eliciting appropriate responses; 5) that roles are defined by the expectations of the actors often backed by sanctions; 6) that roles are learned.

The concept of social role has major features that are highly relevant to the description of nonhuman primate behavior. Other aspects of social role theory based on human language and complex social institutions are, of course, inapplicable to primates other than man. As mentioned above, a role is essentially an abstraction based on repeated and relatively stereotyped patterns of behavior in specific situations. Many individuals perform or display this constellation of behaviors, hence the emphasis is on the normative behavior of many rather than on that of a single individual.

The definitive attribute of a social role is that it always has as a context at least two interacting individuals. If there is a dominant monkey this implies the existance of a subordinate. A leader must be followed by at least one other animal, and for each mother there must be an offspring. Since roles are reciprocal each participant holds a set of expectations relative to the role. There are right times and right places for the performance of a role and there are many obvious as well as many subtle cues that indicate the attributes of a role. Among humans, clothing, decorations, adornments, and names may all serve to signal roles. For humans there may also be a large nonverbal component of postures, gestures, and the manner in which words are spoken. A monkey, such as an adult male baboon, relies upon nonverbal cues to signal roles; for example, his large canines and thick mantle may be shown in display as he asserts a role of dominance. Many morphological features are prominent in social interactions and for many species these features are an essential part of such behavior as courtship, mothering, or leading.

Human subjects, of course, can discuss and clarify much of their social behavior if asked to by the anthropologist. Language makes possible labels that provide clues to why humans behave in certain ways toward one another. The ability to talk about relationships, obligations, restrictions, and other aspects of behavior enables the investigator to perceive regularities and reasons for regularities even when they are not part of the conscious knowledge of the members of the society practicing them.

Nonhuman primate social roles must be defined entirely upon behaviors that can be observed, regularities that reflect the general order of the society. Statements regarding psychological motivations for monkey or ape role behavior are speculative and should not be offered unless they are clearly labeled as speculation. Not all of the many often repeated and relatively stereotyped behaviors of nonhuman primates are associated with role behavior.

Within an overall description of the group, and in addition to the idiosyncratic behaviors of individuals, statements can be made outlining social positions, rank, or roles independent of the individuals occupying them. Some of these roles will be discussed below.

LEARNING SOCIAL ROLES

How an individual monkey or ape learns a social role, how an alpha male (if there is one in a group) learns to behave not only as an adult male but also as the top-ranking animal and how females learn the roles characteristic of adulthood are examples of the questions asked in the study of primate social roles. Clearly there is a strong basis in the biology of a species for members of each sex to respond differently to stimuli and to learn certain behaviors more readily than others. Physiological, hormonal influences on behavior are strong and very important in various phases of development from earliest life. The effect of sex hormones on the developing fetal brain greatly influences whether male female patterns of behavior will predominate later in life (Young, Goy, and Phoenix 1964).

However, learned aspects of behavior, such as a social role, depend to a great extent on the social context of the learning. We know relatively little about which environmental stimuli affect a maturing primate, especially during the early months or days of life. Progress is being made, however, in the investigation of early experience.

Harlow and his associates have investigated the development and importance of some components of the emotional attachments of rhesus monkeys (Harlow 1959*b* 1962*a* and *b*; Harlow and Harlow 1966; Harlow and Zimmerman 1959). An artificial or surrogate mother was designed and then the investigators varied the attributes of the "mother." For example, some of the surrogates were cloth, others were wire; some gave milk, others did not; some moved, others were stationary (Alexander and Harlow 1965; Cross and Harlow 1963; Hansen 1966; Harlow 1962*b*; Seay, Alexander, and Harlow 1964; Seay, Hansen, and Harlow 1962; Harlow and Harlow 1965). These researchers concluded that clinging was the most important factor in the mother-infant bond, but in addition, there were a large number of other important variables.

Rhesus infants were raised under a variety of conditions. Some infants were kept in complete isolation, others with one of the different types of surrogate mother, and still others with their real mother (Harlow and Harlow 1962; Harlow, Harlow, and Hansen 1963; Harlow, Harlow, Dodsworth, and Arling 1966). Some infants were allowed limited and controlled access to other rhesus. Infants allowed to play with other infants even a few minutes each day appeared to benefit greatly from the experience. When the young females grew to sexual maturity and were mated (by no means an easy task in many instances) the relationship with their offspring was observed.

Laboratory experimental manipulation of rhesus early experience demonstrated conclusively that the experiences of an infant, and even those of its mother, strongly affect the infant's later behavior patterns — having effects that vary from obliteration of social abilities to only small deviations from the normal behavior of laboratory rhesus (Mason 1960, 1961*a*, 1961*b*, 1965; Mason and Sponholz 1963). Few, if any of the experiments noted above could have been attempted in a field context. However, the parameters of "normal" behavior can only be determined by observation of animals living in a free-ranging social situation. Only by constant interaction between field and laboratory research can the determinants of social behavior be understood.

The kinds of information that have been collected in traditional studies are insufficient to answer questions concerning the development of nonhuman primate gender roles. It is possible, however, to observe many subtle behaviors that may provide clues to the experiences of a young monkey or ape, experiences that may be very influential in shaping gender and social roles. A multitude of crucial questions remain unanswered, as for example: Where does the infant (from the day of birth) look when in the mother's arms? Which other animals look at the infant? How intent are these glances and what else is the watching animal doing? Which animals touch the infant and how? We need to know not only how long a time an infant clings and how long it is on or off the nipple, but also, how tightly does the mother clasp the infant? How quickly does she respond, by grasping the infant, to what is happening in the social group around her, and to what extent and how does she convey her reactions to the infant? In other words, how does the mother influence the infant and pass information about group events to it via her own reactions?

The observer cannot tell, in detail, how tightly or loosely, how roughly or gently a mother holds her infant, but with practice he might at least attempt to devise a scale of firmness so that some measure could be obtained and perhaps correlated with infant reaction. Qualities of touch and contact are exceedingly important. All such measures must be made from close enough range that the observer is able to see precisely what happens and to hear the softest vocalizations. If the animals are too far away soft vocalizations may be missed completely. A well-planned colony is an excellent location for these types of observations and with the addition of telemetric techniques and filming of interactions it should be possible to measure motor responses to social stimuli, whether the latter are visual, auditory, tactile, or olfactory. Telemetry should also help greatly in defining the types of stimuli to which the maturing young primate is sensitive. Until the time telemetry is practical either in a colony or in the field we must rely upon improving the quality of the kinds of information we are able to see and to record.

It is probably correct to assume that the young female primate learns much from her mother regarding the roles she will assume later in life, although it is not likely that she learns it all from her mother. It is less obvious where the young male learns about male roles before the age when he spends hours with peers in the play group. How much attention the infant male monkey pays to the behavior of adult males in a group remains to be determined, but at least we can start by recording the animals to which the young male appears to respond or attend. It is probably that in the play group the male and the female learn different skills as well as encounter many of the same kinds of experience. Most motor skills and many social relationships are identical for males and females—but it is those experiences that are not shared or that are perceived differently that are of concern in the development of gender roles. Certainly when one watches a troop of monkeys, it does not take long to realize that from an early age the relationships of the young to adults differ by sex.

Observers have tended to regard infants as a rather uniform group of animals and to concentrate upon and emphasize the range of variability in the temperament and status of the care-taking animal, the mother or other female. Although there is a wide latitude of behavior among mothers, and all adults for that matter, there is also a surprising degree of variability in infants — from the first days of life. Some infants are more active than others and respond differently to social and physical stimuli. It is unlikely that even the most dominant mother could bequeath her status to an offspring whose basic temperament was characterized by passivity or withdrawal as customary responses to social interactions, whereas the very aggressive, robust son of a retiring low-ranking female might some day become alpha of the group. These are hypothetical examples, but much more attention should be paid to differences in responsiveness, and to other characteristics of the infants themselves. It is the interaction between mother and infant and the influence from other members of the group that shape the growing infant's patterns of behavior, but it is an interaction, and a two-way one, with the infant contributing far more than has been appreciated.

Some studies of human mother-infant behavior have also emphasized interactional aspects in early development (Escalona 1968; Rheingold 1960, 1961, 1966, 1969). Rheingold (1966:12) writes,

> The infant, by his appearance and behavior, modifies the behavior of other social objects. He not only evokes responses from them but maintains and shapes their responses by reinforcing some and not others. From our individual experiences, we know how effective he can be! He is so effective because he is relatively helpless yet active and because he is so attractive to his beholders. The amount of attention and the number of responses directed to the infant are enormous — out of all porportion to his age, size, and accomplishments. Under ordinary circumstances, in any human group containing an infant, the attention directed toward him is usually considerable.

Hall (this volume) has emphasized the importance of social learning to most aspects of nonhuman primate life. This learning is not restricted to any phase of development and to some degree it continues throughout life. It is most important during immaturity when the growing animal's patterns of social behavior are forming. It is difficult to estimate just how much a young primate learns by observing others in the group, but if our meager evidence is indicative, we have grossly underestimated the importance of such observation. Furthermore, the young primate learns from highly selective observations. Just any other animal in the group will not do as a model for an infant or juvenile. The young monkey or ape must pay particular attention to specific older animals, for what are probably exceedingly complex reasons. This does not mean that the immature cannot learn something from every animal in the troop, only that certain skills develop from contact with specific classes of animals.

The experimental introduction of new foods into the diet of Japanese macaques by the Japan Monkey Centre yielded fascinating information as the channels of learning among group members; certain age-sex classes learned much more from each other than from others. Itani (1958) introduced caramels to the Takasakiyama troop and observed that the infants and juveniles under three years took the candy. He suggested that the younger the monkey the more willingly it adapted to certain new situations, and also that after three years of age Japanese macaques found it much more difficult to adapt to new circumstances.

Kawai (1965) recorded new behaviors for the Koshima troop including "placer mining" of wheat, making a gesture of "give-me-some," and washing sweet potatoes. Tsumori (1967:207) reports that "in 1953 a female first tried potato washing, and by 1962 this behavior had been picked up by all the troop members except newly born infants, 1-year-old infants, and adults more than 12 years old." Tsumori (1967) has summarized these observations and reported on his research and experiments on newly acquired behavior patterns and social interactions.

Van Lawick's observations of several chimpanzee orphans showed beyond doubt that there are many skills the young chimp does not learn if it loses its mother (personal communication; van Lawick-Goodall 1968b). Although other chimpanzees may be termiting, playing, and doing all the things they normally do around the orphan, the youngster does not learn from them.

It is very probable that learning patterns and processes may vary at different ages and in different contexts. This remains to be investigated in detail in the field, where social environments can vary from group to group and where considerable change may be induced in group structure and in behavior patterns by the introduction of various learning experiences. The environment for learning, if it changes much from generation to generation, may influence the short-term structure of the group and the experiences of young. Longer term changes in this environment would affect genetic contributions from one generation to the next, and the interaction between behavior and biology may be seen in these effects. On a shorter term basis, the organization of behavior patterns and social structures approximate the behavior traditions of the group and they change little from one generation to the next if there are no substantial changes in the habitat or in the larger social environment.

ROLES AND SOCIAL DOMINANCE

Let us return briefly to the problem of defining social roles in a non-human primate group. Among many Old World species each individual male and often some of the adult females occupies a rank of dominance relative to most other adults in the group. This ranking may be sharply or vaguely definable and may be more applicable to some social activities than to others. It is not useful to define each animal's rank as a different social role since each possible position in the troop does not have specific characteristics except that, for the most part, they are rankings in a series as long as the number of identifiable positions. It does not mean much to say that such-

and-such an animal is social role number 9 in a ranked hierarchy of 13. Some general categories of behavior, such as leader, have more meaning. The total number of social roles within a troop is not clear, and there may be some disagreement as to which ought to be included. The number of social roles may be related to the size of the group, but not to this alone since complexity of organization is also a very important factor.

Not all species organizations necessarily have the same roles. For example, the hamadryas adult male leader of several females and young might be referred to as a harem overlord (Kummer 1968a), but such a role is not present in the savanna baboon troop. Not all species have a central hierarchy within the ranking of the adult males (Hall and DeVore, this volume), but for those that do, this is a role that may be occupied by several animals simultaneously.

Some social roles may be assumed involuntarily, such as that of motherhood, whereas others are acquired by struggle, such as that of an alpha male. In some groups the role of alpha exists over time regardless of which individual occupies this position. In terms of daily life it matters very much which animal is the alpha; different personalities and temperaments lead to very different kinds of behavior of the top-ranking male, and these behaviors affect the troop. Should he be extremely aggressive and threatening, the troop will probably behave differently from one headed by a more relaxed and nonaggressive individual. It is quite possible that a group with the first type of leading male could control a larger home range relative to adjacent troops than would be possible for a troop with the second type of male in the top position.

Certainly it seems possible to predict the eventual status of some young males by the way they behave during subadulthood. DeVore (1962) referred to one young male in particular, Brash, whose behavior strongly suggested that he would eventually achieve high rank in the baboon troop. Subadult male behavior varies so much that most observers of baboon or macaque groups have been tempted to predict (at least in private) that certain of the young males are "most likely to succeed."

ROLE CONFLICT

Among human societies, role conflict may become a problem. Behaviors associated with one role may be in conflict with those ascribed to another, such as in the roles of mother and wife. What is appropriate to one role may be quite out of place within the other context and conflict results. It is not clear whether or to what extent this is a problem among nonhuman primates, but there are several situations in which these conflicts conceivably could arise. Take, for example, the role of alpha male and that of sexual consort. If the female is very much lower in status, which most females are, the close attention or proximity of a very dominant and perhaps usually aggressive adult male may repress the female's sexual interest or change it into fear, subordination, and retreat. The male must behave in such a manner as to allay her fears if she is to remain in close physical contact for copulation. He may force her to copulate regardless of how fearful she is, but in the long run if this were necessary each time he copulated he would probably be much less successful in contributing to the next generation than if he could reassure the female easily and maintain a consort relationship. Grooming and gestures of placation, such as lip-smacking or grimacing among many species, reduce tension and indicate to the subordinate that the more powerful animal does not intend harm or attack.

It has been suggested in the literature that the roles of motherhood and sexual receptivity are incompatible for the adult female, but there is some question whether this is true. The adult female among most species has already started to wean her latest infant at the time she becomes sexually receptive again and in many species the infant or small juvenile by then is spending a lot of time with peers in play. Instances have been recorded for a number of Old World species (langurs and macaques, personal observation), of an estrous female copulating with an adult male while carrying her infant.

In evolution from monkey, to ape, to man, the trend has been for an increase in the length of the infant's dependence upon its mother. This period varies among different monkey species from a few months (vervets, Struhsaker 1965, 1967b) to a year and a half (baboons, DeVore 1962; langurs, Jay 1965a). A chimpanzee infant relies upon its mother for from three to four years (van Lawick-Goodall 1968b) and a young human for a decade and in many cultures still longer. The length of time an infant is closely associated with its mother is determined to a large extent by the rate of its physical development, and the length of dependency has important implications for the nature of its relationship with its mother as it matures. Frequent association with the mother can continue long after weaning. Among rhesus monkeys on Cayo Santiago the offspring may have a special relationship with its mother long after it reaches maturity and her status and temperament can influence the status of the offspring for many years (Sade 1965, 1966, 1968). In contrast, the langur monkey of North India does not continue a close relationship with its mother after it is weaned and it is often impossible to tell which adult female is the mother of any specific juvenile in the group. Whether or not the presence of a young monkey or ape interferes with the sexual behavior of its mother is clearly the result of many factors, including rate of development, length of dependency, and the behavior patterns characteristic of the species.

It may well be advantageous for a female nonhuman primate not to have young with her while consorting with an adult male since she may be in the midst of aggressive interactions with other males over her attentions.

In other instances adult male baboons have been observed to give kindly attention to infants (personal observation), but this is not common and may be limited to males of certain personalities or temperaments and/or status at the time. Jay (1963*b*, 1965*a*) recorded instances of young juvenile langur males deliberately harassing adult males copulating with their mother. The result was not harmful to the juvenile, or to the mother, or, most decidedly, to the adult male. Clearly, more information is required before it can be established that maternal and sexual roles are incompatible.

It is common occurrence among many macaques and baboon troops for an adult female carrying an infant on her front, to threaten or attack another female or young animal. The infant responds invariably by clinging tighter and trying its best to maintain contact with its mother, even though it appears frightened by the commotion and fast movement. The mother of a newborn may try to refrain from aggressive interactions; this was very noticeable among North Indian langurs, but when the infant was older this author at least was never impressed with the reluctance of a mother to interact. Females who do not interact much in dominance fights without an infant tend not to change their patterns as mothers and, similarly, dominant aggressive females interact regardless of the presence of a young infant after it is a few weeks old.

One feature of the life of nonhuman primates is the infrequent necessity of playing several roles at the same time. The dominant leader is usually just that. He can signal aggressive intentions or peaceful ones, such as a desire to groom, to be groomed, to copulate, or to sit quietly; he is seldom called upon to play very different roles at the same time in an incompatible way. This is also true for the female. When she is a mother and cares for a very small infant she is not sexually receptive; if she is very dominant or if she is active in leading the group, she continues to be. It is difficult to imagine a situation among the monkeys and apes where there are conflicting social roles, as there often are for modern man. There may be such conflicts, but they are not obvious.

Benedict (1969) and others have suggested that social roles generate social order by assuring cooperation among the participants in a society. Certainly the definition of predictable patterns of relationships among the members of a group helps insure regularity of behavior and continuity of performance of the jobs that must be done. Because the definition of a social role is independent of the individual occupying the role, the tasks or attributes of the role continue although the holder of the role may change.

Roles may be more or less rigidly defined depending on the society and the circumstances as well as on the content of the role. Roles are responsive to change just as are social systems, and the effects of the environment and habitat on behavior manifest themselves over time in roles as well as in structure. Since some primate societies have been described as much more rigidly structured than others—as for example, DeVore and Hall's account of baboons (1965) compared to van Lawick-Goodall's of the Gombe chimpanzee (1967, 1968*b*)—it is reasonable to expect that social roles in these societies will vary in their definition, complexity, and attributes. A relationship between the role system and maturation rates has also been suggested (Benedict 1969), but it is difficult at this time to tell whether the length of various stages of development is a causal factor in the stringency of roles or whether both are related to other more pervasive factors in the physical environment of the social group. One thing is clear, the complexity of social behavior increases immensely from monkey, to chimpanzee, to man. The amount of learning that may modify behavior also increases vastly, and the range of subtleties in all phases of social interaction for the chimpanzee is much greater than that for the monkey, regardless of the kind of monkey. Man, of course, surpasses all the nonhuman primates in the variety of his behavior.

DOMINANCE

Let us go back for the moment to a very important aspect of social behavior, dominance, and use it to illustrate some qualities of social roles and their development. The more we know about monkeys and apes, the more difficult it is to give a simple answer to the question, "What is dominance?" The answers we gave in the past are certainly changing. The first few Old World primate species to be studied in the field happened to have quite clear-cut patterns of dominance, power, or priorities—whatever one wished to call the general phenomenon. More studies and especially research on New World monkeys suggested that for many species the unitary early notions of dominance were not meaningful. Dominance is not based on any one element; it is extremely complex, often subtle, and at times very difficult to measure. If dominance rank is understood as expressing priorities accorded different animals, as it usually is, then it is necessary to be quite specific as to the exact situation in which this priority obtains. The priorities of animals differ with respect to different objects and situations—for example, with respect to food, to estrous females, and the leadership of the group when it moves.

Reference has been made to the ability of an animal to control or influence the behavior of animals around him or her. Bernstein (1966) has suggested that this ability to control might be an accurate way to describe aspects of social behavior among some New World monkeys. Here too, the control is specific to certain situations and activities and does not necessarily generalize to all social activities where priority of access is at issue.

In general, Old World adult male primates perform several social functions that include leading the group when it moves, keeping order within the social unit by stopping fights among other animals, and providing defense against strange or threatening animals. The adult females of some species may also perform some of these tasks, but in most species males are more aggressive, stronger, and larger than females. In many social situations male responses, of course, are very different from those observed among females. Although in some species there is some overlap in the activities and strength of males, and females, the male is normally the protector, policeman, and leader, whereas the female is the mother, nurturer, and follower. With the strength and position of a dominant male comes the right both to fight and to deny the right to fight to any other member of the group. It is part of the freedom of the strongest individual to use his prowess to do as he pleases. Through learning and experience these abilities are channeled in ways that are beneficial for maintaining the group and protecting its members. Selection has undoubtedly rapidly eliminated those rare animals whose temperaments were extremes of antisocial aggressiveness.

Among Old World monkeys there are various types of dominance hierarchies. In general, adult male dominance structures, such as those among *Macaca* and *Papio* groups, are more easily defined than are the relationships among adult females. The positions of power of the latter are often much less clearly definable relative to each other. The status changes that occur during the various stages of the reproductive cycle create many ambiguities in status among females and often there is no true hierarchy. Linear dominance structures have been described, as in some North Indian langurs (Jay 1965a), and also more complicated structures, such as the central hierarchies characteristic of the savanna baboons studied by Washburn and DeVore (1963). Among the savanna baboons a few males combine and work together as a coalition against which no single male may succeed. The members of the coalition seldom, if ever, allow themselves to be in a situation where they must act independently of one another. Occasionally an individual member of the central hierarchy may be subordinate to other adult males in the troop that are considerably below the coalition of males in the total troop hierarchy.

As mentioned earlier, an individual adult male may be able to take precedence in one type of situation and not in others. There is a tendency among many kinds of monkeys for the most dominant male to be able to consort with any female, take the best food, and sleep on the most desirable branch. Incidents are recorded from the field, however, indicating that a male may, for example, have the most frequent copulation rate of any in the group and yet be unable to take first choice of food.

A general ranking that includes all adult males in the group may be drawn up after a substantial number of observation hours. As a generalization this ranking is a useful statement about the group, but when the group is not together as a unit and instead has broken up temporarily into smaller subgroupings other males will assume the role of top animal and will give all the proper signals of authority and control. If the most dominant adults are out of sight, even a subadult male may assume the stance and actions of an alpha male—only to act subordinate from the moment he sees a stronger male approaching. For example, in a rhesus group the most dominant adult male generally carries his tail erect, whereas less dominant group members do not, but when the alpha male strolls out of sight it is not uncommon for the animal next in rank to raise his tail. When the dominant animal returns the lower status one's tail lowers quickly, probably to avoid giving the returning male the impression that a challenge has been tendered. Even the young monkey has ample opportunity in peer play groups to experience dominance position over still smaller or weaker young monkeys (Dolhinow and Bishop, this volume). Opportunism is not limited to the human primate.

The impression may have been given so far that the animal in control is the largest or strongest. An association may exist between physical abilities and rank, but personality also plays a very important part in determining status. It takes only a very short period of observation to realize that individual monkeys and apes are extremely different in temperament and personality. Some are quick to respond with threat or aggression to the slightest stimulus from another animal whereas other individuals, in contrast, are best described as placid and reluctant to interact aggressively unless it cannot be avoided. The relation between the temperament of an individual and his or her ranking is important, but unfortunately, it is a relationship that is exceedingly hard to document. In the few field studies that have emphasized individual differences (Jay 1965b; van Lawick-Goodall 1968b) the possibility that personality affects social rank and roles is clear. To assert that rank is determined entirely by temperament is premature, and doubtless as incorrect as most simple explanations. Individual rank is the result of interplay between personality, physical abilities, and experience— including the influence of closely related animals. Affiliations and dominance hierarchies may be based more on kinship ties than we have suspected.

The effect of experience and learning on dominance should not be underestimated. Conditioning is very important and experiences from the early months of life, before weaning, may be very influential in later behavior. The extent and nature of this carryover from early experience into adult life is far from understood, but it is possible that subsequent experiences during the juvenile and subadult stages of development may largely override early experiences. Although the precise effects of early experience are not known, it is probable that among those species in which mother and offspring maintain social contacts for years, and in which siblings continue to support one another after they are independent of the mother an individual may gain a great deal of support over time.

A general tendency has been reported among the rhesus macaques on Cayo Santiago for the offspring of very dominant females to be more dominant than the young of very subordinate females (Sade 1965, 1966, 1968). Certainly among the Gombe chimpanzees the offspring of the well-known female Flo have had advantages from the substantial protection their dominant mother was able to give them when they were young. How lasting these benefits are remains to be determined. Among many monkey species the infants of high-ranking females may annoy adult females of low status with little or no fear of retaliation. Reverse effects may also operate; after early years of dependence on a very subordinate mother the young animal may find it difficult to rise in status. When more is known about individuals that have attained much greater dominance status than the mother, it may be possible to determine whether there are common features of past experiences, individual temperament, exceptional physical attributes, or some combination of factors that can be correlated with eventual status.

Adult status changes over time and is affected sometimes by fortuitous events or factors. The case of Mike, an adult male chimpanzee in the Gombe, is a striking example of a serendipitous rise to power (van Lawick-Goodall, personal communication). Mike accidentally discovered that slapping empty tins down a slope caused a tremendous racket that frightened other adult males. The other males apparently credited the noise and alarm to Mike and he enjoyed immediate rise in status since the other males were frightened of him. This is, admittedly, the most unusual of any recorded change in status, but there are other ways to change status. An individual may join with another or others in a coalition; he may move from one troop to another in which the adult males are less a threat; or, as has been documented, a male may move out of a troop taking with him several others, usually females and young. He may even move away from conspecifics; an adult male rhesus macaque that moved into the langur troop at Kaukori enjoyed the highest dominance rank possible, and the rare encounters he had with other rhesus indicated he was not dominant within his own species (Jay 1965a).

The extent to which the stresses of low rank can effect the general health or temperament of an individual and serve to reinforce his low status is unknown, but this must be considered as a possible factor in rank over long periods of time.

What motivates an animal to dominate another is not clear and since it is impossible for a monkey or ape to answer questions we must content ourselves with speculations based upon many observations of the same animals. It may be puzzling when a high-ranking adult male does not compete for some high-priority object such as a ripe fruit or an estrous female, especially when in the past the other animals seeking the object have been consistently unable to compete successfully against him. It is quite possible that according to mood and personal preference food may not be as important to an animal as copulation, or that the dominant male may not always care to have what may have become the object of a contest, or that the object is one about which there would be absolutely no contest in any case. Because a potentially dominant animal does not always attempt to secure a premium item, it is necessary in order to evaluate dominance, to average out assessments of general status over long periods.

Several adult female langurs in the Kaukori troop seldom interacted with the others to gain a specially good piece of food, right of way on a path, or a favored sleeping place. However, when these females did show any indication that they wanted to have a ripe mango, or to sit where another female was sitting, the other females never offered any sign of contest. For these females, total numbers of "wins and losses" were much less significant than the circumstance and percentage of wins out of attempts.

Rowell (1966b) has suggested that rank is a function of the behavior of relatively subordinate animals in the group. Certainly the behavior of dominant and subordinate is reciprocal, and sometimes the actions of the subordinate stimulate the higher ranking animal to threaten or to attack. In a cage situation these confrontations are probably much intensified because the animals are usually in full sight of each other and unable to escape.

Restlessness in the subordinate often appears to stimulate the dominant animal to aggression, but this will depend upon both the mood and the temperament of the high-ranking animal. Certain types of movement appear to draw attention more than others and the stances of subordinates are frequently accompanied by vocalizations that seem only to serve to call attention to the presence and possible plight of the less dominant animal. Instead of decreasing the irritation of a dominant animal these actions on the part of the lower animal appear often to increase the likelihood of attack.

To use the langur again as a source of illustration, a threat or mild attack often was the response to the feeble attempts of a young or low-dominance female to groom a more dominant animal, male or female. The cautious, hesitating, too-light touch of the nervous subordinate can startle, tickle, or annoy the dominant and elicit a threat rather than the relaxation associated with grooming. In these instances the behavior of the subordinate definitely appears to stimulate the actions of the higher ranking animal. The confident firm motions of casual grooming when dominance is not an issue is easily identified and very different from what has just been described.

The subordinate learns to inhibit his behavior and to keep out of the way of the dominant on many occasions. Sometimes this avoidance may last for years and then the subordinate will challenge a higher ranking animal. Such challenges are rarely directed to an individual much further up the ladder of rank; status is increased gradually and by small steps rather than greatly and suddenly. Of course, a rather dramatic increase in rank is possible if other factors intervene or if the ascending animal is able to win against a much higher animal, perhaps because the latter is ill or weak from other encounters. In any event, the subordinate is not irrevocably conditioned to a low rank and may take advantage of changes to challenge another group member.

The older subadult or young adult male is apt to be very active in challenging any adult that he might be able to dominate. This is a period of social development during which there is most stress and tension for the young animal, and often during these years the young male becomes peripheral to much of the action in the group involving the dominant adult males. This varies with different species but the transition from subadult to adult for the male is likely to be one of the most dangerous and traumatic stages in his life.

Rank is not a stable state that remains unaltered. Alpha or dominant males are not always on the top, nor are they necessarily first in rank for long periods. In some species the role of most dominant or controlling male is always present, but the occupant of the position changes. Social ranks change in the lifetime of an individual, starting with the earliest rankings in play groups, through the adult years, and continue to change with increasing age. The old animal that can no longer fight and is incapacitated by the normal toll of aging, including missing teeth and stiff joints, loses status gradually until he may become very peripheral to troop activities. Only longterm studies will document the changes of rank that occur as individuals mature and age.

SOCIAL COMMUNICATION[3]

From the moment of birth a primate is in constant communication with its mother and soon thereafter with other members of the social group. Not all communication, or the passing of information from one animal to another or others, takes place via stereotyped gestures; much of what animals communicate is conveyed very subtly. Patterns of communication include all sensory modalities, and the passing of information among individ-

---

[3] A very helpful article written by S. A. Altmann (1968c) contains an excellent bibliography on primate communication including Old and New World nonhuman primates. Other general references on communication include S. A. Altmann 1967b, Marler 1961, 1965, and 1968, and Sebeok 1965a, 1965b, 1968. The following films illustrate many aspects of primate behavior. The films are listed under the address where they may be rented or purchased.

University of California Extension Media Center, 2223 Fulton Street, Berkeley, Calif., 94720: "Baboon Behavior" by S. L. Washburn and I. DeVore; "Baboon Ecology" by S. L. Washburn and I. DeVore; "Baboon Social Organization" by S. L. Washburn and I. DeVore; and "Monkeys of Mysore" by Paul Simonds.

Audio Visual Services, 6 Willard Building, Pennsylvania State University, University Park, Pa.: "Howler Monkeys of Baro Colorado Island" by C. R. Carpenter; "Rhesus Monkeys in India" by C. H. Southwick; "Mountain Gorilla" by G. B. Schaller; "Nature and Development of Affection" rhesus monkeys by H. F. Harlow and R. Zimmerman; and "Behavior of the Macaques of Japan: the *Macaca fuscata* of the Koshima and Takasakiyama Colonies" by C. R. Carpenter.

Public Health Service Audiovisual Facility, Chamblee, Ga.: "The Rhesus Monkeys of Santiago Island" by C. and E. Schwartz.

National Geographic Society, Washington, D. C.: "Chimpanzees of the Gombe Reserve" by Jane van Lawick-Goodall.

uals is complex, seldom involving only a single channel such as vision, audition, or touch. Both the degree of subtlety of the message and number of ways in which information is passed among group members increase among primates in the sequence of prosimian, ape, and man. As seen in the preceding chapters, primates live in relatively small social groups and the majority of communication takes place among animals that are close together most of the day. This is the social context for acts of communication, and past as well as present social factors mediate and modify messages.

Patterns of communication are learned in a social context and they are based in the biology of each species. Both the biological basis and the social context are necessary for normal development of the full communication repertoire of any primate; to the extent that the young individual is deprived of any portion of this total context there will be deficits, which may or may not be remediable later in life. Specific gestures or sounds may be determined by morphological features of a species and may not be common to all primates (Hill 1956, 1953–1966; Hill and Booth 1957; Schultz 1961, 1969), just as intricacies of social relationships are often specific to the adaptations of a particular species.

Some communication acts are based on the morphological features of only one sex, usually the male. Adult male langurs can produce a canine grind that is socially meaningful, but females are not able to produce the same sound with their much smaller canines. Also, physical characteristics of adult males may emphasize gestures that are given by both sexes. The yawn threat of the adult male baboon is far more impressive (certainly to the human observer) than is the similar gesture given by an adult female. Pilo-erection of the head and shoulder hair of adult male baboons is much more striking and noticeable than a similar reaction given by an adult female.

As indicated above, species-specific features sometimes enter into communication. In certain species, animals have special structures such as laryngeal pouches, large hyoid bones, and other anatomical features with which they produce or emphasize some sounds. The existence of these structures indicates their importance in the evolution of communication patterns for the species possessing them.

In many species an important communicatory function is played by the natal coat color of newborn and very young infants, which differs strikingly from that of the adults of the same species (Napier and Napier 1967). The infant is treated very differently during the months when it has its natal coat color than when it grows older and develops adult coloration. In general, the young primate receives solicitous caretaking attention during its early weeks. It is undoubtedly a complex combination of the infant's color, motor patterns, and vocalizations that elicits caretaking attention and inhibits aggression. Together all these factors communicate the infant's helplessness and need for protection and nurturing.

Some behaviors, such as gestures or specific social roles, are characteristic of many species of monkeys and apes, and constellations of socially meaningful gestures are often shared in detail by all the species within a genus. The presence of similarities of behavior patterns among different kinds of primates had prompted speculations that they may be innate and under genetic control. However, at a very general level, during development and even later in life, individual primates of many kinds may experience very similar social stimuli, regardless of differences in the details of group life. Every animal must accommodate to others of different rank and temperament, guard itself against danger, control and protect young, signal intentions, and express a wide range of emotions from placation to aggression. These similarities in the social context of maturation could account for behavioral similarities among even distantly related primates. It is more likely that these patterns are phenotypic adaptations to social and physical environments than that they are genetically determined patterns.

A great many stereotyped gestures and vocalizations associated with specific types of interactions are characteristically given by animals of specific status, age, and sex. Although there are many variations and complexities in the daily life of primates, certain basic recurrent situations are common to all and are repeated a great many times in approximately similar ways. Basic categories of interaction include threat and aggression, submission and subordination, alarm and location calls, and actions, such as presenting, associated with sexual behavior.

Some gestures given in other situations are not common to all primates. For example, gestures of placation or reassurance have been reported for only a few species and mainly for chimpanzees, but it is possible that this more subtle kind of gesture has been overlooked in previous studies and will be reported in future observations. Much less frequent use of an apparently similar and much simpler gesture was reported for the north Indian langur; at the end of some interactions the hand of the more dominant animal is placed on the back or side of the lower ranking animal. Ransom (personal communication) recently described a gesture among forest-living baboons in the Gombe that includes reaching back with the foot and placing it on the body of another animal, and this gesture may serve in some manner to express placation or subordination. Leaning against another animal may communicate subtle feelings, and when the tail is entwined with another's or placed along the side of another animal's body or back this deliberate contact may well signal the need for reassurance. This behavior has been observed many times among crab-eating macaques (personal observation) and Simonds (1965) reports some instances among the south Indian bonnet macaques.

Very subtle changes in posture and in quality of movement can signal changes in an animal's mood. These signals are often not the stereotyped, easily recognizable gestures that lead to facile generalization about stages of arousal or invite descriptive statements. Changes in body tension, for example, are good indicators of an animal's involvement in or attentiveness to an external situation, and these cues certainly are noticed by others in the group and responded to appropriately. If an animal is very dominant, other group members do well to be aware of any cue, no matter how subtle, as to his mood. In contrast, few except for the young pay heed to the moods of a very low-ranking adult. In fact, as was mentioned above, the cowering, nervous, or very subordinate animal, by the quality of its movements may appear to stimulate the very aggression it is presumably attempting to avoid by submissive signals. However, many times signals of helplessness and of low status do prevent attack and further punishment by the stronger or higher status animal. One can speculate that the latter situations have been of great importance in evolution, and that the additional traumas initiated when the behavior of the fearful animal stimulates a dominant one to aggression are less important to the success of long-range social adjustment and selection.

In communication behavior among the nonhuman primates, the subordinate often attempts to deemphasize the very features that the stronger maximizes. The frightened or attacked subordinate may press to the ground or assume a posture that makes it appear as small as possible. The tone of the action is withdrawal or active placation. A subordinate adult male may show his teeth in a wide fear grimace, but this is very different from the gesture that he would use against a weaker animal, when he might flash his canines in a wide yawn of threat.

The animal in a dominant role does not always have to do something to assert his or her status relative to other animals. Often the observer is made aware of an animal's status by the reaction of others. A very dominant adult male sitting quietly may be passed by another that pauses, presents, grimaces, and then moves quickly past. Although the dominant may have given no apparent sign that he even noticed the presence of the other, the observer may safely conclude that the episode was based on past experiences during which the presenting animal was not ignored. In such an interaction the role of the one male as dominant is clear, acknowledged, and unambiguous without his having actively to enforce it.

Gestural communication patterns convey more information and are more varied than are vocal patterns of communication. Because of this, if the nonhuman primate is allowed to hear but not to see what is going on around it (if tapes of sounds are played but no image of the actors is given to the monkey or ape), the listening animal knows relatively little of what is happening. On the other hand, a monkey or ape allowed visual contact with others and deprived of audition would be able to respond to interactions. For humans the relative importance of vocal and gestural communication patterns is overwhelmingly in favor of the vocal. Most people have experienced the frustrations that arise in attempting communication with someone who does not speak the same language. The available gestures are exceedingly limited, both in number and in the kinds of concepts and information that they make it possible to share. Only with a formal system of sign language can humans communicate adequately by gesture alone.

In nonhuman primates gestures do not always produce the same response among different species. In fact, the same motor patterns sometimes may have nearly opposite effects on viewers in different species. Consider tail position. A very dominant adult male rhesus monkey and also a baboon being attacked by stronger animals each may hold his tail high in the air, signifying exactly opposite motivations. It must be added, however, that in each of these examples the position of the tail is combined with other important signals, and it is the total complex that carries the message to other animals.

Vocalizations among monkeys and apes are useful chiefly to call attention to the animal making the sounds. After their attention is thus engaged the animals also notice the other signals that are being given, whether in the form of stereotyped gestures, or, perhaps, postural cues. Certain kinds of information need to be transmitted as quickly as possible, as for example a warning to others of potential danger. Each animal within hearing distance of an alarm call responds immediately with actions that minimize the likelihood of danger—that is, with escape behavior. The listener does not wait to catch a glimpse of the sender to receive more information, and it is highly adaptive that such calls produce prompt and effective responses. A social group living in thick forest or in a habitat where visibility is poor depends heavily upon vocalizations that convey information over a distance and are independent of gesture. Although location calls are a common element in most species' repertoires, some kinds of primates such as gibbons use them far more than others. Again, their usefulness will depend in part on the terrain as well as on the spacing patterns characteristic of the species.

The range of variation and complexity of emotions is vast in man as compared to other mammals, including the rest of the primates. The relatively small number of complex emotions that are expressed and one assumes felt in ape and particularly in monkey behavior correspond to the relatively limited number of meanings in their communication system. The biology of those emotions that are apparent is undoubtedly similar for both monkeys and apes and is also similar to our own in comparable situations. Similarity across many taxonomic boundaries, then, comes from the nature of nonhuman primate expressions and their meanings.

Monkey and ape communication cannot ever approach the complexity of routine human communication. For example, one chimpanzee may reassure another, and the details of how this is done are determined by the age, sex, status, and relationships among the actors, but there are no messages among nonhuman primates so complex as those man sends with human language. The chimpanzee cannot say, "I am leaving now and if it is sunny we will meet in the evening at the fig tree in the next valley." It is only the addition of language, the ability to name objects as well as to express subjective conditions, that made possible the vast increase in man's information-sharing. And, language, in turn, is predicated on the human brain; the anatomical basis for some of these skills has been discussed by many authors (Lenneberg 1967; Geschwind 1964; Lancaster 1967, 1968b).

Much is being learned about the relationship between brain and behavior by the telestimulation of various parts of the primate brain (Delgado 1963a, 1963b, 1964, 1965, 1966, 1967; MacLean 1958, 1959, 1962; MacLean and Delgado 1953; Ploog 1964). Ploog and Melnechuk (1969) elicited sounds by electrically stimulating the limbic area, and Robinson (1967a, 1967b) reports that by stimulating parts of the limbic system 500 vocalizations were produced that resembled the sounds of a normal rhesus monkey. He generalized that "the areas from which expressive calls can be elicited are the same limbic areas that generate feeling states that support feeding, sexual behavior, alerting, and the behavioral manifestations of the primitive drives" (in Ploog and Melnechuk 1969:454).

When Robinson attempted to correlate the physiological or behavioral responses that were simultaneously elicited by the electrical stimulation of the limbic system he found that no strong correlations could be made because the sounds were often accompanied by several responses. For example, the same stimulus might result in a sound and also pupil dilation, piloerection, penile erection, and eating behavior. The work of Robinson and others indicates that the neural basis for human language and for nonhuman primate vocalizations differs. The calls elicited when a monkey's brain is stimulated in the limbic system are associated with emotional behaviors, and sounds cannot be elicited by stimulating the areas of a monkey's brain that are related to the cortical areas associated with speech in the human brain.

Human language is influenced by cortical control (Myers 1969), so it is not possible to stimulate words, even though stimulation of certain parts of the cortex can interfere with speech (Penfield and Roberts 1959). Evolution has produced new systems of communication in the human primate.

PROBLEMS IN THE STUDY OF PRIMATE COMMUNICATION

Questions invariably arise as to the meanings of gestures and vocalizations, the functions of acts of communication, and even more basically, as to what constitutes basic units of communication within a social group. (Altmann 1962, 1965, 1967b, 1968a; Marler 1965; Struhsaker 1967a). It is clear from the field studies that, compared to humans, apes and especially monkeys are quite restricted in the kinds of information they are able to communicate. Individual nonhuman primates express a range of emotional states, and the messages they send to each other are assumed to be relevant to these emotions. Great difficulties are raised when the human observer attempts to distinguish and describe the internal states or motivations of an animal that cannot indicate what it is feeling except by motor patterns and sounds unlike those of the human communication system. Generalizations by the observer are possible only after long observation of similar responses that are consistent and repeated. Interactions of nonhuman primates refer to the immediate situation and the present time of action; there is no indication that reference is made to the future, except in the sense of the next few minutes as a continuation of an on-going interaction. Animals learn from experience but cannot refer to events that are in the past.

It is possible to determine the emotional states of signaling animals only by inference from their actions and the responses of others to them (setting aside for the moment the possibilities of monitoring internal reactions with telemetry). Motivations for action and reaction are sometimes apparent and at other times obscure. The more that is known about the individual actors and the history of their relationships, the more likely that the observer will be correct in his interpretations of motivation. Because it is often not possible to know when an animal is paying attention to or is even aware of an interaction in which it is not directly involved, the stimuli for its subsequent actions are often not apparent. There is usually a very gradual build-up toward many kinds of interactions, such as fights, and the observer may soon perceive cues of involvement and be able to stay with the action and anticipate what is going to happen. He can do this only when there has been a mild build-up of tension and a gradual change in the intensity and form of signals among the animals, changes it is assumed correspond to changes in motivation. When the animal suddenly enters an interaction in which others have been participating the observer is likely to be left with the impression of instantaneous involvement—which is undoubtedly not the case.

The social basis or context of communication is far more complex than the actual forms of gestures or vocalizations might lead an observer to think. On one level, all interactions may be described in terms of motor patterns in space and sounds, and this is a necessary first step in any study when the animals are unfamiliar and the species' repertoire unknown. However, other questions arise that are not answered by complex descriptions of the ways animals move or produce sounds. After a short period of observation it becomes clear that past relationships greatly influence an animal's reactions and that it is not possible to understand interactions only on the basis of descriptions of what happens on one occasion. The same acts by the same individuals do not always elicit identical responses in others.

Behavior may appear to be regulated by signals when only the present few minutes are considered, but there are more basic relations, rooted in experience, and these are difficult to document. A severe fight may continue to influence the participants' reactions to each other, and the gestures and vocalizations used to maintain a status quo will vary according to the nature of these experiences.

Many attempts have been made to categorize social responses and to interpret the effects of communication (Altmann 1962, 1967*b*; Bobbitt, Gordon and Jensen 1966; Hall 1963*b*; Marler 1968; Mason 1959; Schaller 1965). Various authors have divided actions and responses into simple dichotomies, such as approach and withdrawal. Gestures and vocalizations then may be divided according to whether they stimulate others to come nearer or to move away, and at one level most elements of social communication may be put in one or other of these categories. Other factors such as the social context and the past relationships of the individual participants may enter into communication to determine either approach or withdrawal. In order for certain animals to approach others, some gestures that have been described as appeasement or placation may have to be given. In a different situation the same gesture may give the other animal permission to withdraw.

Specific gestures or motor patterns may appear in very different contexts and great care must be exercised not to assume that because the same motor patterns are used the motivations for the actions are the same. This always remains to be demonstrated.

The observer must always generalize the meaning of acts of communication on the basis of the reactions of other animals to these acts. Sometimes the general context is useful and sometimes it is not. Some expressions or gestures or vocalizations are similar in similar situations and are clearly related to them. Presenting by the female prior to copulation seems almost universal among monkeys. The motor patterns involved are very similar, as one would expect since all are quadrupedal animals. Given this similarity of anatomy, then, the motor patterns of the prelude to being mounted by the male also would be expected to share common features. If threat is expressed by nonhuman primates by showing the teeth then expressions of threat will be similar among different species.

It is impossible to record any single gesture used in one incident and expect to tell from that what the status of an animal is relative to the group. One cannot generalize that an adult male rhesus carrying his tail high is the most dominant animal in the group, but only that he is likely to be the highest ranking of the animals in immediate sight. It is also not possible to count the frequency with which an animal carries his tail up and expect that this will reflect general status in the troop. It is only in the context of the social setting that generalizations as to status or rank in the social group may be made.

Various attempts have been made to take one unit or gesture and trace its occurrence among many species. Often this involves assumptions as to homologies of meaning that may or may not hold true for those species considered. Similar motor patterns may be used in quite different contexts, as noted above, and if the gesture or action is placed in its larger setting, within a matrix of other gestures or movements, apparent similarities among different species may not be borne out. The value of crosstaxa comparisons lies in their suggestions for research on possible homologies. Comparisons are, in themselves, only suggestive and remain hypothetical until they are investigated experimentally.

The problem of correct interpretation of primate communication is compounded by the fact that the human ear may not perceive differences in sounds that are possibly very meaningful to nonhuman primate listeners. For this reason recordings of vocalizations should be made in the field for subsequent spectrographic analysis, so that the structure of sounds may be differentiated and the differences among calls correlated with the responses of listening monkeys or apes (examples are to be found in Rowell 1962; Goustard 1963; Schaller 1963; Struhsaker 1965; Altmann and Altmann 1965; and Bertrand 1969). Structural similarities among the vocalizations of related species may suggest details of evolutionary relationships.

Similarly, gestures should be recorded on film for later detailed frame-by-frame analysis to determine the component parts of what are often very quick actions. Even after long periods of observation the human observer should never assume that he has seen or been aware of all the details of an act of communication. It is possible that he has, of course, but double-checking must become a standard method of a behavioral study when the means of detailed analysis are available. For example, it is exceedingly difficult and often impossible to measure or determine the communicative significance of many kinds of physical contact, and this will remain a problem until, possibly, telemetry will allow the monitoring of internal changes in response to such contact. Certainly the situations in which some of these actions occur suggest strongly that there are aspects of reassurance involved.

Experimental investigation of communication patterns is necessary, but there are problems in effecting these investigations. The field is a difficult and in some instances an impossible location for the detailed analysis of behaviors. The alternative, a laboratory or colony, also presents complications since many important aspects of social life may not be present. Many situations requiring communication do not occur in captive conditions (predators are lacking, for example), and even when the animals are subjected to frightening or threatening stimuli the alternatives of defense or flight are usually not available. This does not mean that many aspects of social communication can not be investigated under captive conditions, it only underlines the necessity to assess as closely as possible the limitations placed on the animal's behavior by the context of the experiment.

SUMMARY

There are many and varied patterns of primate behavior, some of which characterize many kinds of monkeys and apes and others that are restricted to very closely related species. The single template that can be superimposed on the many varieties of monkeys and apes is one that emphasizes the basically social nature of these animals and of the general processes of maturation and learning that each individual experiences during its lifetime. The manner in which sociability is expressed and the strength of specific bonds may vary tremendously, but they are always present. There are minimal requirements for normal development common to all primates, and each individual animal shares most of its patterns of interindividual relationship and the vocabulary for their expression with each other individual of its species.

Variations on themes of social organization are related at least in part to features of the environment, including the social environment both within a group and in conspecific groups, and to the ecological factors of an area. Depending on the demands of the habitat, whether it is rich or harsh or somewhere between these extremes, the structure and composition of groups will vary to maximize the available food, water, and space, and to minimize competition for necessary items of life. Social structures are sensitive to changes in the habitat, especially over time, and to other external pressures that are not the result only of climate or of vegetation changes.

Every primate social group is characterized by a structure or network of relationships among its members, whether the total group consists of only two animals or a hundred. These structures vary little over time in overall design, although the details of relationships within the structure may demonstrate considerable variability. Within the structure of a group there are more-or-less clearly defined social roles that mediate the activities and relationships necessary to express group stability and well-being. Some of these roles have been discussed and their characteristics outlined.

As an intimate part of the social nature of primates, systems of communication have developed that enable the animals to convey among themselves a great deal of information about their emotional states at the time of the interaction. The forms of communication systems may vary so greatly that a member of one species may not be able to communicate very much with an individual of another taxon, but there appears to be a common set of shared experiences, emotions, and social situations that are familiar to all nonhuman primates. A message is seldom carried by one mode, but rather, many modalities combine to increase the richness and message content possible in social signals.

With new techniques that enable the observer to monitor internal reactions it will be possible to fill in many of the lacunae in our understanding of primate behavior and these techniques will certainly guide in the experimental investigation of behavior patterns. Until that time we must rely upon continuous, careful, sensitive recording of the external observable events in primate social behavior.

Alexander, B. K., and H. F. Harlow, 1965, "Social Behavior of Juvenile Rhesus Monkeys Subjected to Different Rearing Conditions During the First Six Months of Life," *Zool, Jb. Physiol.,* 71:489-508.

Altman, J., 1966, *Organic Foundations of Animal Behavior.* New York: Holt, Rinehart and Winston, Inc.

Altmann, S. A., 1962, "A Field Study of the Sociobiology of Rhesus Monkeys, *Macaca mulatta," Ann. N. Y. Acad. Sci.,* 102(2):338-435.

_____, 1965, "Sociobiology of Rhesus Monkeys. II: Stochastics of Social Communication," *Journal of Theoretical Biology,* 8:490-522.

_____, ed., 1967a, *Social Communication Among Primates.* Chicago: University of Chicago Press.

_____, 1967b, "Structure of Social Communication," in *Social Communication Among Primates,* S. A. Altmann, ed. Chicago: University of Chicago Press, pp. 325-362.

_____, 1968a, "Sociobiology of Rhesus Monkeys. III: The Basic Communication Network," *Behaviour,* 32(1-3):17-32.

_____, 1968b, "Sociobiology of Rhesus Monkeys. IV: Testing Mason's Hypothesis of Sex Differences in Affective Behavior," *Behaviour,* 32(1-3):49-69.

_____, 1968c, "Primates," in *Animal Communication: Techniques of Study and Results of Research,* T. A. Sebeok, ed. Bloomington: Indiana University Press, pp. 466-522.

_____, and J. Altmann, 1965, "Vocal Communication in Wild Baboons," *Amer. Zool.,* 5:694.

Andrew, R. J., 1963a, "Evolution of Facial Expression," *Science,* 142:1034-1041.

_____, 1963b, "The Origin and Evolution of the Calls and Facial Expressions of Primates," *Behaviour,* 20:1-109.

Ayer, A. Ananthanarayana, 1948, *The Anatomy of Semnopithecus entellus.* Madras, India: Indian Publishing House, Ltd.

Azuma, S., and A. Toyoshima, 1962, "Progress Report of the Survey of Chimpanzees in Their Natural Habitat, Kabogo Point Area, Tanganyika," *Primates,* 3(2).

_____, and _____, 1965, "Chimpanzees at Kabogo Point," in *Sociological Studies of Monkeys and Apes,* S. Kawamura and J. Itani, eds. Tokyo: Chuokoron.

Bandura, A., and R. H. Walters, 1963, *Social Learning and Personality Development.* New York: Holt, Rinehart and Winston, Inc.

Barlow, G., 1968, "Ethological Units of Behavior," in *The Central Nervous System and Fish Behavior,* D. Ingle, ed. Chicago: University of Chicago Press, pp. 217–232.

Barnett, S. A., 1963, *The Rat: A Study in Behavior.* Chicago: Aldine.

Bastock, M., D. Morris, and M. Moynihan, 1953, "Some Comments on Conflict and Frustration in Animals," *Behaviour,* 6:66–84.

Bauchop, T., and R. W. Martucci, 1968, "Ruminant-like Digestion of the Langur Monkey," *Science,* 161:698–700.

Beach, F. A., 1945, "Current Concepts of Play and Animals," *American Naturalist,* 79:523–541.

———, 1965, *Sex and Behavior.* New York: Wiley.

Benedict, B., 1969, "Role Analysis in Animals and Men," *Man,* 4(2):203–214.

Berkson, G., W. A. Mason, and S. V. Saxon, 1963, "Situations and Stimulus Effects on Stereotyped Behaviors of Chimpanzees," *J. Comp. Physiol. Psychol.,* 56(4): 786–792.

Berlyne, D. E., 1960, *Conflict, Arousal, and Curiosity.* New York: McGraw–Hill.

Bernstein, I. S., 1966, "Analysis of a Key Role in a Capuchin (*Cebus albifrons*) group," *Tulane Studies Zool.* 13:49–54.

———, and W. A. Mason, 1962, "The Effects of Age and Stimulus Conditions on the Emotional Responses of Rhesus Monkeys: Responses to Complex Stimuli," *J. Genet. Psych.* 101:279–298.

Bertrand, M., 1969, *The Behavioral Repertoire of the Stumptail Macaque,* Basel, Switzerland: S. Karger. *Bibliotheca Primatologica,* No. 11.

Bingham, H. C., 1932, "Gorillas in a Native Habitat," *Carnegie Inst. Wash. Publ.,* 426:1–66.

Bishop, A., 1962, "Control of the Hand in Lower Primates," *Ann. N. Y. Acad. Sci.,* 102(II):316–337.

Blanford, W. T., 1888–1891, *The Fauna of British India: Mammalia.* London: Taylor and Francis.

Bobbitt, R. A., B. N. Gordon, G. D. Jensen, 1966, "Development and Application of an Observational Method: Continuing Reliability Tests," *J. Psychol.* 63: 83–88.

Bolwig, N., 1959, "A Study of the Behaviour of the Chacma Baboon, *Papio ursinus,*" *Behaviour,* 14:136–163.

———, 1961, "An Intelligent Tool-using Baboon," *S. Afr. J. Sci.,* 57:147–152.

Booth, C. P., 1968, "Taxonomic Studies of *Cercopithecus mitis* Wolf (East Africa)," *National Geogr. Society Research Reports, 1963 Projects,* pp. 37–51.

Brain, C. K., 1968, "Observations of the Behaviour of Vervet Monkeys, *Cercopithecus aethiops,*" *Proc. Zool. Soc. S. Africa Symp. on African Mammals.*

Brown, J. L., and R. W. Hunsperger, 1963, "Neuroethology and the Motivation of Agonistic Behavior," *Animal Behavior,* 11:439–448.

Buettner-Janusch, J., ed., 1962, "The Relatives of Man," *Ann. N. Y. Acad. Sci.,* 102:181–514.

———, ed., 1963-1964, *Evolutionary and Genetic Biology of Primates.* New York: Academic Press.

Burridge, Kenelm, O. L., 1957, "A Tangu Game," *Man,* 51:88–89.

Butler, R. A., 1954, "Incentive Conditions Which Influence Visual Exploration," *J. Exp. Psychol.,* 48:19–23.

Carpenter, C. R., 1934, "A Field Study of the Behavior and Social Relations of Howling Monkeys," *Comp. Psychol. Monogr.,* 10(48):1–168.

———, 1935, "Behavior of Red Spider Monkeys in Panama," *J. Mammal.,* 16:171–180.

———, 1940, "A Field Study in Siam of the Behavior and Social Relations of the Gibbon, *Hylobates lar,*" *Comp. Psychol. Monogr.,* 16(5):1–212.

_____, 1942, "Sexual Behavior of Free Ranging Rhesus Monkeys, *Macaca mulatta*," *J. Comp. Psychol.*, 33:113–142.

_____, 1952, "Social Behavior of Nonhuman Primates," *Colloq. Int. Cent. Nat. Rech. Sci (Paris)*, 34:227–246.

_____, 1964, *Naturalistic Behavior of Nonhuman Primates*. University Park: Pennsylvania State University Press.

Carthy, J. D., and F. J. Ebling, eds., 1964, *The Natural History of Aggression*. New York: Academic Press.

Champion, F. W., 1934, *The Jungle in Sunlight and Shadow*. New York: Scribner.

Clark, W. E., Le Gros, 1936, "The Problem of the Claw in Primates," *Proc. Zool. Soc. Lond.*, 1936:1–24.

_____, 1959, *The Antecedents of Man*. Chicago: Quadrangle Books (subsequent editions 1960, 1962).

_____, 1962, *The Antecedents of Man*, rev. ed. Edinburgh, Scotland: Edinburgh University Press.

Cobb, S., 1958, *Foundations of Neuropsychiatry*. Baltimore: Williams and Wilkins.

Collias, N. E., 1944, "Aggressive Behavior Among Vertebrate Animals," *Physio. Zool.*, 17:83–123.

_____, 1962, "Social Development in Birds and Mammals," in *Roots of Behavior*, E. L. Bliss, ed. New York: Harper, pp. 264–273.

Conaway, C. H., and C. B. Koford, 1964, "Estrous Cycles and Mating Behavior in a Free-ranging Band of Rhesus Monkeys," *J. Mammal.*, 45:577–588.

Corner, G. W., 1942, *The Hormones in Human Reproduction*. Princeton, N. J.: Princeton University Press.

Crook, J. H., 1966, "Gelada Baboon Herd Structure and Movement: A Comparative Report," *Symp. Zool. Soc. Lond.*, 18:237–258.

_____, ed., 1970a, *Social Behaviour in Birds and Mammals: Essays on the Social Ethology of Animals and Man*. London and New York: Academic Press.

_____, 1970b, "The Socio-ecology of Primates," in *Social Behaviour in Birds and Mammals: Essays on the Social Ethology of Animals and Man*, J. H. Crook, ed. London and New York: Academic Press, pp. 103–166.

_____, and S. Gartlan, 1965, "Evolution of Primate Societies," *Nature* (Lond.) 20:1200–1203.

Cross, H. A., and H. F. Harlow, 1963, "Observation of Infant Monkeys by Female Monkeys," *Percept. Mot. Skills*, 16:11–15.

Cunningham, D. D., 1904, *Some Indian Friends and Acquaintances, A Study of the Ways of Birds and Other Animals Frequenting Indian Streets and Gardens*. New York: Dutton.

Delgado, J. M. R., 1963a, "Effect of Brain Stimulation on Task-free Situations," in *The Physiological Basis of Mental Activity*, R. N. Peón, ed., New York: Elsevier Publishing Co., pp. 260–280.

_____, 1963b, "Cerebral Heterostimulation in a Monkey Colony," *Science*, 141:161–163.

_____, 1964, "Free Behavior and Brain Stimulation," *Internatl. Rev. Neurobiol.*, 6:349–449.

_____, 1965, "Sequential Behavior Induced Repeatedly by Stimulation of the Red Nucleus in Free Monkeys," *Science*, 148:1361–1363.

_____, 1966, "Aggressive Behavior Evoked by Radio Stimulation in Monkey Colonies," *Amer. Zool.*, 6:669–681.

_____, 1967, "Social Rank and Radio-Stimulated Aggressiveness in Monkeys," *J. Nerv. Mont. Dis.*, 144:383–390.

de Reuck, A., and J. Knight, eds., 1966, *Conflict in Society*. Boston: Little, Brown.

DeVore, I., 1962, "The Social Behavior and Organization of Baboon Troops," unpublished doctoral thesis, University of Chicago.

DeVore, I., 1963, "Mother-infant Relations in Free-ranging Baboons," *Maternal Behavior in Mammals,* H. L. Rheingold, ed. New York: Wiley, pp. 305–335.

——, ed., 1965a, *Primate Behavior: Field Studies of Monkeys and Apes.* New York: Holt, Rinehart and Winston, Inc.

——, 1965b, "Male Dominance and Mating Behavior in Baboons," in *Sex and Behavior,* F. A. Beach, ed. New York: Wiley, pp. 266–289.

——, and K. R. L. Hall, 1965, "Baboon Ecology," in *Primate Behavior: Field Studies of Monkeys and Apes,* I. DeVore, ed. New York: Holt, Rinehart and Winston, Inc., pp. 20–52.

——, and S. L. Washburn, 1961, "Baboon Behavior" (16 mm sound-color film), Berkeley, California: University Extension, University of California.

Dobzhansky, T., 1962, *Mankind Evolving.* New Haven and London: Yale University Press.

Dolhinow, P., and N. Bishop, 1970, "The Development of Motor Skills and Social Relationships Among Primates Through Play," *Minn. Symp. on Child Psych.,* 4:141–198.

Donisthorpe, J., 1958, "A Pilot Study of the Mountain Gorilla *(Gorilla gorilla beringei)* in South-west Uganda, February to September 1957," *S. Afr. J. Sci.,* 54(8):195–217.

Eccles, J. C., M. Ito, and J. Szentagothai, 1967, *The Cerebellum as a Neuronal Machine.* New York: Springer Verlag.

Eimerl, S., and I. DeVore, 1965, *The Primates.* New York: Time, Inc.

Ellefson, J. O., 1967, "A Natural History of Gibbons in the Malay Peninsula *(Hylobates lar),*" unpublished doctoral thesis, University of California, Berkeley.

——, 1968, "Territorial Behavior in the Common White-handed Gibbon, *Hylobates lar* Linn.," in *Primates: Studies in Adaptation and Variability,* P. C. Jay, ed. New York: Holt, Rinehart and Winston, Inc., pp. 180–199.

Emlen, J. T., and G. Schaller, 1960, "Distribution and Status of the Mountain Gorilla *(Gorilla gorilla beringei)* – 1959," *Zoologica,* 45(1):41–52.

Erickson, G. E., 1963, "Brachiation in the New World Monkeys," *Symp. Zool. Soc. Lond.,* "The Primates," 10:135–164.

Escalona, S. K., 1968, *The Roots of Individuality: Normal Patterns of Development in Infancy,* Chicago: Aldine.

Etkin, W., 1964, "Theories of Socialization and Communication," in *Social Behavior and Organization Among Vertebrates,* W. Etkin, ed. Chicago: University of Chicago Press, pp. 167–205.

Fiedler, W., 1956, "Übersicht über das System der Primates," in *Primatologia,* Vol. 1, H. Hofer, A. H. Schultz, and D. Stark, eds. New York: S. Karger, pp. 1–266.

Frank, L. K., 1958, "Tactile Communication," *ETC: A Review of General Semantics.* San Francisco: Autumn.

Freeman, D., 1964, "Human Aggression in Anthropological Perspective," in *The Natural History of Aggression,* J. D. Carthy and F. J. Ebling, eds. New York: Academic Press, pp. 109–119.

Frisch, J. E., 1959, "Research on Primate Behavior in Japan," *Amer. Anthrop.,* 61:584–596.

——, 1968, "Individual Behavior and Intertroop Variability in Japanese Macaques," in *Primates: Studies in Adaptation and Variability,* P. C. Jay, ed. New York: Holt, Rinehart and Winston, Inc., pp. 243–252.

Furuya, Y., 1961, "The Social Life of Silver Leaf Monkeys *(Trachypithecus cristatus), Primates,* 3:41–60.

Gartlan, J. S., 1966, "Ecology and Behavior of the Vervet Monkey, Lolui Island, Lake Victoria, Uganda," Ph.D. Thesis, University of Bristol.

_____, and C. K. Brain, 1968, "Ecology and Social Variability in *Cercopithecus aethiops* and *C. mitis*," in *Primates: Studies in Adaptation and Variability*, P. C. Jay, ed. New York: Holt, Rinehart and Winston, Inc., pp. 253–292.

Geschwind, N., 1964, "The Development of the Brain and the Evolution of Language," *Monograph Series in Language and Linguistics*, 17:155–169.

Gilbert, C., and J. Gillman, 1951, "Pregnancy in the Baboon *(Papio ursinus),*" *S. Afr. J. Med. Sci.*, 16:115–124.

Gillman, J. and C. Gilbert, 1946, "The Reproductive Cycle of the Chacma Baboon, *Papio ursinus* with Special Reference to the Problems of Menstrual Irregularities as Assessed by the Behaviour of the Sex Skin," *S. Afr. J. Med. Sci.*, 11 (Biological Supplement):1–54.

Goffman, E., 1959, *The Presentation of Self in Everyday Life*, New York: Doubleday.

_____, 1961, *Encounters*, Indianapolis: Bobbs-Merrill.

Goodall, J., 1962, "Nest Building Behavior in the Free-ranging Chimpanzee," *Ann. N. Y. Acad. Sci.*, 102:455–467.

_____, 1963, "Feeding Behaviour of Wild Chimpanzees: A Preliminary Report," *Symp. Zool. Soc. London*, 10:39–47.

_____, 1964, "Tool-using and Aimed Throwing in a Community of Free-living Chimpanzees," *Nature*, 201:1264–1266.

_____, 1965, "Chimpanzees of the Gombe Stream Reserve," in *Primate Behavior: Field Studies of Monkeys and Apes*, I. DeVore, ed. New York: Holt, Rinehart and Winston, Inc., pp. 425–473.

Goodman, M., 1968, "Phylogeny and Taxonomy of the Catarrhine Primates from Immunodiffusion Data," *Taxonomy and Phylogeny of Old World Primates with References to the Origin of Man*, B. Chiarelli, ed., Turin: Rosenberg and Sellier, pp. 95–107.

Goustard, M., 1963, "La Structure Sociale d'une Colonie de *Macaca irus*," *Ann. Sci. Natur. Zool.*, Ser. 12, 3:287–322.

Haddow, A. J., 1952–1953, "Field and Laboratory Studies on an African Monkey, *Cercopithecus ascanius schmidti* Matschie," *Proc. Zool. Soc. Lond.*, 122(II):297–394.

Hall, E. T., 1959, *The Silent Language*. New York: Doubleday.

_____, 1966, *The Hidden Dimension*. New York: Doubleday.

Hall, K. R. L., 1962, "The Sexual, Agonistic, and Derived Social Behaviour Patterns of the Wild Chacma Baboon, *Papio ursinus*," *Proc. Zool. Soc. London*, 139:283–327.

_____, 1963a, "Observational Learning in Monkeys and Apes," *Brit. J. Psychol.*, 54:201–226.

_____, 1963b, "Some Problems in the Analysis and Comparison of Monkey and Ape Behavior," in *Classification and Human Evolution*, S. L. Washburn, ed., Viking Fund Pub. No. 37. New York: Wenner-Gren Foundation, pp. 273–300.

_____, 1963c, "Tool-using Performances as Indicators of Behavioral Adaptability," *Current Anthropology*, 4:479–494.

_____, 1964, "Aggression in Monkey and Ape Societies," in *The Natural History of Aggression*, J. D. Carthy and F. J. Ebling, eds. New York: Academic Press, pp. 51–64.

_____, 1965a, "Behavior and Ecology of the Wild Patas Monkeys, *Erythrocebus patas*, in Uganda," *J. Zool. Soc. London*, 148:15–87.

_____, 1965b, "Social Organization of the Old World Monkeys and Apes," *Symp. Zool. Soc. London*, 14:265–289.

_____, 1968, "Social Learning in Monkeys," in *Primates: Studies in Adaptation and Variability*, P. C. Jay, ed. New York: Holt, Rinehart and Winston, Inc., pp. 383–397.

Hall, K. R. L., and I. DeVore, 1965, "Baboon Social Behavior," in *Primate Behavior: Field Studies of Monkeys and Apes,* I. DeVore, ed. New York: Holt, Rinehart and Winston, Inc., pp. 53–110.

————, and J. S. Gartlan, 1965, "Ecology and Behaviour of the Vervet Monkey, *Cercopithecus aethiops,* Lolui Island, Lake Victoria," *Proc. Zool. Soc. London,* 145:37–56.

————, and M. J. Goswell, 1964, "Aspects of Social Learning in Captive Patas Monkeys," *Primates,* 5(3–4):59–70.

————, and B. Mayer, 1967, "Social Interactions in a Group of Captive Patas Monkeys *(Erythrocebus patas),*" *Folia Primat.,* 5:213–236.

Hallowell, A. I., 1955, "Cultural Factors in Spatial Orientation," *Culture and Experience,* Philadelphia: University of Pennsylvania Press, pp. 184–202.

Hamburg, D. A. 1963, "Emotion in the Perspective of Human Evolution," in *Expression of the Emotions in Man,* P. Knapp, ed., New York: International Universities, pp. 300–317.

Hansen, E. W., 1966, "The Development of Maternal and Infant Behavior in the Rhesus Monkey," *Behaviour* 27:107–149.

————, and W. A. Mason, 1962, "Socially Mediated Changes in Lever-responding of Rhesus Monkey," unpublished doctoral thesis, University of Wisconsin.

Harlow, H. F., 1959a, "Basic Social Capacity of Primates," in *The Evolution of Man's Capacity for Culture,* J. N. Spuhler, ed. Detroit: Wayne State University Press, pp. 40–52.

————, 1959b, "Love in Infant Monkeys," *Scientific American,* 200:68–74.

————, 1960, "Primary Affectional Patterns in Primates," *Amer. J. Ortho-Psychiat.,* 30:676–684.

————, 1962a, "The Heterosexual Affectional System in Monkeys," *Am. Psychologist,* 17:1–9.

————, 1962b, "Development of the Second and Third Affectional Systems in Macaque Monkeys," in *Research Approaches to Psychiatric Problems,* T. T. Tourlentes, S. L. Pollack, and H. E. Himwich, eds. New York: Grune and Stratton, pp. 209–229.

————, 1964, "Early Social Deprivation and Later Behavior in the Monkey," in *Unfinished Tasks in the Behavioral Sciences,* A. Abrams, H. H. Garner, and J. E. P. Toman, eds. Baltimore: Williams and Wilkins, pp. 154–173.

————, and M. K. Harlow, 1962, "The Effect of Rearing Conditions on Behavior," *Bulletin of the Menninger Clinic,* 26(5):213–224.

————, and ————, 1965, "The Affectional System," in *Behavior of Nonhuman Primates, Vol. II,* A. M. Schrier, H. F. Harlow, and F. Stollnitz, eds. New York: Academic Press, pp. 287–334.

————, and ————, 1966, "Learning to Love," *American Scientist,* 54:244–272.

————, ————, R. O. Dodsworth, and G. L. Arling, 1966, "Maternal Behavior of Rhesus Monkeys Deprived of Mothering and Peer Association in Infancy," *Proceedings of the Am. Philos. Soc.,* 110:58–66.

————, ————, and E. W. Hansen, 1963, "The Maternal Affectional System and Rhesus Monkeys," in *Maternal Behavior in Mammals,* H. L. Rheingold, ed. New York: Wiley, pp. 254–281.

————, and P. H. Settlage, 1934, "Comparative Behavior of Primates, VII. Capacity of Monkeys to Solve Patterned String Tests," *J. Comp. Psychol.* 18:432–435.

————, and R. R. Zimmerman, 1958, "The Development of Affectional Responses in Infant Monkeys, *Proc. Amer. Phil. Soc.,* 5:501–509.

————, and ————, 1959, "Affectional Responses in the Infant Monkey," *Science,* 130:421–432.

Harris, G., 1964, "Sex Hormones, Brain Development and Brain Function," *Endocrin.,* 75:627–648.

Harrisson, B., 1962, *Orangutan.* London: Collins.

Heape, W., 1894, "The Menstruation of *Semnopithecus entellus,*" *Phil. Trans. Roy. Soc. London,* B. Vol. 185:411–471.

Hill, W. C. O., 1936, "Supplementary Observations on the Purple-faced Leaf Monkey (genus *Kasi*)," *Ceylon J. of Sci. B.,* 20(1):115–133.

———, 1937, "The Type of *Semnopithecus thersites,*" *Ceylon J. Sci. B.,* 20(2): 207–209.

———, 1956, "Behaviour and Adaptations of the Primates," *Proc. Roy. Soc. Edinburgh,* 66B:94–110.

———, 1953–1966, *Primates: Comparative Anatomy and Taxonomy.* Edinburgh: University Press, and New York: Interscience.

———, and A. N. Booth, 1957, "Voice and Larynx in African and Asiatic Colobidae," *J. Bombay Nat. Hist. Soc.,* 54:309–321.

Hinde, R. A., 1965, "Rhesus Monkey Aunts," *Determinants of Infant Behavior. III.* Proceedings of the Third Tavistock Study Group on Mother-Infant Interaction held at the house of the CIBA Found., Lond., Sept., 1963, B. M. Foss, ed. London: Methuen, Ltd.

———, and T. E. Rowell, 1962, "Communication by Postures and Facial Expressions in the Rhesus Monkey *(Macaca mulatta),*" *Proc. Zool. Soc. Lond.,* 138(I):1–21.

———, ———, and Y. Spencer-Booth, 1964, "Behaviour of Socially Living Rhesus Monkeys in Their First Six Months," *Proc. Zool. Soc. Lond.,* 143:609–649.

———, and N. Tinbergen, 1958, "The Comparative Study of Species-specific Behavior," in *Behavior and Evolution,* A. Roe and G. G. Simpson, eds. New Haven: Yale University Press, pp. 251–268.

Hingston, R. W. G., 1920, *A Naturalist in Himalaya.* London: Witherby.

Hofer, H., 1962, "Über die Interpretation der Ältesten Fossilen Primatengehirne," *Bibl. Primat. Fasc.,* 1:1–31.

Hooton, E. A., 1942, *Man's Poor Relations.* New York: Doubleday.

Hutt, C., 1966, "Exploration and Play in Children," *Symposium of the Zoological Society of London,* 18:61–81.

Hutton, T. H., 1867, "On the Geographical Range of *Semnopithecus entellus,*" *Proc. Zool. Soc. Lond.,* pp. 944–952.

Huxley, J., 1964, *Evolution: The Modern Synthesis.* New York: Wiley.

Imanishi, K., 1957a, "Identification: A Process of Socialization in the Subhuman Society of *Macaca fuscata,*" *Primates,* 1–1:1–29.

———, 1957b, "Social Behavior in Japanese Monkeys, *Macaca fuscata,*" *Psychologia,* 1:47–54.

———, 1960, "Social Organization of Subhuman Primates in Their Natural Habitat," *Cur. Anthro.* 1:393–407.

———, 1963, "Social Behavior in Japanese Monkeys, *Macaca fuscata,*" in *Primate Social Behavior,* C. H. Southwick, ed. Princeton, N. J.: Van Nostrand, pp. 68–81.

———, 1964, "The Individual in the Society of Japanese Monkeys," *Japan Quart.,* 11:293–360.

———, and S. A. Altmann, 1965, eds., *Japanese Monkeys: A Collection of Translations,* selected by K. Imanishi, edited by S. A. Altmann, publ. by the editor.

Itani, J., 1954, *The Monkeys of Mt. Takasaki.* Tokyo: Kobunsha.

———, 1958, "On the Acquisition and Propagation of a New Food Habit in the Troop of Japanese Monkeys at Takasakiyama," *Primates,* 1–2:84–98.

———, 1959, "Paternal Care in the Wild Japanese Monkey, *Macaca fuscata fuscata,*" *Primates,* 2:61–93.

———, 1961, "The Society of Japanese Monkeys," *Japan Quart.,* 8:421–430.

Itani, J., 1963, "Vocal Communication of the Wild Japanese Monkey," *Primates,* 4:11–66.

———, K. Tokuda, Y. Furuya, K. Kano, and Y. Shin, 1963, "The Social Construction of Natural Troops of Japanese Monkeys in Takasakiyama," *Primates,* 4(3):1–42.

Izawa, K. and J. Itani, 1966, "Chimpanzees in Kasakati Basin, Tanganyika (I) Ecological Study in the Rainy Season. 1963–1964," in *Kyoto Univ. African Studies,* K. Imanishi, ed., Vol. 1:73–149.

Jay, P. C., 1962, "Aspects of Maternal Behavior Among Langurs," *Ann. N. Y. Acad. Sci.,* 102(II):468–476.

———, 1963a, "The Indian Langur Monkey (Presbytis entellus)," in *Primate Social Behavior,* C. H. Southwick, ed. Princeton, N. J.: Van Nostrand, pp. 114–123.

———, 1963b, "Mother-infant Relations in Langurs," in *Maternal Behavior in Mammals,* H. Rheingold, ed. New York: Wiley, pp. 282–304.

———, 1965a, "The Common Langur of North India," in *Primate Behavior: Field Studies of Monkeys and Apes,* I. DeVore, ed. New York: Holt, Rinehart and Winston, Inc., pp. 197–249.

———, 1965b, "Field Studies," in *Behavior of Nonhuman Primates,* A. M. Schrier, H. F. Harlow, and F. Stollnitz, eds. New York: Academic Press, pp. 525–591.

———, ed., 1968a, *Primates: Studies in Adaptation and Variability.* New York: Holt, Rinehart and Winston, Inc.

———, 1968b, "Primate Field Studies and Human Evolution," in *Primates: Studies in Adaptation and Variability,* P. C. Jay, ed. New York: Holt, Rinehart and Winston, Inc., pp. 487–503.

———, and D. Lindburg, 1965, "The Indian Primate Ecology Project (September 1964–June 1965)," unpublished manuscript.

Jensen, G. D., R. A. Bobbitt, and B. N. Gordon, 1966a, "The Development of mutual independence in Mother-infant pigtailed monkeys, *Macaca nemestrina,*" in *Social Interaction Among Primates,* S. A. Altmann, ed. Chicago: University of Chicago Press, pp. 43–53.

———, ———, and ———, 1966b, "Sex Differences in Social Interaction Between Infant Monkeys and Their Mothers," *Recent Advances in Biological Psychiatry,* 9:283–293.

Jolly, A., 1966, *Lemur Behavior: A Madagascar Field Study.* Chicago: University of Chicago Press.

Kaufman, I. C., and L. A. Rosenblum, 1966, "A Behavioral Taxonomy for *Macaca nemestrina* and *Macaca radiata:* Based on Longitudinal Observation of Family Groups in the Laboratory," *Primates,* 7(2):205–258.

Kawai, M., 1958, "On the Rank System in a Natural Group of Japanese Monkeys," *Primates,* 1(2):84–98.

———, 1965, "Newly-Acquired Pre-cultural Behavior of the Natural Troop of Japanese Monkeys on Koshima Island," *Primates,* 6(1):1–30.

———, and H. Mizuhara, 1959, "An Ecological Study on the Wild Mountain Gorilla *(Gorilla gorilla beringei),*" *Primates,* 2(1):1–42.

Kellogg, W. N., 1968, "Communication and Language in the Home-raised Chimpanzee," *Science,* 162:423–427.

Koford, C. B., 1963a, "Group Relations in an Island Colony of Rhesus Monkeys," in *Primate Social Behavior,* C. H. Southwick, ed. Princeton, N. J.: D. Van Nostrand, pp. 136–152.

———, 1963b, "Rank of Mothers and Sons in Bands of Rhesus Monkeys," *Science,* 141:356–357.

_____, 1965, "Population Dynamics of Rhesus on Cayo Santiago," in *Primate Behavior: Field Studies of Monkeys and Apes*, I. DeVore, ed. New York: Holt, Rinehart and Winston, Inc., pp. 160–174.

_____, 1966, "Population Changes in Rhesus Monkeys, 1960–1965," *Tul. Studies in Zool.*, 12:1–7.

Köhler, W., 1925, *The Mentality of Apes*. New York: Harcourt.

Kourilsky, R., A. Soulairac, and P. Grapin, 1965, *Adaptation et Aggressivité*. Paris: Presses Universitaires de France.

Kummer, H., 1956, "Rang-kriterien bei Mantelpavianen," *Rev. Suiss. Zool.*, 63: 288–297.

_____, 1957, "Soziales Verhalten einer Mantelpavian-Gruppe," *Schweiz-Zeitsch. Psychol.*, No. 33.

_____, 1967, "Tripartite Relations in Hamadryas Baboons," *Social Communication Among Primates*, S. A. Altmann, ed. Chicago: University of Chicago Press, pp. 63–71.

_____, 1968a, "Social Organization of Hamadryas Baboons: A Field Study," *Bibliotheca Primatologica* 6:1–189.

_____, 1968b, "Two Variations in the Social Organization of Baboons," in *Primates: Studies in Adaptation and Variability*, P. C. Jay, ed. New York: Holt, Rinehart and Winston, Inc., pp. 293–312.

_____, and F. Kurt, 1963, "Social Units of a Free-living Population of Hamadryas Baboons," *Folia Primat.*, 1:4–19.

_____, and _____, 1965, "A Comparison of Social Behavior in Captive and Wild Hamadryas Baboons," in *The Baboon in Medical Research*, H. Vogtborg, ed. Austin: University of Texas Press, pp. 65–80.

Lancaster, J. B., 1965, "Chimpanzee Tool Use," paper presented at the Southwestern Anthropological Association annual meeting, Los Angeles, California.

_____, 1967, "Primate Communication Systems and the Emergence of Human Language," unpublished doctoral thesis, University of California, Berkeley.

_____, 1968a, "On the Evolution of Tool-using Behavior," *American Anthropologist*, 70:56–66.

_____, 1968b, "Primate Communication Systems and the Emergence of Human Language," in *Primates: Studies in Adaptation and Variability*, P. C. Jay, ed. New York: Holt, Rinehart and Winston, Inc., pp. 439–457.

_____, and R. B. Lee, 1965, "The Annual Reproductive Cycle in Monkeys and Apes," in *Primate Behavior: Field Studies of Monkeys and Apes*, I. DeVore, ed. New York: Holt, Rinehart and Winston, Inc., pp. 486–513.

Laughlin, W. S., 1968, "Hunting: an Integrating Biobehavior System and Its Evolutionary Importance," in *Man the Hunter*, R. B. Lee and I. DeVore, eds. Chicago: Aldine, pp. 304–320.

Lee, R. B., and I. DeVore, eds., 1968, *Man the Hunter*. Chicago: Aldine.

Lenneberg, E. H., 1967, *Biological Foundations of Language*. New York: Wiley.

Leuba, C., 1955, "Toward Some Integration of Learning Theories: The Concept of Optimal Stimulation," *Psychol. Rep.*, 1:27–33.

Levine, S., and R. F. Mullins, Jr., 1966, "Hormonal Influences on Brain Organization in Infant Rats," *Science*, 152:1585–1592.

Lewin, K., 1935, *A Dynamic Theory of Personality: Selected Papers*. New York: McGraw-Hill.

Lewis, O. J., 1969, "The Hominoid Wrist Joint," *American Journal of Physical Anthropology*, 30:251–267.

Lindburg, D. G., 1967, "A Field Study of the Reproductive Behavior of the Rhesus Monkey," unpublished doctoral thesis, University of California, Berkeley.

Loizos, C., 1966, "Play in Mammals," in *Play, Exploration, and Territory in Mammals,* P. A. Jewell, ed. London: Academic Press, pp. 1–9.

———, 1967, "Play Behavior in Higher Primates: A Review," in *Primate Ethology,* D. Morris, ed. Chicago: Aldine, pp. 176–218.

Lorenz, K. Z., 1964, "Ritualized Fighting," *The Natural History of Aggression,* J. D. Carthy and F. J. Ebling, eds. New York: Academic Press, pp. 39–50.

———, 1966, *On Aggression.* New York: Harcourt.

McCann, 1928, "Notes on the Common Indian Langur *(Pithecus entellus),"* *J. Bomb. Nat. Hist. Soc.,* 33:192–194.

———, 1933, "Observations on Some of the Indian Langurs," *J. Bomb. Nat. Hist. Soc.,* 36:618–628.

McGill, T. E., ed., 1965, *Readings in Animal Behavior.* New York: Holt, Rinehart and Winston, Inc.

MacLean, P. D., 1958, "The Limbic System with Respect to Self-Preservation and the Preservation of the Species," *J. Nerv. Ment. Dis.,* 127:1–11.

———, 1959, "The Limbic System with Respect to Two Basic Life Principles," in *Central Nervous Systems and Behavior,* M. A. Brazier, ed. Trans. 2nd Conf. New York: Josiah Macy, Jr. Foundation.

———, 1962, "New Findings Relevant to the Evolution of Psychosexual Functions of the Brain," *J. Nerv. Ment. Dis.,* 135:289–301.

———, 1963, "Phylogenesis," in *Expression of the Emotions in Man,* P. Knapp, ed. New York: International Universities, pp. 16–35.

———, and J. M. R. Delgado, 1953, "Electrical and Chemical Stimulation of Fronto-temporal Portion of Limbic System in the Waking Animal," *Electroenceph. Clin. Neurophysiol.,* 5:91–100.

MacRoberts, M. 1965, "Gibralter Macaques," paper presented at the Southwestern Anthropological Assoc. annual meeting, Los Angeles, Calif.

Mann, A. E., 1968, "The Paleodemography of *Australopithecus,"* unpublished doctoral thesis, University of California, Berkeley.

Marler, P., 1961, "The Logical Analysis of Animal Communication," *J. Theoret. Biol.,* 1:295–317.

———, 1965, "Communication in Monkeys and Apes," in *Primate Behavior: Field Studies of Monkeys and Apes,* I. DeVore, ed. New York: Holt, Rinehart and Winston, Inc., pp. 544–584.

———, 1968, "Aggregation and Dispersal: Two Functions in Primate Communication," in *Primates: Studies in Adaptation and Variability,* P. C. Jay, ed. New York: Holt, Rinehart and Winston, Inc., pp. 420–438.

Masland, R. L., 1968, "Some Neurological Processes Underlying Language," *Annals of Otology, Rhinology and Laryngology,* 77:787–804.

Mason, W. A., 1959, "Development of Communication between Young Rhesus Monkeys," *Science,* 130:712–713.

———, 1960, "The Effects of Social Restriction on the Behavior of Rhesus Monkeys, I. Free Social Behavior," *Journal of Comparative and Physiological Psychology,* 53:582–589.

———, 1961*a,* "The Effects of Social Restriction on the Behavior of Rhesus Monkeys, II. Tests of Gregariousness," *Journal of Comparative and Physiological Psychology,* 54:287–290.

———, 1961*b,* "The Effects of Social Restriction on the Behavior of Rhesus Monkeys, III. Dominance Tests," *Journal of Comparative and Physiological Psychology,* 54:694–699.

———, 1964, "Sociability and Social Organization in Monkeys and Apes," *Recent Advances in Experimental Psychology,* L. Berkowitz, ed., 1:277–305.

———, 1965, "The Social Development of Monkeys and Apes," in *Primate Be-*

havior: Field Studies of Monkeys and Apes, I. DeVore, ed. New York: Holt, Rinehart and Winston, Inc., pp. 514–543.

———, 1968, "Use of Space by Callicebus Groups," in *Primates: Studies in Adaptation and Variability,* P. C. Jay, ed. New York: Holt, Rinehart and Winston, Inc., pp. 200–216.

———, and R. R. Sponholz, 1963, "Behavior of Rhesus Monkeys Rasied in Isolation," *J. Psychiat. Res.,* 1:299–306.

Mayr, E., 1963, *Animal Species and Evolution.* Cambridge, Mass.: Harvard University Press.

Merfield, F. G., and H. Miller, 1956, *Gorilla Hunter.* New York: Farrar, Straus.

Miminoshvili, D. I., G. O. Magakian, and G. I. Kokaia, 1960, "Attempts to Obtain a Model of Hypertension and Coronary Insufficiency in Monkeys," *Theoretical and Practical Problems of Medicine and Biology in Experiments on Monkeys,* I. A. Utkin, ed. New York: Pergamon, 103–121.

Mirsky, I. A., R. E. Miller, and J. V. Murphy, 1958, "The Communication of Affect in Rhesus Monkeys," *J. Amer. Psychoanal. Assoc.,* 6:433–441.

Miyadi, B., 1965, "Social Life of Japanese Monkeys," *Science in Japan,* A. H. Livermore, ed. Washington, D. C., A.A.A.S., pp. 315–334.

Mowrer, O. H., 1960, *Learning Theory and the Symbolic Process.* New York: Wiley.

Murphy, J. V., R. E. Miller, and I. A. Mirsky, 1955, "Interanimal Conditioning in the Monkey," *J. Comp. Physiol. Psychol.,* 48:211–214.

Myers, R. E., 1969, "Discussion of the Preceding Paper," in *Comparative and Evolutionary Aspects of the Vertebrate Central Nervous System,* J. M. Petras and C. R. Noback, eds., *Annals of the N. Y. Acad. Sci.,* 167(1):289–292.

Nadel, S. F., 1957, *The Theory of Social Structure.* Glencoe, Ill.: Free Press.

Nalbandov, A. V., 1964, *Reproductive Physiology.* San Francisco: W. H. Freeman.

Napier, J. R., 1962, "The Evolution of the Hand," *Scientific American,* 207:56–62.

———, 1963, "The Locomotor Functions of Hominids," in *Classification and Human Evolution,* S. L. Washburn, ed., Viking Fund Publications in Anthropology, No. 37. New York: Wenner-Gren Foundation, pp. 178–189.

———, and N. Barnicot, eds., 1963, "The Primates," *Symp. Zool. Soc. London, No. 10.*

———, and P. R. Davis, 1959, "The Fore-limb Skeleton and Associated Remains of *Proconsul africanus,*" *Fossil Mammals of Africa, No. 16,* London: British Museum (Natural History).

———, and P. H. Napier, 1967, *A Handbook of Living Primates.* London: Academic Press.

Neville, M. K., 1966, "A Study of the Free-Ranging Behavior of Rhesus Monkeys," unpublished doctoral thesis, Harvard University.

Nishida, T., 1968, "The Social Group of Wild Chimpanzees in the Mahali Mountains," *Primates,* 9:167–224.

Nissen, H. W., 1931, "A Field Study of the Chimpanzee," *Comp. Psychol. Monogr.,* 8(1):1–122.

Noback, C. R., and N. Moskowitz, 1963, "The Primate Nervous System: Functional and Structural Aspects in Phylogeny," in *Evolutionary and Genetic Biology of Primates, Vol. I,* J. Buettner-Janusch, ed. New York: Academic Press, pp. 131–177.

Nolte, A., 1955*a*, "Freilandbeobachtunger über das Verhalten *Macaca radiata* in Südindien," *Z. f. Tierpsychol.,* 12:77–87.

———, 1955*b*, "Field Observations on the Daily Routine and Social Behavior of Common Indian Monkeys, with Special Reference to the Bonnet Monkey *(Macaca radiata* Geoffroy)," *J. Bombay Nat. Hist. Soc.,* 53:177–184.

Osborn, R., 1963, "Observations on the Behaviour of the Mountain Gorilla," *Symp. Zool. Soc. Lond.*, 10:29–37.

Owen, R., 1859, *On the Classification and Geographical Distribution of the Mammalia.* London: G. W. Parker.

Penfield, W., and T. Rasmussen, 1957, *The Cerebral Cortex of Man.* New York: Macmillan.

———, and L. Roberts, 1959, *Speech and Brain Mechanism.* Princeton, N. J.: Princeton University Press.

Petter, J. J., 1962a, "Ecological and Behavioral Studies of Madagascar Lemurs in the Field," *Ann. N. Y. Acad. Sci.*, 102(2):267–281.

———, 1962b, "Recherches sur l'Écologie et l'Ethologie des Lémuriens Malagaches," *Mémoires du Muséum National d'Historie Naturelle, Série A* (Zoologie), 27(1):1–146.

———, 1965, "The Lemurs of Madagascar," in *Primate Behavior: Field Studies of Monkeys and Apes,* I. DeVore, ed. New York: Holt, Rinehart and Winston, Inc., pp. 292–319.

Phillips, W. W. A., 1935, *Manual of the Mammals of Ceylon.* London: Dulan and Company.

Piveteau, Jean, 1957, *Traité de Paléontologie, Tome VII* (Primates, Paléontologie Humaine). Paris: Masson et Cie.

Ploog, D. W., 1964, "Vom Limbischen System Gesteuertes Verhalten," *Der Nervenarzt,* 4:166–174.

———, and T. Melnechuk, 1969, "Primate Communication: A Report Based on an NRP Work Session," *Neurosciences Research Program Bulletin* 7(5):419–510.

Pocock, R. I., 1931, "The Mammal Survey of the Eastern Ghats, Report on the Monkeys," *J. Bomb. Nat. Hist. Soc.*, 35(1):51–59.

———, 1934, "The Monkeys of the Genera *Pithecus* (*presbytis*) and *Pygathrix* Found to the East of the Bay of Bengal," *P. Z. S. Lond.*, 1934:895–961.

———, 1939, *Fauna of British India, Mammals,* 2d ed. Vol. 1. London: Taylor and Francis.

Poirier, F. E., 1968a, "Nilgiri Langur Home Range Change," *Primates,* 9(1–2):29–43.

———, 1968b, "Nilgiri Langur (*Presbytis johnii*) Mother-infant Dyad," *Primates,* 9(1–2):45–68.

———, 1969, "The Nilgiri Langur (*Presbytis johnii*) Troop: Its Composition, Structure, Function and Change," *Folia Primat.*, 10:20–47.

Poole, T. B., 1966, "Aggressive Play in Polecats," *Symposium of the Zoological Society of London,* 18:23–44.

Prater, S. H., 1948, *The Book of Indian Mammals.* Bombay: The Bombay Natural History Society.

Prechtl, H. F. R., 1965, "Problems of Behavioral Studies in the Newborn Infant," in *Advances in the Study of Behavior,* D. S. Lehrman, R. A. Hinde, and E. Shaw, eds. New York: Adademic Press, pp. 75–98.

Rapoport, A., 1966, "Models of Conflict: Cataclysmic and Strategic," in *Conflict in Society,* A. de Reuck and J. Knight, eds. Boston: Little, Brown and Co., pp. 259–287.

Reynolds, V., 1966, "Open Groups in Hominid Evolution," *Man* (n.s.) 1:441–452.

———, and F. Reynolds, 1965, "Chimpanzees in the Budongo Forest," in *Primate Behavior: Field Studies of Monkeys and Apes,* I. DeVore, ed. New York: Holt, Rinehart and Winston, Inc., pp. 368–424.

Rheingold, H. L., 1960, "The Measurement of Maternal Care," *Child Development,* 31:565–575.

————, 1961, "The Effect of Environmental Stimulation upon Social and Exploratory Behavior in the Human Infant," in *Determinants of Infant Behavior*, B. M. Foss, ed. London: Methuen, pp. 143–177.

————, 1966, "The Development of Social Behavior in the Human Infant," *Monographs, Society for Research in Child Development*, Vol. 31, No. 5, pp. 1–17.

————, 1969, "The Social and Socializing Infant," in *Handbook of Socialization Theory and Research*, D. A. Goslin, ed. Chicago: Rand McNally and Co., pp. 779–790.

Ripley, S., 1965, "The Ecology and Social Behavior of the Ceylon Gray Langur," unpublished doctoral thesis, University of California, Berkeley.

————, 1967, "Intertroop Encounters Among Ceylon Gray Langurs (*Presbytis entellus*)," in *Social Communication Among Primates*, S. A. Altmann, ed. Chicago: University of Chicago Press, pp. 237–253.

Roberts, J. M., Malcolm J. Arth, and Robert R. Bush, 1959, "Games in Culture," *American Anthropologist*, 61(4):598–605.

————, and B. Sutton-Smith, 1962, "Child Training and Game Involvement," *Ethology*, 1(2): 166–185.

Robinson, B. W., 1967a, "Neurological Aspects of Evoked Vocalizations," in *Social Communication Among Primates*, S. A. Altmann, ed. Chicago: University of Chicago Press, pp. 135–147.

————, 1967b, "Vocalizations Evoked from Forebrain in *Macaca mulatta*," *Physiol. Behav.*, 2:345–354.

Roe, A., and G. G. Simpson, eds., 1958, *Behavior and Evolution*. New Haven, Conn.: Yale University Press.

Romer, A. S., 1966, *Vertebrate Paleontology*, 3d ed., Chicago: University of Chicago Press.

Rosenberg, B. J., and B. Sutton-Smith, 1960, "A Revised Conception of Masculine-feminine Differences in Play Activities," *Journal of Genetic Psychology*, 96:165–170.

Rosenblum, L. A., 1961, "The Development of Social Behavior in the Rhesus Monkey," unpublished doctoral thesis, University of Wisconsin.

————, and H. F. Harlow, 1963, "Generalization of Affectional Responses in Rhesus Monkeys," *Percept. Mot. Skills*, 16:561–564.

Rowell, T. E., 1962, "Agonistic Noises of the Rhesus Monkey (*Macaca mulatta*)," *Symp. Zool. Soc. Lond.*, 8:91–96.

————, 1966a, "Forest-living Baboons in Uganda," *J. Zool. Soc. Lond.*, 149:344–364.

————, 1966b, "Hierarchy in the Organization of a Captive Baboon Group," *Anim. Behav.*, 14(4): 420–443.

————, 1967a, "A Quantitative Comparison of the Behaviour of a Wild and a Caged Baboon Group," *Anim. Behav.*, 15:499–509.

————, 1967b, "Variability in the Social Organization of Primates," in *Primate Ethology*, D. Morris, ed. Chicago: Aldine, pp. 219–235.

————, and R. A. Hinde, 1962, "Vocal Communication by the Rhesus Monkey (*Macaca mulatta*)," *Proc. Zool. Soc. Lond.*, 138:279–294.

Russell, D. E., P. Louis, and D. E. Savage, 1967, "Primates of the French Early Eocene," *University of California Publications in Geological Sciences, Vol. 73*. Berkeley and Los Angeles: University of California Press.

Sabater Pi, G., 1960, "Beitrag zur Biologie des Flachland Gorillas," *Z. f. Saugetierkunde*, 25(3):133–141.

Sade, D. S., 1964, "Seasonal Cycle in Size of Testes of Free-ranging *Macaca mulatta*," *Folia Primat.*, 2:171–180.

Sade, D. S., 1965, "Some Aspects of Parent-offspring and Sibling Relations in a Group of Rhesus Monkeys, with a Discussion of Grooming," *Amer. J. Phys. Anthrop.*, 23:1–17.

———, 1966, "Ontogeny of Social Relations in a Free-ranging Group of Rhesus Monkeys," unpublished doctoral thesis, University of California, Berkeley.

———, 1968, "Inhibition of Son-mother Mating Among Free-ranging Rhesus Monkeys," *Science and Psychoanalysis*, 12:18–38.

Sarich, V. 1969 "Human Origins: An Immunological View," *Triangle* the Sandoz Journal of Medical Science, 9:55–60.

———, and A. C. Wilson, 1967a, "An Immunological Time Scale for Hominid Evolution," *Science*, 158:1200–1203.

———, and ———, 1967b, "Rates of Albumin Evolution in Primates," *Proceedings of the National Academy of Sciences*, 58:142–148.

Schaller, G., 1961, "The Orang-utan in Sarawak," *Zoologica*, 46(2):73–82.

———, 1963, *The Mountain Gorilla: Ecology and Behavior.* Chicago: University of Chicago Press.

———, 1965, "Field Procedures," in *Primate Behavior: Field Studies of Monkeys and Apes*, I. DeVore, ed. New York: Holt, Rinehart and Winston, Inc., pp. 623–629.

———, 1967, *The Deer and the Tiger: A Study of Wildlife in India.* Chicago: University of Chicago Press.

Scheflin, A. E., 1964, "The Significance of Posture in Communication Systems," *J. Psychiat.*, 27:316–331.

Schein, M. W., and E. B. Hale, 1959, "The Effect of Early Social Experience on Male Sexual Behavior of Androgen Injected Turkeys," *Animal Behaviour*, 7:189–200.

Schleidt, W., M. Schleidt, and M. Magg, 1960, "Störung der Mutter-Kind-Baziehung bei Truthühnern durch Gehorverlust," *Behaviour*, 16:3–4.

Schneirla, T. C., 1946, "Problems in the Biopsychology of Social Organization," *J. Abnorm. Soc. Psychol.*, 41:385–402.

Schultz, A. H., 1934, "Some Distinguishing Characters of the Mountain Gorilla," *J. Mammal.*, 15(1):51–61.

———, 1936, "Characters Common to Higher Primates and Characters Specific for Man," *Quarterly Review of Biology*, 11:259–283, 425–455.

———, 1938, "The Relative Weight of the Testes in Primates," *Anat. Rec.*, 72:387–394.

———, 1961, "Some Factors Influencing the Social Life of Primates in General and of Early Man in Particular," in *Social Life of Early Man*, S. L. Washburn, ed. Chicago: Aldine, pp. 58–90.

———, 1969, *The Life of Primates.* London: Weidenfeld and Nicolson Natural History.

Scott, J. P., 1958, *Aggression.* Chicago: University of Chicago Press.

———, 1962, "Hostility and Aggression in Animals," in *Roots of Behavior*, E. L. Bliss, ed. New York: Harper, pp. 167–178.

Seay, B., 1966, "Maternal Behavior in Primiparous and Multiparous Rhesus Monkeys," *Folia Primat.*, 4:146–168.

———, B. K. Alexander, and H. F. Harlow, 1964, "Maternal Behavior of Socially Deprived Rhesus Monkeys," *J. Abnorm. Soc. Psychol.*, 69:345–354.

———, E. Hansen, and H. F. Harlow, 1962, "Mother-infant Separation in Monkeys," *J. Child Psychol. Psychiat.*, 3:123–132.

Sebeok, T. A., 1965a, "Animal Communication," *Science*, 147:1006–1014.

———, 1965b, "Comments on Language and Communication," in *The Origin of*

Man, P. L. DeVore, ed. New York: Wenner-Gran Foundation, pp. 86–88.

_____, 1968, *Animal Communication: Techniques of Study and Results of Research*, Bloomington: Indiana University Press.

Shirek-Ellefson, J., 1967, "Visual Communication in *Macaca irus*," unpublished doctoral thesis, University of California, Berkeley.

Simonds, P. E., 1963, "Ecology of Macaques," unpublished doctoral thesis, University of California, Berkeley.

_____, 1965, "The Bonnet Macaque in South India," *Primate Behavior: Field Studies of Monkeys and Apes*, I. DeVore, ed. New York: Holt, Rinehart and Winston, Inc., pp. 175–196.

Simons, E. L., 1959, "An Anthropoid Frontal Bone from the Fayum Oligocene of Egypt: the Oldest Skull Fragment of a Higher Primate," *American Museum Novitates* No. 1976. New York: American Museum of Natural History.

_____, 1967, "The Earliest Apes," *Scientific American*, 217:28–35.

Simpson, G. G., 1945, "The Principles of Classification and a Classification of Mammals," *Bull. Am. Museum Nat. His.*, 85:1–350.

_____, 1949, *The Meaning of Evolution*. New Haven, Conn., Yale University Press.

_____, 1953, *The Major Features of Evolution*. New York: Simon and Schuster.

_____, 1962, "Primate Taxonomy and Recent Studies of Nonhuman Primates," *Ann. N. Y. Acad. Sci.*, 102:497–514.

_____, 1963, "The Meaning of Taxonomic Statement," in *Classification and Human Evolution*, S. L. Washburn, ed., Viking Fund Publications in Anthropology, No. 37. New York: Wenner-Gren Foundation, pp. 1–31.

_____, 1964, *This View of Life*. New York: Harcourt.

_____, 1965, *The Geography of Evolution*. Philadelphia and New York: Chilton.

Singh, S. D., 1969, "Urban Monkeys," *Scientific American*, 221(1):108–115.

Sluckin, W., 1965, *Imprinting and Early Learning*. Chicago: Aldine.

Smith, W. J., and E. B. Hale, 1959, "Modification of Social Rank in the Domestic Fowl," *J. Comp. Physiol. Psychol.*, 52:373–375.

Sollas, W. J., 1905, *The Age of the Earth*. London: T. F. Unwin.

Sommer, R., 1969, *Personal Space*. Englewood Cliffs, N. J.: Prentice-Hall.

Southwick, C. H., ed., 1963, *Primate Social Behavior*. Princeton, N. J.: Van Nostrand.

_____, M. A. Beg, and M. R. Siddiqi, 1961a, "A Population Survey of Rhesus Monkeys in Villages, Town, and Temples of Northern India," *Ecology*, 42:538–547.

_____, _____, and _____, 1961b, "A Population Survey of Rhesus Monkeys in Northern India: II. Transportation Routes and Forest Areas," *Ecology*, 42:698–710.

_____, _____, and _____, 1965, "Rhesus Monkeys in North India," in *Primate Behavior: Field Studies of Monkeys and Apes*, I. DeVore, ed. New York: Holt, Rinehart and Winston, Inc., pp. 111–159.

Spencer-Booth, Y., R. A. Hinde, and M. Bruce, 1965, "Social Companions and the Mother-infant Relationship in Rhesus Monkeys," *Nature*, 208:301.

Sperry, R. W., and M. S. Gazzaniga, 1967, "Language Following Surgical Disconnection of the Hemispheres," in *Brain Mechanisms Underlying Speech and Language*, F. L. Darley, ed. New York and London: Grune and Stratton, pp. 108–121.

Stott, K., and C. J. Silsar, 1961, "Observations of the Maroon Leaf Monkey in North Borneo," *Mammalia.*, 25(20):184–189.

Struhsaker, T., 1965, "Behavior of the Vervet Monkey, *Cercopithecus aethiops*," unpublished doctoral thesis, University of California, Berkeley.

Struhsaker, T., 1967a, "Auditory Communication Among Vervet Monkeys (*Cerco-pithecus aethiops*)," in *Social Communication Among Primates*, S. A. Altmann, ed. Chicago: University of Chicago Press, pp. 281–324.

———, 1967b, "Behavior of Vervet Monkeys (*Cercopithecus aethiops*)," *University of California Publications in Zoology*, 82:1–64.

———, 1967c, "Social Structure Among Vervet Monkeys (*Cercopithecus aethiops*)," *Behaviour*, 29(2–4):83–121.

Sugiyama, Y., 1964, "Group Composition, Population Density and Some Sociological Observations of Hanuman Langurs (*Presbytis entellus*)," *Primates*, 5(3–4):7–37.

———, 1965a, "Behavioral Development and Social Structure in Two Troops of Hanuman Langurs (*Presbytis entellus*)," *Primates*, 6(2):213–247.

———, 1965b, "On the Social Change of Hanuman Langurs *(Presbytis entellus)* in Their Natural Condition," *Primates*, 6(3–4):381–418.

———, 1966a, "An Artificial Social Change in a Hanuman Langur Troop," *Primates*, 7(1):41–72.

———, 1966b, "Social Organization of Hanuman Langurs," in *Social Communication Among Primates*, S. A. Altmann, ed. Chicago: University of Chicago Press, pp. 221–236.

———, and K. Yoshiba, 1965, "Home Range, Mating Season, Male Group and Inter-troop Relations in Hanuman Langurs (*Presbytis entellus*)," *Primates*, 6(1):73–106.

Sutton-Smith, B., and J. M. Roberts, 1964, "Ruberics of Competitive Behavior," *Journal of Genetic Psychology*, 105:13–37.

Szalay, F. S., 1968, "The Beginnings of Primates," *Evolution*, 22:19–36.

Thorington, R., 1967, "Feeding and Activity of Cebus and Saimiri in a Colombian Forest," in *Progress in Primatology*, D. Starck, R. Schneida, H. J. Kuhn, eds. Stuttgart: Gustav Fischer, pp. 180–184.

Thorpe, W. H., 1963, "Antiphonal Singing in Birds as Evidence for Avian Auditory Reaction Time," *Nature*, 197:774–776.

Tokuda, K., 1961–1962, "A Study on the Sexual Behavior in the Japanese Monkey Troop," *Primates*, 3(2):1–40.

Tsumori, A., 1967, "Newly Acquired Behavior and Social Interactions of Japanese Monkeys," in *Social Communication Among Primates*, S. A. Altmann, ed. Chicago: University of Chicago Press, pp. 207–219.

Tuttle, R. H., 1967, "Knuckle-walking and the Evolution of Hominid Hands," *Am. J. Phys. Anthrop.*, 26:171–206.

Ullrich, W., 1961, "Zur Biologie und Soziologie der Colobusaffen (*Colobus guereza caudasus* Thomas 1885)," *Der Zoologische Garten*, 25(6):305–368.

van Hooff, J. A. R. A. M., 1962, "Facial Expressions in the Higher Primates," *Symp. Zool. Soc. Lond.*, 8:97–125.

———, 1967, "The Facial Displays of the Catarrhine Monkeys and Apes," in *Primate Ethology*, D. Morris, ed. Chicago: Aldine, pp. 7–68.

van Lawick-Goodall, J., 1967, *My Friends the Wild Chimpanzees*. Washington, D.C.: National Geographic Society.

———, 1968a, "Expressive Movements and Communication in Chimpanzees," in *Primates: Studies in Adaptation and Variability*, P. C. Jay, ed. New York: Holt, Rinehart and Winston, Inc., pp. 313–374.

———, 1968b, "The Behavior of Free-living Chimpanzees in the Gombe Stream Reserve," *Animal Behaviour Monographs*, Vol. 1, Part 3:161–311.

Van Valen, L., 1965, "Treeshrews, Primates and Fossils," *Evolution*, 19:137–151.

Washburn, S. L., 1944, "The Genera of Malaysian Langurs," *J. Mammal.*, 25(3):289–294.

————, 1957, "Ischial Callosities as Sleeping Adaptations," *Amer. J. Phys. Anthrop.*, 15(2):269–276.

————, ed., 1963*a*, *Classification and Human Evolution*, Viking Fund Publication in Anthropology No. 37, New York: Wenner-Gren Foundation.

————, 1963*b*, "Behavior and Human Evolution," in *Classification and Human Evolution*, S. L. Washburn, ed. Chicago: Aldine Publishing Co., pp. 190–203.

————, 1966, "Conflict in Primate Society," in *Conflict in Society*, A. de Reuck and J. Knight, eds. Boston: Little, Brown and Co., pp. 3–15.

————, 1968*a*, "Speculations on the Problem of Man's Coming to the Ground," *Changing Perspectives on Man*, B. Rothblatt, ed. Chicago: University of Chicago Press, pp. 191–206.

————, 1968*b*, *The Study of Human Evolution*, Condon Lectures, Oregon State System of Higher Education. Eugene: University of Oregon Press.

————, and I. DeVore, 1963, "Baboon Ecology and Human Evolution," F. Clark Howell and Francois Boulière, eds., *African Ecology and Human Evolution*, Viking Fund Publications in Anthropology No. 36, New York: Wenner-Gren Foundation.

————, and D. A. Hamburg, 1965*a*. "The Implications of Primate Research," in *Primate Behavior: Field Studies of Monkeys and Apes*, I. DeVore, ed. New York: Holt, Rinehart and Winston, Inc., pp. 607–622.

————, ————, 1965*b*, "The Study of Primate Behavior," in *Primate Behavior: Field Studies of Monkeys and Apes*, I. DeVore, ed. New York: Holt, Rinehart and Winston, Inc., pp. 1–13.

————, and P. Jay, 1965, "The Evolution of Human Nature," unpublished paper presented at the American Anthropological Association Meeting, Denver, Nov. 19, 1965.

————, ————, and J. B. Lancaster, 1965, "Field Studies of Old World Monkeys and Apes," *Science*, 150:1541–1547.

Watson, J. B., 1914, "Imitation in Monkeys," *Psychol. Bull.*, 5:169–178.

Welker, W. I., 1961, "An Analysis of Exploratory and Play Behavior in Animals," in *Functions of Varied Experience*, D. W. Friske and S. R. Maddi, eds. Homewood, Ill.: Dorsey Press, pp. 175–226.

Wilson, A. C., and V. M. Sarich, 1969, "A Molecular Time Scale for Human Evolution," *Proc. Nat. Acad. Sci.*, 63:1088–1093.

Wilson, A. P., 1968, "Social Behavior of Free-ranging Rhesus Monkeys with an Emphasis on Aggression," unpublished doctoral thesis, University of California, Berkeley.

————, and S. H. Vessey, 1968, "Behavior of Free-ranging Castrated Rhesus Monkeys," *Folia Primat.*, 9:1–14.

Wroughton, R. C., 1918, "Summary of Indian Mammal Survey," *J. Bomb. Nat. Hist. Soc.*, 25(4):551–563.

————, 1921, "Summary of the Results from the Indian Mammal Survey," *J. Bomb. Nat. Hist. Soc.*, 27(3):520–521.

Wynne-Edwards, V. C., 1965, "Self-regulating Systems in Populations of Animals," *Science*, 147:1543–1548.

Yamada, M., 1963, "A Study of Blood-relationship in the Natural Society of the Japanese Macaque," *Primates*, 4:43–65.

Yerkes, R. M., 1943, *Chimpanzees: A Laboratory Colony*. New Haven, Conn.: Yale University Press.

————, and A. W. Yerkes, 1929, *The Great Apes*. New Haven, Conn.: Yale University Press.

Yoshiba, K., 1967, "An Ecological Study of Hanuman Langurs, *Presbytis entellus*," *Primates*, 8:127–154.

Yoshiba, K., 1968. "Local and Intertroop Variability in Ecology and Social Behavior of Common Indian Langurs," in *Primates: Studies in Adaptation and Variability,* P. C. Jay, ed. New York: Holt, Rinehart and Winston, Inc., pp. 217–242.

Young, J. Z., 1950, *The Life of Vertebrates.* Oxford: Clarendon Press.

Young, W., R. Goy, and C. Phoenix, 1964, "Hormones and Sexual Behavior," *Science,* 143:212–218.

Zapfe, H., 1958, "The Skeleton of *Pliopithecus (Epipliopithecus) vindobonensis* Zapfe and Hürzeler," *American Journal of Physical Anthropology,* 16:441–455.

Zihlman, A. L., 1967, "Human Locomotion: A Reappraisal of the Functional and Anatomical Evidence," unpublished doctoral thesis, University of California, Berkeley.

Zuckerman, S., 1932, *The Social Life of Monkeys and Apes.* London: Routledge.

_____, 1933, *Functional Affinities of Man, Monkeys and Apes.* London: Routledge.

INDEX